T0358594

QUANTUM STATISTICAL MECHANICS

Selected Works of N N Bogolubov

QUANTUM STATISTICAL MECHANICS

Selected Works of N N Bogolubov

N N Bogolubov, Jr

Moscow State University, Russia

 **World Scientific**

NEW JERSEY · LONDON · SINGAPORE · BEIJING · SHANGHAI · HONG KONG · TAIPEI · CHENNAI

Published by

World Scientific Publishing Co. Pte. Ltd.

5 Toh Tuck Link, Singapore 596224

USA office: 27 Warren Street, Suite 401-402, Hackensack, NJ 07601

UK office: 57 Shelton Street, Covent Garden, London WC2H 9HE

Library of Congress Cataloging-in-Publication Data
Bogoliubov, N. N. (Nikolai Nikolaevich), 1909–1992, author.
    Quantum statistical mechanics : selected works of N.N. Bogolubov / N.N. Bogolubov, Jr., Moscow State University, Russia.
    pages cm
    Includes bibliographical references and index.
    ISBN 978-9814612517 (hardcover : alk. paper)
    1. Quantum statistics. 2. Statistical mechanics. I. Bogoliubov, N. N. (Nikolai Nikolaevich), author. II. Title.
    QC174.4.B65 2014
    530.13'3--dc23

                                              2014015288

British Library Cataloguing-in-Publication Data
A catalogue record for this book is available from the British Library.

Printed in Singapore

FOREWORD

Part I of the Selected Works by N.N. Bogolubov contains some of his papers on statistical mechanics, a field in which he has obtained a number of fundamental results.

The name of Bogolubov has been inseparably linked with the creation of the modern theory of non-ideal quantum macrosystems. His formulation for such important physical phenomena as superfluidity and superconductivity formed the basis of this theory. During the 1940s Bogolubov produced a series of papers dealing with these problems. He developed the method of approximate second quantization which has been considered to be one of the basic tools of quantum statistics. The new method has made possible, in particular, the discovery of a very important physical phenomenon, the stabilization of a condensate in non-ideal gases at temperatures close to zero.

The phenomenon of superfluidity was discovered in 1938 by one of the most prominent Soviet physicists, Academician P.L. Kapitsa. It was found that at temperatures close to absolute zero the viscosity of ^{2}He was equal to zero. A new type of energy spectrum was discovered whose investigation became the main task in the study of properties of matter at low temperatures. However, the dynamical nature of the spectrum was obscure for a long time. It was not clear whether this phenomenon could be interpreted within the usual quantum mechanical scheme for the pair interaction of individual particles.

In 1947 in his classic work "On the Theory of Superfluidity", Bogolubov gave a brilliant physical explanation for the phenomenon of superfluidity. He brought out the dominant role of the interaction of correlated pairs of particles with opposite momenta in the formation of the ground state, whereas an ideal gas does not possess this property. In this way, Bogolubov constructed a special transformation of Bose amplitudes, which was thereafter called the Bogolubov transformation. As a result of these investigations, a consistent microscopic theory of superfluidity was built that describes the energy spectrum for a superfluid system and explains the relation between the superfluid and normal states.

A valuable contribution has been made by Bogolubov to the theory of superconductivity. He showed that the same type of excitations that occur in the superfluids also occur in superconductors, in which a decisive part is played by the interaction of the electrons with lattice oscillations. During construction of the microscopic theory of superconductivity it was found in 1957 that the above-mentioned mathematical methods were also useful for studying this phenomenon. It has been established that between superfluidity and superconductivity there is a deep physical and mathematical analogy. In brief, it can be said that superconductivity is superfluidity of electrons in metal.

Bogolubov has investigated in detail the hydrodynamic stage in the evolution of classic many-particle systems. One of the works devoted to this field, "Hydrodynamics Equations in Statistical Mechanics", published in 1948, is reproduced in this volume. Later he generalized the method of constructing kinetic equations for quantum systems and applied it to studying superfluid liquids.

Studies of degeneracy in systems led Bogolubov in 1961 to the formulation of the method of quasi-averages in his work "Quasi-Averages in problems of statistical mechanics". This method has proved to be a universal tool for systems whose ground states become unstable under small perturbations. The subsequent development of statistical mechanics and quantum field theory demonstrated the fruitfulness of the concept of quasi-averages and of the idea contained in them of spontaneous symmetry breaking. Bogolubov himself successfully applied this method, for example, to problems of superfluid hydrodynamics in the paper "On the Hydrodynamics of a Superfluid Liquid", published in 1963.

The influence of the studies on spontaneous symmetry breaking in macroscopic systems on elementary particle physics and quantum field theory was emphasized by S. Weinberg in his Nobel address, in which he reported that at the beginning of the 1960s he became acquainted with an idea that first appeared in solid theory and was then introduced into particle physics by people who worked in both of these branches of physics. This was the idea of symmetry breaking, according to which the Hamiltonian and the commutation relations of quantum theory can possess an exact symmetry but, nevertheless, the physical states need not correspond to the representations of the symmetry. In particular, it may happen that the symmetry of the Hamiltonian is not the symmetry of the vacuum. The subsequent systematic development of these ideas in quantum filed theory has led to the construction of the theory of electromagnetic and weak interactions, for which S.L. Glashow, A. Salam and S. Weinberg received the Nobel Price in 1979.

In connection with the formulation of the concept of quasi-averages, Bogolubov also proved a fundamental theorem on $1/q^2$ singularities, according to which elementary excitations with energy, which vanishes in the long-wave limit, arise in systems with spontaneous symmetry breaking. In other words, there are massless excitations — quanta of photon or phonon type — whose exchange leads to an interaction of infinite range. Soon after this, a similar result was obtained in quantum field theory.

Bogolubov's concept of quasi-averages has also provided the foundation for the modern theory of phase transitions. It should be mentioned that for problems of statistical mechanics Bogolubov's studies on the theories of superfluidity, superconductivity and quasi-averages produced the basis for the development of the methods of variational inequalities and majorizing estimates for systems of many interacting particles with spontaneously broken symmetry.

Part II is devoted to methods for studying model Hamiltonians for problems in quantum statistical mechanics. In this part methods are proposed for solving certain problems in statistical physics which contain four-fermion interaction. It has been possible, via "approximating (trial) Hamiltonians", to distinguish a whole class of exactly soluble systems.

An essential difference between the two types of problem with positive and negative four-fermion interaction is discovered and examined. The determination of exact solutions for the free energies, single-time and many-time correlation functions, T-products and Green's functions is treated for each type of problem.

I express my sincere thanks to Academician N.N. Bogolubov for valuable remarks. The material of this part was the subject of theoretical seminars and lectures at Moscow State University. I consider it my pleasant duty to thank participants in the theoretical seminars at the V.A. Steklov Mathematical Institute and the Theoretical Physics Laboratory of Joint Institute for Nuclear Research at Dubna for their interest and encouragement for discussion. Some of the results of Part II were discussed in a theoretical physics seminars at the E. T. H., Zurich. The author is thankful to Dr. Alan A. Dzhioev (BLTP, JINR, Dubna, Russia) for the help in presentation of the manuscript for publication and to Dr. Denis Blackmore (Department of Mathematic Science and Center for Applied mathematic and statistics, New Jersey Institute of Technology, USA) for checking the English of the manuscript.

N.N. Bogolubov, Jr.

CONTENTS

ix

PART I

CHAPTER 1

ON THE THEORY OF SUPERFLUIDITY

In the present paper an attempt is made to develop a consistent molecular theory for the phenomenon of superfluidity without any assumptions regarding the structure of the energy spectrum.

With this goal in mind it is more natural to proceed from the model of a non-ideal Bose gas with weak interaction between particles. Similar attempts to account for superfluidity with the aid of the phenomenon of degeneracy of an ideal Bose gas have already been undertaken by L. Tisza and F. London, but these encountered active criticism. It was noted, for instance, that Helium II does not resemble an ideal gas at all, since its molecules strongly interact with each other. By the way, this objection can not be considered as crucial. Indeed, if the goal is to develop a molecular but not a phenomenological theory proceeding only from the common microscopic equations of quantum mechanics, then it is quite clear that any attempts to evaluate the properties of real liquids are hopeless. A molecular theory of superfluidity can be expected only, at least at the first stage, to account for the phenomenon itself proceeding from a simplified model.

In fact, the real objection to the above-mentioned criticism consists in the fact that, in a degenerate ideal Bose gas, particles in the ground state cannot behave as superfluid ones, since nothing can prevent them from exchanging momenta with excited particles in collisions and thereby undergoing friction when moving through the fluid.

In the present paper we shall try to overcome this main difficulty and show that under certain conditions in a weakly non-ideal Bose gas the so called degenerate condensate can move without friction at a sufficiently low velocity relative to elementary excitations. It should be emphasized that in our theory these elementary excitations are a collective effect and cannot be identified with separate molecules. The necessity of considering collective

3

elementary excitations instead of individual molecules was first stressed by Landau in his well known paper on the theory of superfluidity.

Let us consider a system of N identical monoatomic molecules enclosed in a macroscopic volume V and obeying the Bose statistics. As usual, we suppose that its Hamiltonian is of the form

$$H = \sum_{(i \leq i \leq N)} T(\boldsymbol{p}_i) + \sum_{(1 \leq i \leq j \leq N)} \Phi(|\boldsymbol{q}_i - \boldsymbol{q}_j|)$$

where

$$T(\boldsymbol{p}_i) = \frac{|\boldsymbol{p}_i|^2}{2m} = \sum_{(1 \leq \alpha \leq 3)} \frac{(p_i^\alpha)^2}{2m}$$

is the kinetic energy of the i-the molecule and $\Phi(|\boldsymbol{q}_i - \boldsymbol{q}_j|)$ is the potential energy of the pair (i, j).

Then we take advantage of the secondary quantization method to write the basic equation in the form

$$i\hbar \frac{\partial \psi}{\partial t} = -\frac{\hbar^2}{2m} \Delta \psi + \int \Phi(|\boldsymbol{q} - \boldsymbol{q}'_j|) \psi^\dagger(\boldsymbol{q}') \psi(\boldsymbol{q}') \, d\boldsymbol{q}' \, \psi \tag{1}$$

where

$$\psi = \sum_f a_f \varphi_f(\boldsymbol{q}), \quad \psi^\dagger = \sum_f a_f^\dagger \varphi_f^*(\boldsymbol{q}).$$

Here a_f and $a_f^\dagger$ are conjugate operators with the usual commutation relations

$$a_f a_{f'} - a_{f'} a_f = 0,$$

$$a_f a_{f'}^\dagger - a_{f'}^\dagger a_f = \Delta_{f,f'} = \begin{cases} 0, & \boldsymbol{f} \neq \boldsymbol{f}' \\ 1, & \boldsymbol{f} = \boldsymbol{f}' \end{cases}$$

and $\{\varphi_f(\boldsymbol{q})\}$ is an orthonormal

$$\int \varphi_f^*(\boldsymbol{q}) \varphi_{f'}(\boldsymbol{q}) \, d\boldsymbol{q} = \Delta_{f,f'}$$

complete set of functions.

To simplify the problem, hereafter we shall make use of the set of eigenfunctions of the momentum operator for a single particle

$$\varphi_f(\boldsymbol{q}) = \frac{1}{V^{1/2}} \exp\left\{i\frac{\boldsymbol{qf}}{\hbar}\right\}, \quad \boldsymbol{qf} = \sum_{(1\leq\alpha\leq3)} f^\alpha q^\alpha.$$

For this set the operator $N_f = a^\dagger a_f$ corresponds to the number of particles with momentum $\boldsymbol{f}$. For a finite value of the volume V, the vector $\boldsymbol{f}$ is apparently quantized. For instance, under the usual periodic boundary conditions

$$f^\alpha = \frac{2\pi n^\alpha \hbar}{l}$$

where n^1, n^2, n^3 are integers and l is the edge length of a cube of volume V.

However, since here we shall deal with thermodynamics, i.e. bulk properties, we should always imply the limit transition when the walls of a vessel recede to infinity $V \to \infty$, $N \to \infty$ keeping the specific volume $v = V/N$ constant. Therefore, we shall finally proceed to a continuous spectrum replacing sums of the form

$$\sum_f F(\boldsymbol{f})$$

by the integrals

$$\frac{V}{(2\pi\hbar)^3} \int F(\boldsymbol{f})\, d\boldsymbol{f}.$$

Equations (1) are exact equations for the problem of N bodies. To succeed in studying the motion of the system of molecules under consideration we should make an approximation based on the assumption that the interaction energy is small. In accordance with this assumption we shall consider the potential $\Phi(\boldsymbol{r})$ to be proportional to a small parameter ε. Which dimensionless ratio can be taken for ε will be clarified below. Now we only note that the assumption made corresponds, strictly speaking, to neglect of the finiteness of the molecular radius, since here we do not take into account the intensive increase of $\Phi(\boldsymbol{r})$ for small $\boldsymbol{r}$ that ensures the impermeability of the molecules. By the way, as we shall see below, the results which will be obtained can be generalized to the case when one takes into account that the radius is finite.

Proceeding to formulating the approximation we note that, if there is no interaction, i.e. if the parameter ε exactly equals zero, then at zero temperature we could put $N_0 = N$, $N_f = 0$, $(\boldsymbol{f} \neq 0)$. However, in the considered case of a small ε and weakly excited states of the gas, these relations are satisfied approximately in the sense that the major part of the molecules possess momenta close to zero. Of course, the choice of zero momentum as a limit value for particles in the ground state corresponds to a specific choice of framework, namely, one in which the condensate is at rest.

The above speculations are the basis for the following method for approximate solution of Equation (1):

1) In view of the fact that $N_0 = a_0^\dagger a_0$ is quite large compared with unity, the expression

$$a_0 a_0^\dagger - a_0^\dagger a_0 = 1$$

must be small compared with a_0 and $a_0^\dagger$ themselves. Therefore, we consider a_0 and $a_0^\dagger$ as ordinary numbers[a] neglecting their non-commutativity.

2) Putting

$$\psi = \frac{a_0}{\sqrt{V}} + \vartheta, \quad \vartheta = \frac{1}{\sqrt{V}} \sum_{(f \neq 0)} a_f \exp\left\{i\frac{\boldsymbol{fq}}{\hbar}\right\}$$

we consider ϑ as a so called correction term of the first order and in Equation (1) we neglect terms beginning with a quadratic in ϑ, which corresponds to taking into account the weak excitations.

We then obtain basic approximate equations in the form

$$i\hbar\frac{\partial \vartheta}{\partial t} = -\frac{\hbar^2}{2m}\Delta\vartheta + \frac{N_0}{V}\Phi_0\vartheta + \frac{N_0}{V}\int \Phi(|\boldsymbol{q}-\boldsymbol{q}'|)\vartheta(\boldsymbol{q}')\,d\boldsymbol{q}'$$

$$+ \frac{a_0^2}{V}\int \Phi(|\boldsymbol{q}-\boldsymbol{q}'|)\vartheta(\boldsymbol{q}')\,d\boldsymbol{q}',$$

$$i\hbar\frac{\partial a_0}{\partial t} = \frac{N_0}{V}\Phi_0 a_0, \tag{2}$$

where

$$\Phi_0 = \int \Phi(|\boldsymbol{q}|)\,d\boldsymbol{q}.$$

To proceed from the operator wave function ϑ to the creation and annihilation operators a_f and $a_f^\dagger$ we use the Fourier expansion

$$\Phi(|\boldsymbol{q}-\boldsymbol{q}'|) = \sum_f \frac{1}{V} \exp\left\{\frac{i}{\hbar}\big(\boldsymbol{f}(\boldsymbol{q}-\boldsymbol{q}')\big)\right\}\nu(\boldsymbol{f}). \tag{3}$$

---

[a]A similar remark was used by Dirac in his monograph "Foundation of Quantum Mechanics" at the end of Section 3 named "Waves and Bose-Einstein particles."

Due to the radial symmetry of the potential, the coefficients of the expansion

$$\nu(\boldsymbol{f}) = \int \Phi(|\boldsymbol{q}|) \exp\left\{-i\frac{\boldsymbol{f}\boldsymbol{q}}{\hbar}\right\} d\boldsymbol{q}$$

depend on upon the length $|\boldsymbol{f}|$ of the vector $\boldsymbol{f}$. We substitute Equation (1) into Equation (2) to obtain

$$i\hbar\frac{\partial a_f}{\partial t} = \left\{T(\boldsymbol{f}) + E_0 + \frac{N_0}{V}\nu(\boldsymbol{f})\right\}a_f + \frac{a_0^2}{V}\nu(\boldsymbol{f})a_{-f},$$

$$E_0 = \frac{N_0}{V}\Phi_0.$$

If we put

$$a_f = \exp\left(\frac{E_0}{i\hbar}t\right)b_f, \quad a_0 = \exp\left(\frac{E_0}{i\hbar}\right)b, \tag{4}$$

we can also write

$$i\hbar\frac{\partial b_f}{\partial t} = \left\{T(\boldsymbol{f}) + \frac{N_0}{V}\nu(\boldsymbol{f})\right\}b_f + \frac{b^2}{V}\nu(\boldsymbol{f})b_{-f}^{\dagger},$$

$$-i\hbar\frac{\partial b_{-f}^{\dagger}}{\partial t} = \frac{(b^*)^2}{V}\nu(\boldsymbol{f})b_f + \left\{T(\boldsymbol{f}) + \frac{N_0}{V}\nu(\boldsymbol{f})\right\}b_{-f}^{\dagger}. \tag{5}$$

On solving this system of two equations with constant coefficients, we see that the dependence of the operators b_f and $b_f^{\dagger}$ on time is expressed by a linear combination of exponents of the form

$$\exp\left(\pm\frac{E(\boldsymbol{f})}{i\hbar}\right)$$

where

$$E(\boldsymbol{f}) = \left(2T(\boldsymbol{f})\frac{N_0}{V}\nu(\boldsymbol{f}) + T^2(\boldsymbol{f})\right)^{1/2}. \tag{6}$$

Now we note that if

$$\nu(0) = \int \Phi(|\boldsymbol{q}|) d\boldsymbol{q} > 0, \tag{7}$$

then in the considered case of sufficiently small ε the expression in Equation (6) under the square root sign is positive. Thus. the operators b_f and $b_f^{\dagger}$ turn out to be periodic functions of time. If, conversely, the inequality $\nu(0) < 0$ holds, then for small momenta this expression is negative, and therefore the quantity $E(\boldsymbol{f})$ becomes complex. Consequently, the operators b_f and $b_f^\dagger$ contain a real exponential increasing with time, so that the states with small N_f turn out to be unstable.

To ensure stability of weakly excited states, we shall consider below only such types of interactions among the molecules for which the inequality (7) is valid. It is interesting to note that inequality (7) is simply the condition for thermodynamic stability of a gas at absolute zero. Indeed, at zero temperature the free energy coincides with the internal energy, while the leading term in the expression of the internal energy is

$$E = \frac{N^2}{2V} \int \Phi(|\boldsymbol{q}|)\, d\boldsymbol{q},$$

since the correction terms, for instance, the average kinetic energy, are proportional to higher powers of ε. Hence, the pressure p is given by

$$p = -\frac{\partial E}{\partial V} = \frac{N^2}{2V^2} \int \Phi(|\boldsymbol{q}|)\, d\boldsymbol{q} = \frac{\rho^2}{2m^2} \int \Phi(|\boldsymbol{q}|)\, d\boldsymbol{q}$$

where $\rho = Nm/V$ is the mass density of the gas. Therefore, inequality (7) is equivalent to the condition of thermodynamic stability

$$\frac{\partial p}{\partial \rho} > 0.$$

Note, finally, that since here we take into account only the leading terms, we can write with the same degree of accuracy

$$E(\boldsymbol{f}) = \left(2T(\boldsymbol{f})\frac{N}{V}\nu(\boldsymbol{f}) + T^2(\boldsymbol{f})\right)^{1/2} = \left(\frac{|\boldsymbol{f}|^2 \nu(\boldsymbol{f})}{m\nu} + \frac{|\boldsymbol{f}|^4}{4m^2}\right)^{1/2}, \qquad (6')$$

instead of Equation (6). Thus, for small momenta[b]

$$E(\boldsymbol{f}) = \left(\frac{\nu(0)}{m\nu}\right)^{1/2}|\boldsymbol{f}|(1 + \dots) = \left(\frac{\partial p}{\partial \rho}\right)^{1/2}|\boldsymbol{f}|(1 + \dots) \qquad (8)$$

where $\dots$ denotes terms vanishing together with $\boldsymbol{f}$.

Let us now agree to take any square root we encounter with a positive sign. Then for small momenta

$$E(\boldsymbol{f}) = c|\boldsymbol{f}|(1 + \dots) \qquad (9)$$

---

[b]If we write the corresponding frequency $E(\boldsymbol{f})/\hbar$ and take the limit $\hbar \to 0$, $\boldsymbol{f}/h = k$, we obtain the classical A.A. Vlasov formula for the dependence of frequency on wave number.

where c is the velocity of sound at zero temperature. Conversely, for sufficiently large momenta one can expand $E(\boldsymbol{f})$ in powers of ε and write

$$E(\boldsymbol{f}) = \frac{|\boldsymbol{f}|^2}{2m} + \frac{1}{\nu}\nu(\boldsymbol{f}) + \dots.$$

Since the quantity $\nu(\boldsymbol{f})$ vanishes with increasing $|\boldsymbol{f}|$, for sufficiently large momenta the energy $E(\boldsymbol{f})$ approaches the kinetic energy of a single molecule $T(\boldsymbol{f})$.

Let us now return to Equation (5) and introduce new mutually conjugate operators ξ_f and $\xi_f^\dagger$ instead of b_f and $b_f^\dagger$ by the relations

$$\xi_f = \left(b_f - L_f b_{-f}^\dagger\right)\left(1 - |L_f|^2\right)^{1/2},$$

$$\xi_f^\dagger = \left(b_f^\dagger - L_f^* b_{-f}\right)\left(1 - |L_f|^2\right)^{1/2}, \tag{10}$$

where L_f are numbers defined as follows

$$L_f = \frac{Vb^2}{N_0^2 \nu(\boldsymbol{f})}\left\{E(\boldsymbol{f}) - T(\boldsymbol{f}) - \frac{N_0}{V}\nu(\boldsymbol{f})\right\}.$$

We have

$$|L_f|^2 = \left(\frac{N_0}{V}\nu(\boldsymbol{f})\right)^2\left(E(\boldsymbol{f}) + T(\boldsymbol{f}) + \frac{N_0}{V}\nu(\boldsymbol{f})\right)^{-2},$$

$$1 - |L_f|^2 = 2E(\boldsymbol{f})\left(E(\boldsymbol{f}) + T(\boldsymbol{f}) + \frac{N_0}{V}\nu(\boldsymbol{f})\right)^{-1}. \tag{11}$$

If we reverse Equations (10), we find

$$b_f = \left(\xi_f + L_f \xi_{-f}^\dagger\right)\left(1 - |L_f|^2\right)^{1/2},$$

$$b_f^\dagger = \left(\xi_f^\dagger + L_f^* \xi_{-f}\right)\left(1 - |L_f|^2\right)^{1/2}. \tag{12}$$

We substitute these relations into Equations (5) to obtain

$$i\hbar\frac{\partial\xi_f}{\partial t} = E(\boldsymbol{f})\xi_f, \quad -i\hbar\frac{\partial\xi_f^\dagger}{\partial t} = E(\boldsymbol{f})\xi_f^\dagger. \tag{13}$$

It is not difficult to check directly that the new operators satisfy the same commutation relations as the operators a_f and $a_f^\dagger$ do

$$\xi_f \xi_{f'} - \xi_{f'}\xi_f = 0, \quad \xi_f \xi_{f'}^\dagger - \xi_{f'}^\dagger\xi_f = \Delta_{f,f'}. \tag{14}$$

This implies that the excited states of the considered system of molecule can be thought of as an ideal gas of so called elementary excitations, or quasiparticles, with energies depending on momenta according the formula $E = E(\boldsymbol{f})$. Just as the molecules have been described by the operators a_f and $a_f^\dagger$, these quasiparticles are described by the operators ξ_f and $\xi_f^\dagger$, and therefore they obey the Bose statistics. The operator

$$n_f = \xi_f^\dagger \xi_f$$

represents the number of quasiparticles with momentum $\boldsymbol{f}$.

The above remark will be absolutely clear when we consider the total energy

$$H = H_{\text{kin}} + H_{\text{pot}}$$

where

$$H_{\text{kin}} = \frac{\hbar}{2m} \int \psi^\dagger(\boldsymbol{q}) \Delta \psi(\boldsymbol{q}) \, d\boldsymbol{q},$$

$$H_{\text{pot}} = \frac{1}{2} \int \Phi(|\boldsymbol{q} - \boldsymbol{q}'|) \psi^\dagger(\boldsymbol{q}) \psi^\dagger(\boldsymbol{q}') \psi(\boldsymbol{q}) \psi(\boldsymbol{q}') \, d\boldsymbol{q} \, d\boldsymbol{q}'$$

$$= \frac{1}{2V} \sum_f \nu(\boldsymbol{f}) \int \exp\left\{ i \frac{\boldsymbol{f}(\boldsymbol{q} - \boldsymbol{q}')}{\hbar} \right\} \psi^\dagger(\boldsymbol{q}) \psi^\dagger(\boldsymbol{q}') \psi(\boldsymbol{q}) \psi(\boldsymbol{q}') \, d\boldsymbol{q} \, d\boldsymbol{q}'.$$

For the kinetic energy we have

$$H_{\text{kin}} = \sum_f T(\boldsymbol{f}) a_f^\dagger a_f = \sum_f T(\boldsymbol{f}) b_f^\dagger b_f.$$

We calculate the potential energy in accordance with the accepted approximation. Namely, in the expression

$$\psi^\dagger(\boldsymbol{q}) \psi^\dagger(\boldsymbol{q}') \psi(\boldsymbol{q}) \psi(\boldsymbol{q}') = \left(\frac{a_0^\dagger}{\sqrt{V}} + \vartheta^\dagger(\boldsymbol{q}) \right)$$

$$\times \left(\frac{a_0^\dagger}{\sqrt{V}} + \vartheta^\dagger(\boldsymbol{q}') \right) \left(\frac{a_0}{\sqrt{V}} + \vartheta(\boldsymbol{q}) \right) \left(\frac{a_0}{\sqrt{V}} + \vartheta(\boldsymbol{q}') \right)$$

we omit the terms beginning with the cubic in ϑ and $\vartheta^\dagger$. Then we have

$$H_{\text{pot}} = \Phi_0 \left\{ \frac{1}{2} \frac{N_0^2}{V} + \frac{N_0}{V} \sum_{f \neq 0} b_f^\dagger b_f \right\} + \frac{b^2}{2V} \sum_{f \neq 0} \nu(\boldsymbol{f}) b_f^\dagger b_{-f}^\dagger$$

$$+ \frac{(b^*)^2}{2V} \sum_{f \neq 0} \nu(\boldsymbol{f}) b_f b_{-f} + \frac{N_0}{V} \sum_{f \neq 0} \nu(\boldsymbol{f}) b_f^\dagger b_f.$$

Since the following relations hold here

$$\sum_{f \neq 0} b_f^\dagger b_f = \sum_{f \neq 0} N_f = N - N_0,$$

in the accepted approximation we have

$$\frac{1}{2} \frac{N_0^2}{V} + \frac{N_0}{V} \sum_{f \neq 0} b_f^\dagger b_f = \frac{1}{2} \frac{N^2}{V},$$

and therefore

$$H = \frac{N_0}{2V} \Phi_0 + \frac{b^2}{2V} \sum_{f \neq 0} \nu(\boldsymbol{f}) b_f^\dagger b_{-f}^\dagger + \frac{(b^*)^2}{2V} \sum_{f \neq 0} \nu(\boldsymbol{f}) b_f b_{-f}$$

$$+ \frac{N_0}{V} \sum_{f \neq 0} \nu(\boldsymbol{f}) b_f^\dagger b_f + \sum_f T(\boldsymbol{f}) b_f^\dagger b_f.$$

We express the operators b_f and $b_f^\dagger$ in terms of the operators ξ_f and $\xi_f^\dagger$ to find finally

$$H = H_0 + \sum_{f \neq 0} E(\boldsymbol{f}) n_f, \quad n_f = \xi_f^\dagger \xi_f, \tag{15}$$

where

$$H_0 = \frac{1}{2} \frac{N^2}{V} + \Phi_0 \sum_{f \neq 0} \frac{1}{2} \left(E(\boldsymbol{f}) - T(\boldsymbol{f}) - \frac{N_0}{N} \nu(\boldsymbol{f}) \right)$$

$$= \frac{1}{2} \frac{N^2}{V} + \frac{V}{2(2\pi\hbar)^3} \int \left\{ E(\boldsymbol{f}) - T(\boldsymbol{f}) - \frac{N_0}{N} \nu(\boldsymbol{f}) \right\} d\boldsymbol{f}. \tag{16}$$

Thus, the total energy of non-ideal gas under consideration consists of the energy of the ground state H_0 and the sum of individuals energies of separate quasiparticles. The quasiparticles do not, apparently, interact with each other and form an ideal Bose gas.

It is easy to see that the absence of interaction between the quasiparticles is due to the approximation used, where the terms beginning with a cubic in ξ_f and $\xi_f^\dagger$ have been omitted. Therefore, the result obtained is relevant only for weakly excited states. Had we taken into account either the omitted cubic terms in the expression for energy or, respectively, the quadratic terms in Equations (13) as a small perturbation, we would find a weak interaction between quasiparticles which is responsible for approaching the state of statistical equilibrium in the system.

Proceeding to a study of the state of statistical equilibrium, we prove that the total momentum of quasiparticles $\sum_f \boldsymbol{f} n_f$ is conserved. With this aim in view we consider the components of the total momentum of the system of molecules. We have

$$\sum_{(1 \leq i \leq N)} p_i^\alpha = \int \psi^\dagger(\boldsymbol{q}) \left\{ -i\hbar \frac{\partial \psi^\dagger(\boldsymbol{q})}{\partial q^\alpha} \right\} d\boldsymbol{q} = \sum_f f^\alpha a_f^\dagger a_f = \sum_f f^\alpha b_f^\dagger b_f,$$

and hence, due to transformation formulae (12), we see that

$$\sum_{(1 \leq i \leq N)} p_i^\alpha = \sum_f f^\alpha \frac{(\xi_f^\dagger + L_f^* \xi_{-f})(\xi_f + L_f \xi_{-f}^\dagger)}{1 - |L_f|^2}.$$

But in view of the fact that L_f and L_f^* are invariant with regard to the replacement of $\boldsymbol{f}$ by $-\boldsymbol{f}$, we can write

$$\sum_f f^\alpha \frac{L_f^* \xi_{-f} \xi_f}{1 - |L_f|^2} = \sum_f f^\alpha \frac{L_f \xi_f^\dagger \xi_{-f}^\dagger}{1 - |L_f|^2} = 0,$$

$$\sum_f f^\alpha \frac{|L_f|^2 \xi_{-f} \xi_{-f}^\dagger}{1 - |L_f|^2} = \sum_f f^\alpha \frac{|L_f|^2 (\xi_{-f}^\dagger \xi_{-f} - 1)}{1 - |L_f|^2} = -\sum_f f^\alpha \frac{|L_f|^2}{1 - |L_f|^2} \xi_f^\dagger \xi_f,$$

and therefore

$$\sum_{(1 \leq i \leq N)} p_i^\alpha = \sum_f f^\alpha n_f.$$

Thus, the total momentum of the system of molecules is equal to that of the system of quasiparticles. Since the total momentum of the system of molecule is conserved, the sum

$$\sum_f \boldsymbol{f} n_f$$

is really an integral of motion.

It is easy to see that the total number of quasiparticles $\sum_f n_f$ is not conserved. They can be created and annihilated. Therefore, with the aid of the usual reasoning we see that in the state of statistical equilibrium the average occupation numbers $\bar{n}_f$ ($\boldsymbol{f} \neq 0$) are given by

$$\bar{n}_f = \left\{ A \exp\left(\frac{E(\boldsymbol{f}) - \boldsymbol{f}\boldsymbol{u}}{\theta} \right) - 1 \right\}^{-1}, \quad A = 1, \tag{17}$$

where θ is the temperature and $\boldsymbol{u}$ is an arbitrary vector. By the way, the length of this vector must be limited from above. Indeed, since the average occupation numbers must be positive, for all $\boldsymbol{f} \neq 0$ the following inequality must hold

$$E(\boldsymbol{f}) > \boldsymbol{f}\boldsymbol{u}$$

which yield the inequality

$$E(\boldsymbol{f}) > |\boldsymbol{f}|\,|\boldsymbol{u}|.$$

But by virtue of the properties of the function $E(\boldsymbol{f})$ established above, the ratio

$$E(\boldsymbol{f})/|\boldsymbol{f}|$$

is a continuous positive function of $|\boldsymbol{f}|$ which takes the value $c > 0$ at $|\boldsymbol{f}| = 0$ and grows with $|\boldsymbol{f}|/2m$ as $|\boldsymbol{f}| \to \infty$. Therefore, the ratio considered has a strictly positive minimum value. Thus, the condition for positivity of the numbers $\bar{n}_f$ is equivalent to the inequality

$$|\boldsymbol{u}| \leq \min\{E(\boldsymbol{f})/|\boldsymbol{f}|\}. \tag{18}$$

If for small momenta the function $E(\boldsymbol{f})$ had decreased in proportion to the square root of the momentum but no to the momentum itself, as the kinetic energy of a molecule does, the right hand side of the inequality obtained would be equal to zero and zero would be the only possible value for $\boldsymbol{u}$. However, in the case considered the $\boldsymbol{u}$ may be arbitrary, the only restriction being that its length must be sufficiently short.

Note that formula (17) describes a distribution of momenta of quasiparticles in the gas such that it moves as an entity with the velocity $\boldsymbol{u}$. First we have chosen a framework such that in the ground state the condensate, i.e. the system of molecules, is at rest. Had we proceeded to a framework in which in the ground state the gas of quasiparticles is at rest, we would, conversely, observe the motion of the condensate with velocity $\boldsymbol{u}$. Since this relative motion is stationary motion in the state of statistical equilibrium in the absence of external fields, we see that it is not accompanied by friction and is, therefore, the phenomenon of superfluidity.[c]

We have already seen that at small momenta the energy of the quasiparticle is approximately equal to $c|\boldsymbol{f}|$ where c is the velocity of sound. Therefore, at small momenta the quasiparticle is nothing but a phonon. With increasing momentum when the kinetic energy $T(\boldsymbol{f})$ becomes large in comparison with the coupling energy of the molecule, the energy of the quasiparticle transforms continuously into the individual energy $T(\boldsymbol{f})$ of the molecule.

Thus, it is not possible to speak of the subdivision of the quasiparticles into two different sorts, phonons and rotons.

Now we consider the distribution of momenta in a system of molecules for the state of statistical equilibrium. We introduce a function $w(\boldsymbol{f})$ by requiring that $Nw(\boldsymbol{f})\,d\boldsymbol{f}$ is the average number of particles with momenta from the elementary volume $d\boldsymbol{f}$ in the momentum space. This function is, obviously, normalized in the sense that

$$\int w(\boldsymbol{f})\,d\boldsymbol{f} = 1. \tag{19}$$

Now let $F(\boldsymbol{f})$ be an arbitrary continuous function of momentum. Then the average value of the dynamical variable

$$\sum_{(i\leq i\leq N)} F(\boldsymbol{p}_i)$$

is

$$N\int F(\boldsymbol{f})w(\boldsymbol{f})\,d\boldsymbol{f}. \tag{20}$$

---

[c]If we take a framework in which the condensate moves with velocity $\boldsymbol{u}$, it is not difficult to see that the energy of the considered system of molecules is

$$H = \sum_f \{E(\boldsymbol{f}) - \boldsymbol{f}\boldsymbol{u}\}n_f + H_0 + M\boldsymbol{u}^2/2.$$

In view of Landau's speculations from the above mentioned paper this expression makes the property of superfluidity obvious. Indeed the appearance of elementary excitations is energetically unfavorable, since it is accomplished by increase of energy.

On the other hand, this average is equal to

$$\sum_f F(\boldsymbol{f})\overline{N}_f = \frac{V}{(2\pi\hbar)^3}\int F(\boldsymbol{f})\overline{N}_f\,d\boldsymbol{f}. \tag{21}$$

Thus, if we compare Equations (20) and (21), we find

$$w(\boldsymbol{f}) = \frac{\nu}{(2\pi\hbar)^3}\overline{N}_f = \frac{\nu}{(2\pi\hbar)^3}\overline{b_f^\dagger b_f}.$$

Then expressing the operators b_f and $b_f^\dagger$ in terms of ξ_f and $\xi_f^\dagger$ we obtain

$$w(\boldsymbol{f}) = \frac{V}{(2\pi\hbar)^3}\left(1 - |L_f|^2\right)^{-1}\overline{(\xi_f^\dagger + L_f^*\xi_{-f})(\xi_f + L_f^*\xi_{-f}^\dagger)}$$

$$= \frac{\nu}{(2\pi\hbar)^3}\frac{\bar{n}_f + |L_f|^2(\bar{n}_{-f} + 1)}{1 - |L_f|^2}, \tag{22}$$

where in view of Equation (17)

$$\bar{n}_f = \left\{\exp\left(\frac{E(\boldsymbol{f}) - \boldsymbol{f}\boldsymbol{u}}{\theta}\right) - 1\right\}^{-1}. \tag{23}$$

The obtained expression (22) for the probability density is valid only for $\boldsymbol{f} \neq 0$. Therefore due to the normalization condition (19), the general expression for the probability density for the molecule momenta is

$$w(\boldsymbol{f}) = C\delta(\boldsymbol{f}) + \frac{\nu}{(2\pi\hbar)^3}\frac{\bar{n}_f + |L_f|^2(\bar{n}_{-f} + 1)}{1 - |L_f|^2} \tag{24}$$

where $C\delta(\boldsymbol{f})$ is the Dirac delta-function and C is the number determined by the equality

$$C = 1 - \frac{\nu}{(2\pi\hbar)^3}\int\frac{\bar{n}_f + |L_f|^2(\bar{n}_{-f} + 1)}{1 - |L_f|^2}\,d\boldsymbol{f}. \tag{25}$$

The value of C is, obviously, equal to N_0/N, since CN is the average number of molecules with zero momentum.

In the formulae obtained, on the basis of Equation (11) we have

$$\frac{|L_f|^2}{1 - |L_f|^2} = \frac{\left(\dfrac{N_0}{V}\nu(\boldsymbol{f})\right)^2}{2E(\boldsymbol{f})\left\{E(\boldsymbol{f}) + T(\boldsymbol{f}) + \dfrac{N_0}{V}\nu(\boldsymbol{f})\right\}},$$

$$\frac{1}{1 - |L_f|^2} = \frac{E(\boldsymbol{f}) + T(\boldsymbol{f}) + \frac{N_0}{V}\nu(\boldsymbol{f})}{2E(\boldsymbol{f})}, \tag{26}$$

and consequently at zero temperature the momentum probability density is

$$w(\boldsymbol{f}) = C\delta(\boldsymbol{f}) + \frac{\nu}{(2\pi\hbar)^3} \frac{\left(\frac{N_0}{V}\nu(\boldsymbol{f})\right)^2}{2E(\boldsymbol{f})\left\{E(\boldsymbol{f}) + T(\boldsymbol{f}) + \frac{N_0}{V}\nu(\boldsymbol{f})\right\}}$$

where

$$1 - C = \frac{\nu}{(2\pi\hbar)^3} \int \frac{\left(\frac{N_0}{V}\nu(\boldsymbol{f})\right)^2}{2E(\boldsymbol{f})\left\{E(\boldsymbol{f}) + T(\boldsymbol{f}) + \frac{N_0}{V}\nu(\boldsymbol{f})\right\}}\, d\boldsymbol{f}. \tag{27}$$

Thus, at zero temperature as well, only a part of the molecules possess exactly zero momentum. The rest are continuously distributed over the whole spectrum of momenta.

In view of above remarks, the approximate method used is applicable only while the following inequality holds

$$(N - N_0) = 1 - C \ll 1.$$

Therefore, to ensure the smallness of the integral (27), the interaction between the molecules should be sufficiently small.

Now let us determine how the smallness of the interaction should be understood. We put

$$\Phi(r) = \Phi_m\, F\!\left(\frac{r}{r_0}\right)$$

where $F(\rho)$ is a function such that it and its derivatives take values of the order of unity for $\rho \sim 1$ and vanish rapidly for $\rho \to \infty$. Then we have

$$\nu(\boldsymbol{f}) = \Phi_m\, r_0^3 \omega\!\left(\frac{|\boldsymbol{f}|r_0}{\hbar}\right)$$

where $\omega(x)$ is a function taking values of ~ 1 for $x \sim 1$ and vanishing rapidly as $x \to \infty$.

If in equation (27) we proceed to dimensionless variables and reduce the three-dimensional integral to a one-dimensional one, we find

$$\frac{N - N_0}{N} = \frac{\nu}{r_0^3}\, \eta\, \frac{1}{(2\pi)^2} \int\limits_0^\infty \frac{\eta\omega^2(x)x\,dx}{\alpha(x)\{x\alpha(x) + x^2 + \eta\omega(x)\}} \tag{28}$$

where

$$\alpha(x) = \left(x^2 + 2\eta\omega(x)\right)^{1/2},$$

$$\eta = \frac{r_0^3 N_0}{V}\, \Phi_m\left(\frac{\hbar^2}{2mr_0^2}\right)^{-1} \sim \frac{r_0^3}{\nu}\, \Phi_m\left(\frac{\hbar^2}{2mr_0^2}\right)^{-1}.$$

It is not difficult to see that for small η the integral in the right-hand side of Equation (28) is of the order of $\sqrt{\eta}$ and the condition for applicability of the method considered is given by the inequality

$$\eta \ll 1, \quad (\nu/r_0^3)\eta^{3/2} \ll 1,$$

that is

$$\frac{r_0^3}{\nu}\, \Phi_m \ll \frac{\hbar^2}{2mr_0^2}, \quad \left(\frac{r_0^3}{\nu}\right)^{1/3}\Phi_m \ll \frac{\hbar^2}{2mr_0^2}. \tag{29}$$

For temperatures different from zero, similar consideration of the general formula (24) will result in an auxiliary condition for the weakness of the interaction requiring that the temperature should be low compared with the temperature of the λ-point.

We see that the condition for the smallness of interaction in the form of expression (29) automatically excludes the possibility for taking into account short-range repulsive forces, since for that it would be necessary to consider intensive growth of the function $\Phi(r)$ as $r \to \infty$. However, it is not difficult to modify the results obtained here in order to extend them over the more realistic case of a gas of low density with molecules of a finite radius.

Indeed, in our final formulae, the potential $\Phi(r)$ enters only the expression

$$\nu(\boldsymbol{f}) = \int \Phi(|\boldsymbol{q}|) \exp\left\{-i\frac{\boldsymbol{fq}}{\hbar}\right\} d\boldsymbol{q} \tag{30}$$

proportional to the amplitude of the Born probability of pair collision. Since at low density the interaction between molecules is mainly realized via pair collisions, expression (30) should be replaced[d] by an expression proportional to the amplitude of the exact probability of pair collision. In other words, we should put

$$\nu(\boldsymbol{f}) = \int \Phi(|\boldsymbol{f}|)\varphi(\boldsymbol{q}, \boldsymbol{f}) \, d\boldsymbol{q}, \tag{31}$$

where $\varphi(\boldsymbol{q}, \boldsymbol{f})$ is the solution of the Schrödinger equation for the relative motion of a pair of molecules

$$-\frac{\hbar^2}{m} \Delta\varphi + \{\Phi(|\boldsymbol{q}|) - E\}\varphi = 0$$

which behaves like $\exp\{-i\boldsymbol{f}\boldsymbol{q}/\hbar\}$ at infinity.

The replacement of Equation (30) by Equation (31) in the expression for $E(\boldsymbol{f})$ will result in formulae valid for gases of low density. Therefore, the condition for existence of superfluidity $\nu(0) > 0$ will be written, for instance, in the form

$$\int \Phi(|\boldsymbol{f}|)\varphi(|\boldsymbol{q}|) \, d\boldsymbol{q} > 0 \tag{32}$$

where $\varphi(|\boldsymbol{q}|)$ is the spherically symmetric solution of the equation

$$-\frac{\hbar^2}{m} \Delta\varphi + \Phi(|\boldsymbol{q}|)\varphi = 0$$

approaching unity at infinity.

In order to connect the inequality (32) with the condition for the thermodynamic stability, as this has been done above, we calculate the leading term in the expansion of the gas free energy at zero temperature in powers of density. Since at zero temperature the free energy coincides with the internal energy, we have the following expression for the energy per molecule

$$\mathscr{E} = \bar{T} + \frac{1}{2\nu} \int \Phi(|\boldsymbol{q}|)g(|\boldsymbol{q}|) \, d\boldsymbol{q} \tag{33}$$

where $\bar{T}$ is the average kinetic energy of a molecule, $g(r)$ is the molecular distribution function, approaching unity as $r \to \infty$.

On the other hand, but the virial theorem, the pressure p can be determined from the formula

$$p\nu = \frac{2}{3}\bar{T} - \frac{1}{6} \int \Phi'(|\boldsymbol{q}|) \, |\boldsymbol{q}| \, g(|\boldsymbol{q}|) \, d\boldsymbol{q}. \tag{34}$$

---

[d]I was kindly informed of this important fact by L.D. Landau.

Now note that the leading term in the expression of the molecular distribution function at zero temperature in powers of the density is, obviously, equal to $\varphi^2(|\boldsymbol{q}|)$. Therefore, omitting in Equations (33) and (34) the terms proportional to the square root of the density, we find

$$\mathscr{E} = \bar{T} = \frac{1}{2\nu} \int \Phi(|\boldsymbol{q}|)\varphi^2(|\boldsymbol{q}|) \, d\boldsymbol{q},$$

$$p\nu = \frac{2}{3}\bar{T} - \frac{1}{6} \int \Phi'(|\boldsymbol{q}|)\varphi^2(|\boldsymbol{q}|) \, d\boldsymbol{q}.$$

Then, taking into account that

$$p\nu = -\nu \frac{\partial \mathscr{E}}{\partial \nu}$$

we obtain the equation for determining the leading term in the expression for $\bar{T}$. After calculations we find

$$\mathscr{E} = \frac{1}{2\nu} \int \Phi(|\boldsymbol{q}|)\varphi(|\boldsymbol{q}|) \, d\boldsymbol{q} = \frac{\nu(0)}{2\nu}, \qquad p = \frac{\nu(0)}{2\nu^2}.$$

Thus, in the considered case of a gas of low density, the condition for the existence of superfluidity (32) is equivalent to the usual condition for thermodynamic stability of a gas at zero temperature, i.e.,

$$\frac{\partial p}{\partial \nu} < 0.$$

As an example, let us consider a model where the molecules are ideal rigid spheres of diameter r_0, so that

$$\Phi(r) = +\infty, \quad r < r_0,$$

$$\Phi(r) = +o, \quad r > r_0.$$

After simple calculations we find

$$\nu(0) = 4\pi \frac{\hbar^2 r_0}{m}.$$

If we suppose that there is weak attraction between spheres and put

$$\Phi(r) = +\infty, \quad r < r_0,$$

$$\Phi(r) = \varepsilon\Phi_0(r) < 0, \quad r > r_0,$$

where ε is a small parameter, we obtain, up to the terms of the order of ε^2,

$$\nu(0) = 4\pi \frac{\hbar^2 r_0}{m} + 4\pi \int_{r_0}^{\infty} r^2 \Phi(r)\, dr.$$

Thus, in this model the appearance of superfluidity is due to the relationship between the repulsion and attraction forces. The repulsion forces promote superfluidity, while the attraction forces hinder it.

Finally, it should be emphasized that it is apparently possible to proceed to consideration of realistic fluids in the frame of the theory developed here, if such semi-phenomenological concepts as the free energy for weakly nonequilibrium states is used.

CHAPTER 2

QUASI-AVERAGES IN PROBLEMS OF STATISTICAL MECHANICS

Part A.   QUASI-AVERAGES

1.   Green's Functions, Defined with Regular Averages; Additive Conservation Laws and Selection Rules

In modern statistical mechanics all newly developed methods involve obtaining an understanding and use of the methods of the quantum field theory.

The introduction of Green's functions is very fruitful, since, for example, with their help it is possible to generalize diagrammatic perturbation methods in statistical mechanics and to perform partial summation of expressions. We shall, first of all, discuss the definition of Green's functions. As is known these functions are expressed as linear forms in the average values

$$\langle \dots \Psi^\dagger(t_j, x_j) \dots \Psi(t_s, t_s) \dots \rangle \tag{1.1}$$

with coefficients made up of products of the step function $\theta(t_i - t_k)$. We will use the following notation: $x = (\vec{r}, \sigma)$ represents all the space coordinates $(\vec{r})$ and the series of discrete indices (σ), characterizing the spin of the particle, their type, etc; $\Psi(t, x)$, $\Psi^\dagger(t, x)$ – represent field operators in the Heisenberg picture. These operators can be expressed in "quasi-discrete" summation

$$\Psi(t, x) = \frac{1}{\sqrt{V}} \sum_{(k)} a_{k\sigma}(t)\, e^{i(\vec{k} \cdot \vec{r})};$$

$$\Psi^\dagger(t, x) = \frac{1}{\sqrt{V}} \sum_{(k)} a_{k\sigma}^\dagger(t)\, e^{-i(\vec{k} \cdot \vec{r})} \tag{1.2}$$

where $a_{k\sigma}^\dagger$ is the creation operator and $a_{k\sigma}$ is the destruction operator, which satisfy the usual Bose or Fermi commutation relations. In these sums $k^\alpha = 2\pi n_\alpha/L$; $\alpha = 1, 2, 3$; n_α – integer; $V = L^3$ – the volume of the system.

The definition of the Green's functions is independent of the nature of the system. They are linear forms of the averages of the type (1.1). The question of defining the Green's function reduces to the definition of expression (1.1). Usually they are defined as averages with respect to the Gibbs Grand Canonical ensemble, in accordance with which there always appears the usual statistical mechanical limit $V \to \infty$. That is,

$$\langle \ldots \Psi^\dagger(t_j, x_j) \ldots \Psi(t_s, t_s) \ldots \rangle$$

$$= \lim_{V \to \infty} \frac{\mathrm{Tr}\left(\{\ldots \Psi^\dagger(t_j, x_j) \ldots \Psi(t_s, t_s) \ldots\} e^{-\frac{H}{\theta}}\right)}{\mathrm{Tr}\, e^{-\frac{H}{\theta}}} \qquad \cdot \quad (1.3)$$

where H is the total Hamiltonian of the system, and includes terms with chemical potential due to the conservation of the number of particles.

Let us agree to call the average values (1.1), which are defined by the relationship (1.3), the *regular averages*, and the corresponding Green's functions, the *Green's functions, constructed from regular averages*.

Let us now draw our attention to the well known fact that the additive laws of conservation lead to selection rules for regular averages and also for Green's functions.

For example we have the conservation law for the total number of particles

$$N = \sum_{(k,\sigma)} a_{k\sigma}^\dagger a_{k\sigma} = \sum_{(\sigma)} \int \Psi^\dagger \Psi \, d\vec{r},$$

so that $[H, N] = 0$, where H is the total Hamiltonian of the system (including the term μN, where μ is the chemical potential. Whenever $H = U^\dagger H U$, where $U = e^{i\varphi N}$ and φ is an arbitrary real number, then the Hamiltonian is invariant under the gradient transformation of the 1st kind:

$$a_{k\sigma} \to U^\dagger a_{k\sigma} U = e^{i\varphi} a_{k\sigma}.$$

Therefore we have

$$\mathrm{Tr}\left(\{\ldots a_{k\sigma}^{\dagger}(t)\ldots a_{k'\sigma'}(t')\ldots\}e^{-\frac{H}{\theta}}\right)$$

$$= \mathrm{Tr}\left(\{\ldots a_{k\sigma}^{\dagger}(t)\ldots a_{k'\sigma'}(t')\ldots\}U^{\dagger}e^{-\frac{H}{\theta}}U\right)$$

$$= \mathrm{Tr}\left(U\{\ldots a_{k\sigma}^{\dagger}(t)\ldots a_{k'\sigma'}(t')\ldots\}U^{\dagger}e^{-\frac{H}{\theta}}\right)$$

$$= e^{-i\varphi n}\,\mathrm{Tr}\left(\{\ldots a_{k\sigma}^{\dagger}(t)\ldots a_{k'\sigma'}(t')\ldots\}e^{-\frac{H}{\theta}}\right)$$

where n is the difference between the numbers of a and $a^{\dagger}$ operators in the products $\ldots a_{k\sigma}^{\dagger}(t)\ldots a_{k'\sigma'}(t')\ldots$. From this, on the basis of the definition (1.3) we find

$$(1 - e^{-i\varphi n})\langle\ldots a_{k\sigma}^{\dagger}(t)\ldots a_{k'\sigma'}(t')\ldots\rangle = 0$$

and thus

$$\langle\ldots a_{k\sigma}^{\dagger}(t)\ldots a_{k'\sigma'}(t')\ldots\rangle = 0$$

if in the given product the number of creation operators is not equal to the number of destruction operators.

Since the Green's functions are expressed as linear forms of regular averages these same selection rules also hold for Green's functions. For example,

$$\langle T(\ldots a_{k\sigma}^{\dagger}(t)\ldots a_{k'\sigma'}(t')\ldots)\rangle = 0, \quad \text{if } n \neq 0.$$

Let us examine the selection rules derived from conservation of total momentum. The total momentum operator is

$$\vec{\mathscr{P}} = \sum_{(k,\sigma)} a_{k\sigma}^{\dagger}a_{k\sigma}.$$

The law of conservation of total momentum gives

$$\mathrm{Tr}\left\{[\vec{\mathscr{P}}, \mathscr{U}]\,e^{-\frac{H}{\theta}}\right\} = \mathrm{Tr}\left\{\mathscr{U}\left[e^{-\frac{H}{\theta}}, \vec{\mathscr{P}}\right]\right\} = 0.$$

Let us take

$$\mathscr{U} = \ldots \Psi^{\dagger}(t_j, \vec{r}_j + \vec{\xi}, \sigma_j)\ldots\Psi(t_s, \vec{r}_s + \vec{\xi}, \sigma_s)\ldots$$

and note that

$$\sum_\alpha \frac{1}{i} \frac{\partial}{\partial \vec{r}_\alpha} \langle \mathscr{U} \rangle = \langle [\vec{\mathscr{P}}, \mathscr{U}] \rangle = \frac{\mathrm{Tr}\left\{ [\vec{\mathscr{P}}, \mathscr{U}] e^{-\frac{H}{\theta}} \right\}}{\mathrm{Tr}\, e^{-\frac{H}{\theta}}} = 0.$$

Therefore, the average (1.1) do not change under the translation

$$\vec{r}_j \to \vec{r}_j + \vec{\xi}$$

where $\vec{\xi}$ is an arbitrary vector. Saying this differently the regular averages (1.1) must be spatially homogeneous. Let us now make use of the momentum representation (1.2). We get

$$\sum_\alpha \frac{\partial}{\partial \vec{r}_\alpha} \langle \dots \Psi^\dagger(t_j, \vec{r}_j, \sigma_j) \dots \Psi(t_s, \vec{r}_s, \sigma_s) \dots \rangle$$

$$= \frac{1}{V^{n/2}} \sum_{(\dots k_\nu, \sigma_\nu \dots)} \left(\sum_\alpha \vec{k}_\alpha \right) \langle \dots a^\dagger_{-k_j, \sigma_j}(t_j) \dots a_{k_s, \sigma_s}(t_s) \dots \rangle e^{i(\vec{k}_1 \vec{r}_1 + \dots + \vec{k}_n \vec{r}_n)} = 0$$

using translational invariance. This leads to the selection rule

$$\langle \dots a^\dagger_{-k_j, \sigma_j}(t_j) \dots a_{k_s, \sigma_s}(t_s) \dots \rangle = 0, \quad \text{if } \vec{k}_1 + \dots + \vec{k}_n \neq 0.$$

Such relationships are also satisfied for the Green's functions. We have for example

$$\langle T(\dots a^\dagger_{-k_j, \sigma_j}(t_j) \dots a_{k_s, \sigma_s}(t_s) \dots) \rangle = 0, \quad \text{if } \vec{k}_1 + \dots + \vec{k}_n \neq 0.$$

An analogous situation arises when the laws of conservation of total spin and other additive dynamical variables are taken into account.

The selection rules become more descriptive if one introduces a diagrammatic presentation of perturbation theory. For the purpose of formulating perturbation theory the total Hamiltonian is divided into two parts, H_0 and H_1,

$$H = H_0 + H_1$$

where the expansion is carried out "in powers of H_1". As a rule, a Hamiltonian is picked for H_0 which corresponds to the "ideal gas" without interaction, and all interactions are included in H_1. Note that by separating H_0 it is guaranteed that the above mentioned conservation laws for additive variables are also satisfied for the "dynamical system in zeroth approximation" characterized by the Hamiltonian H_0. In such a separation of H_0, exactly the same selection rules are obtained for the hierarchy of the exact Green's functions as for the hierarchy of of the zeroth approximation Green's functions. Let us consider the diagrammatic presentation using either the general Feynman diagrams for the case of zero temperature or the corresponding diagrams of Matsubara, C. Bloch for $\theta > 0$. In both cases the diagrams are characterized by lines and loops. In each one of the loops the conservation laws are satisfied, and all allowed lines likewise satisfy these laws.

Assume for example, that we have a system of particles with spin $\sigma = \pm 1/2$ with the total number of particles conserved. Also, assume that there exist total momentum conservation laws a conservation law for the total z component of the spin. Then all "contractions" i.e. all the zeroth approximation Green's functions of the type

$$\overline{a_{p\sigma}^{\dagger} a_{p'\sigma'}^{\dagger}}; \quad \overline{a_{p\sigma} a_{p'\sigma'}}; \quad \overline{a_{p\sigma} a_{p'\sigma'}^{\dagger}}; \quad \overline{a_{p\sigma}^{\dagger} a_{p'\sigma'}}$$

when $\vec{p} \neq \vec{p}'$ or $\sigma \neq \sigma'$, are exactly equal to zero.

The only allowed lines of the particles under consideration will be the lines which correspond to the contractions

$$\overline{a_{p\sigma} a_{p\sigma}^{\dagger}}; \quad \overline{a_{p\sigma}^{\dagger} a_{p\sigma}}$$

conserving $\vec{p}$ and σ.

The same situation exists for the "wide", or "summed" lines correspond to the exact Green's functions. The exact Green's functions of type

$$\ll a_{p\sigma}^{\dagger}, a_{p'\sigma'}^{\dagger} \gg, \quad \ll a_{p\sigma}, a_{p'\sigma'} \gg, \quad \ll a_{p\sigma}, a_{p'\sigma'}^{\dagger} \gg, \quad \ll a_{p\sigma}^{\dagger}, a_{p'\sigma'} \gg$$

when $\vec{p} \neq \vec{p}'$ or $\sigma \neq \sigma'$, are all equal to zero. The only allowed lines will be the "wide" lines characterized by the Green's functions $\ll a_{p\sigma}, a_{p\sigma}^{\dagger} \gg$, $\ll a_{p\sigma}^{\dagger}, a_{p\sigma} \gg$ conserving $\vec{p}$ and σ.

As demonstrated, the selection rules considerably simplify the topological structure of the diagrams and the actual calculations.

2. Degeneracy of the Statistical Equilibrium States; Introduction of Quasi-averages

In applying the diagram technique one cannot forget that it is only a handy presentation of the ordinary perturbation theory, and that one encounters similar difficulties and sometimes complex problems about the convergence of the resultant expansion.

At the present time the convergence can be proven for only a number of simplest models. For more realistic problems one can only assume the presence of a certain correspondence between the real solutions and the resultant formal expansions. Such formal expansions will be used, in particular, for the formation of approximate solutions. A very effective procedure here is the partial summation of the (in some sense) "main" terms, which is easily performed with the help of the diagram technique. If the perturbation is "weak enough" and is characterized by a small parameter, the approximate solution take the form of asymptotic expressions. When the smallness parameter of the perturbation has a zero value, the "correctness" of the partial summations can be ascertained. Although the mathematics of these procedures is not fully justified, nevertheless, in many important problems one can obtain physically correct results, not only for the asymptotic formulas, but also for results pertaining to the qualitative properties of exact solutions. However, in a number of cases, for example in the theory of super conductivity and in the theory crystalline state, the ordinary diagram technique does not lead to physically correct results. In our opinion it would be not enough to limit the expansion by referring to such formal reasons as the lack of convergence, complexity of the analytical structure pertaining to the small parameter, etc. It follows that we have to look for physically basic, constructive solutions of the newly appearing problems.

Let us now turn our attention to the well known quantum mechanical problems of degeneracy. When investigating the problem of finding eigenfunctions in quantum mechanics one discovers that perturbation theory in its regular form, developed for non-degenerate cases, cannot be directly applied to problems having degeneracy. It is essential to modify it first. In problems of statistical mechanics we always have a case of degeneracy due to presence of the additive conservation laws.

However, at first glance it might seem that the degeneracy is not important and can in practice be neglected. Actually in these quantum mechanical problems linear combination of different eigenfunctions can correspond to one eigenvalue of the energy. The eigenfunction in this case contains undetermined constants.

In statistical mechanics the average value of any dynamical variable $\mathscr{U}$

is always unambiguously determined:

$$\langle \mathscr{U} \rangle = \frac{\mathrm{Tr}\left(\mathscr{U}\, e^{-\frac{H}{\theta}}\right)}{\mathrm{Tr}\, e^{-\frac{H}{\theta}}}.$$

It follows that the *Green's functions constructed from regular averages must likewise be unambiguously determined*. From this it might seem that when studying statistical equilibrium, for example, with the help of the diagram technique, one need not take into account the presence of degeneracy. However, in reality, the situation is not so simple. In order to form an intuitive feeling for the nature of the problem here let us look at the case of an ideal isotropic ferromagnetic.

For the definition let us assume a dynamical system, characterized by the Heisenberg model Hamiltonian

$$H = -\frac{1}{2} \sum_{(f_1, f_2)} I(f_1 - f_2)(\vec{S}_{f_1} \cdot \vec{S}_{f_2}) \tag{2.1}$$

where (f) represents space points, corresponding to the sites of the crystalline lattice (occupying volume V), $\vec{S}_f$ is the spin vector with the usual commutation rules, $I(f_1 - f_2)$ is a non-negative number. For example we may assume that $I(f_1 - f_2)$ is greater that zero when the sites f_1, f_2 are "nearest neighbors".

For this dynamical system each of the components of the total spin vector $\vec{S} = \sum_{(f)} \vec{S}_f$ is an integral of the motion. We also have

$$S_x S_y - S_y S_x = iS_z,$$
$$S_y S_z - S_z S_y = iS_x,$$
$$S_z S_x - S_x S_z = iS_y.$$

From this it follows that

$$i\,\mathrm{Tr}\left(S_z e^{-\frac{H}{\theta}}\right) = i\,\mathrm{Tr}\left((S_x S_y - S_y S_x)e^{-\frac{H}{\theta}}\right).$$

But, in so far as S_x commutes with H, we obtain

$$\mathrm{Tr}\left(S_y S_x e^{-\frac{H}{\theta}}\right) = \mathrm{Tr}\left(S_x e^{-\frac{H}{\theta}} S_y\right) = \mathrm{Tr}\left(S_x S_y e^{-\frac{H}{\theta}}\right)$$

and thus

$$\mathrm{Tr}\left(S_z \mathrm{e}^{-\frac{H}{\theta}}\right) = 0.$$

Similarly we find

$$\mathrm{Tr}\left(S_x \mathrm{e}^{-\frac{H}{\theta}}\right) = 0, \quad \mathrm{Tr}\left(S_y \mathrm{e}^{-\frac{H}{\theta}}\right) = 0.$$

Introducing the magnetization vector,

$$\vec{\mathscr{M}} = \mu \frac{1}{V} \sum_{(f)} \vec{S}_f = \mu \frac{1}{V} \vec{S}$$

we have

$$\mathrm{Tr}\left(\vec{\mathscr{M}} \mathrm{e}^{-\frac{H}{\theta}}\right) = 0$$

and therefore:

$$\langle \vec{\mathscr{M}} \rangle = \lim_{V \to \infty} \frac{\mathrm{Tr}\left(\vec{\mathscr{M}} \mathrm{e}^{-\frac{H}{\theta}}\right)}{\mathrm{Tr}\, \mathrm{e}^{-\frac{H}{\theta}}} = 0. \tag{2.2}$$

The regular average of the vector $\vec{\mathscr{M}}$ is equal to zero. This corresponds to the isotropy of this dynamical system with respect to the spin rotation group.

Let us note that expression (2.2) is correct for all temperatures θ, and *in particular, for temperatures below the Curie point.*

Let us now investigate specifically this last case. As is known, when the magnitude of the magnetization vector is different form zero its direction can be taken arbitrary. In this sense the *statistical equilibrium state in this system is degenerate.*

Now let us include an external magnetic field $B\vec{e}$ ($B > 0$, $\vec{e}^2 = 1$) changing the Hamiltonian (2.1) to the Hamiltonian

$$H_{B\vec{e}} \to H + B(\vec{e} \cdot \vec{\mathscr{M}})V. \tag{2.3}$$

Then, taking into account the characteristic property of isotropic ferromagnetic when the temperature is below the Curie point, we can see

$$\langle \vec{\mathscr{M}} \rangle = \vec{e} M_B.$$

Furthermore, M_B will approach the limit, different from zero when the intensity B of the external magnetic field approaches zero. From the formal point of view we have here an "instability" of regular averages. When the term $B(\vec{e} \cdot \vec{\mathscr{M}})V$ is added to the Hamiltonian (2.1) with an infinitesimally small[e] B the average $\langle \vec{\mathscr{M}} \rangle$ obtains a limit which is different from zero, for example,

$$\vec{e}m; \quad \text{where,} \quad m = \lim_{B \to 0} M_B.$$

Let us now introduce the concept of "quasi-averages" for a dynamical system with the Hamiltonian (2.1).

Take any dynamical variable A, which is a linear combination of the products

$$S_{f_1}^{\alpha_1}(t_1) \dots S_{f_r}^{\alpha_r}(t_r)$$

and define the quasi-average $\prec A \succ$ of this variable

$$\prec A \succ = \lim_{B \to 0} \langle A \rangle_{B\vec{e}}$$

where $\langle A \rangle_{B\vec{e}}$ is the regular average of A with the Hamiltonian $H_{B\vec{e}}$. In this manner the presence of degeneracy in the problem is reflected in the dependence of the quasi-averages on the arbitrary direction $\vec{e}$.

It is not difficult to see that

$$\langle A \rangle = \int \prec A \succ d\vec{e}. \tag{2.4}$$

Now, it is understood that for the description of the case under consideration, (the degenerate statistical equilibrium state), the quasi-averages are more convenient, more "physical" than the regular averages. These latter express the same quasi-averages, only they are averaged in all directions of $\vec{e}$.

Further note that the regular averages

$$\langle S_{f_1}^{\alpha_1}(t_1) \dots S_{f_r}^{\alpha_r}(t_r) \rangle$$

must be *invariant* with respect to spin rotation group. The corresponding quasi-averages

$$\prec S_{f_1}^{\alpha_1}(t_1) \dots S_{f_r}^{\alpha_r}(t_r) \succ \tag{2.5}$$

will posses only the property of *covariance*; when there is a rotation of the spin a similar rotation must be made on the vector $\vec{e}$ so that the expression (2.5) does not change.

---

[e]When we talk about an infinitesimally small B we always mean that *first* the statistical mechanical limit $V \to \infty$ is carried out *and then* B approaches zero.

In such a way the quasi-averages will not have the selection rules which for regular averages depended upon their invariance with respect to the spin rotation group. The arbitrary direction $\vec{e}$, which is the direction of the magnetization vector, characterizes the degeneracy of the statistical equilibrium state under consideration. In order to remove the degeneracy the direction $\vec{e}$ must be fixed. We will pick the λ-axis for this direction. Then all the quasi-averages will become definite numbers. Exactly the same type of averages are encountered in the theory of ferromagnetism.

In other words, we can remove the degeneracy of the statistical equilibrium state with respect to spin rotation group by including in the Hamiltonian H the additional invariant member BM_zV with an infinitely small B.

Let us now look at another example of degeneracy, this time turning to the theory of the crystalline state. Consider a dynamical system with spinless particles having a binary interaction characterized by a Hamiltonian of the ordinary type

$$H = \sum_{(p)} \left(\frac{p^2}{2m} - \mu \right) a_p^\dagger a_p + \frac{1}{2V} \sum_{(p_1, p_2, p_1', p_2')} a_{p_1}^\dagger a_{p_2}^\dagger a_{p_1'} a_{p_2'}$$
$$\times \, \delta(\vec{p}_1 + \vec{p}_2 - \vec{p}_1' - \vec{p}_2') \nu(|\vec{p}_1 - \vec{p}_1'|) p \qquad (2.6)$$

in which $\delta(\vec{p})$ is the discrete δ function, $\nu(p)$ is the Fourier transform of the interaction potential energy $\phi(r)$ of a pair of particles. Assume that this type of interaction is such that our dynamical system must be in a crystalline state when the temperature is low enough $\theta < \theta_{cr}$. Consider the observed particle density $\rho(\vec{r})$, which evidently, must be a periodic function of $\vec{r}$ with the period of the crystal lattice. It would be natural to consider that $\rho(\vec{r})$ is equal to the regular average of the operator density $\Psi^\dagger(\vec{r})\Psi(\vec{r})$

$$\rho(\vec{r}) = \langle \Psi^\dagger(\vec{r})\Psi(\vec{r}) \rangle = \frac{1}{V} \sum_{(q)} \left\{ \sum_{(k)} \langle a_k^\dagger a_{k+q} \rangle \right\} e^{i(\vec{q} \cdot \vec{r})}.$$

This, however, is not true.

Actually, in the present case the total momentum

$$\vec{p} = \sum_{(k)} \vec{k} a_k^\dagger a_k$$

is conserved and thus, as mentioned in 1. we have the selection rule

$$\langle a_k^\dagger a_{k+q} \rangle = 0, \quad \text{if} \quad q \neq 0,$$

from which follows

$$\langle \Psi^\dagger(\vec{r})\Psi(\vec{r})\rangle = \frac{1}{V}\sum_{(k)}\langle a_k^\dagger a_{k+q}\rangle = \frac{N}{V} = \text{const.}$$

In such a way the value of the regular average of the operator density cannot be equal to the periodic function $\rho(\vec{r})$. It is clear that this situation is brought about by the conservation of momentum in the statistical equilibrium state considered. Actually the crystal lattice, as a whole, can be arbitrary placed in space. In particular, our Hamiltonian possesses translational invariance, and thus the lattice can be arbitrary translated.

No special position of the crystal lattice is preferred in space, and when we take a regular average we thereby average over all possible positions of this lattice. In order to remove the degeneracy and introduce the quasi-average we must include in the Hamiltonian the term

$$\varepsilon \int U(\vec{r})\Psi^\dagger(\vec{r})\Psi(\vec{r})\,d\vec{r}; \quad \varepsilon > 0, \quad \varepsilon \to 0 \tag{2.7}$$

corresponding to the infinitely small external filed $_\varepsilon U(\vec{r})$. We will denote the resulting Hamiltonian by H_ε. As $U(\vec{r})$ we will take the periodic function of $\vec{r}$ with the periodicity of the lattice in such a way that the external field $_\varepsilon U(\vec{r})$ removes the degeneracy, *thus fixing the position of our crystal in space.*

In as much as we are inherently investigating only the physically stable cases, it is clear that the inclusion of the infinitely small external field can only slightly change the physical properties of the dynamical system under consideration. In as much as the position of the crystal is now fixed in space, taking the regular average of the density operator $\Psi^\dagger(\vec{r})\Psi(\vec{r})$ with the Hamiltonian H_ε (with infinitesimally small ε) actually produces the average for the system with the initial Hamiltonian H, *but without the extra average* over the position of the whole crystal lattice in space.

In this way we will obtain the observed density distribution of particles $\rho(\vec{r})$. Let us formally define quasi-averages by placing

$$\prec \ldots \Psi^\dagger(t_j,\,\vec{r}_j)\ldots\Psi(t_s,\,\vec{r}_s)\ldots\succ = \lim_{\varepsilon\to 0}\langle\ldots\Psi^\dagger(t_j,\,\vec{r}_j)\ldots\Psi(t_s,\,\vec{r}_s)\ldots\rangle_{H_\varepsilon}.$$

Thus, as was just indicated

$$\prec \Psi^\dagger(\vec{r})\Psi(\vec{r})\succ = \rho(\vec{r}).$$

Noting, that

$$\prec \Psi^\dagger(\vec{r})\Psi(\vec{r}) \succ = \frac{1}{V} \sum_{(q)} \left\{ \sum_{(k)} \prec a_k^\dagger a_{k+q} \succ \right\} e^{i(\vec{q}\cdot\vec{r})}$$

we see that the quasi-averages

$$\prec a_k^\dagger a_{k'} \succ; \quad k' \neq k \tag{2.8}$$

cannot *all* be equal zero.

In this way the selection rules, determined by the law of conservation of the total momentum, *are not satisfied for these quasi-averages*.

Now note that the quasi-averages generally depend upon a series of arbitrary parameters, for example, upon an arbitrary vector $\vec{\xi}$. Actually, if we replace the function $U(\vec{r})$ by an equally acceptable function $U(\vec{r}+\vec{\xi})$ then it is not difficult to show that the quasi-average (2.8) becomes

$$\prec a_k^\dagger a_{k'} \succ e^{i(\vec{k}-\vec{r}\,')\vec{\xi}}.$$

The quasi-average become well defined, when we *fix* the function $U(\vec{r})$. Up to this point we have investigated cases involving the degeneracy of the statistical equilibrium state, connected with the law of conservation of the total spin or the total momentum. In both cases the degeneracy can be removed and adequate physical quasi-averages can be introduced by including the appropriate infinitesimally small external field.

Let us now turn to those cases when the degeneracy is connected with the law of conservation of the total number of particles. Let us start with elementary example of condensation of a Bose-Einstein ideal gas. In order to conveniently extract the condensate we shall take the ideal has Hamiltonian in the form

$$H = -\mu a_0^\dagger a_0 + \sum_{|k|>\varepsilon} \left(\frac{k^2}{2m} - \mu \right) a_k^\dagger a_k, \quad \varepsilon > 0. \tag{2.9}$$

Here we shall let ε approach zero *after* taking the limit $V \to \infty$. We shall find that the average number for certain momentum states

$$N_0 = \left\{ \exp\left(-\frac{\mu}{\theta}\right) - 1 \right\}^{-1},$$

$$N_k = \left\{ \exp\left(\frac{1}{\theta}\left(\frac{k^2}{2m} - \mu\right)\right) - 1 \right\}^{-1}, \quad |k| > \varepsilon.$$

will become large. From this it is seen that $\mu < 0$. Expressing the total number of particles by N we obtain

$$\frac{N}{V} = \frac{N_0}{V} + \frac{1}{V} \sum_{|k|>\varepsilon} \left\{ \exp\left(\frac{1}{\theta}\left(\frac{k^2}{2m} - \mu\right)\right) - 1 \right\}^{-1},$$

(2.10)

$$\mu = -\theta \ln\left(1 + \frac{1}{N_0}\right).$$

Let us consider the Bose-Einstein condensation, where $n_0 = \lim\limits_{V\to\infty} \dfrac{N_0}{V}$ is the thermodynamic limit of $\dfrac{N_0}{V}$ and is different from zero. In this case, when taking the limit in the expression (2.10), we shall find

$$n = \lim_{V\to\infty} \frac{N}{V} = n_0 + \frac{1}{(2\pi)^3} \int_{|k|>\varepsilon} \frac{d\vec{k}}{\exp\{k^2/2m\theta\} - 1}.$$

Here, letting the "cut-off momentum" ε approach zero, we shall finally obtain

$$n = \lim_{V\to\infty} \frac{N}{V} = n_0 + \frac{1}{(2\pi)^3} \int \frac{d\vec{k}}{\exp\{k^2/2m\theta\} - 1}.$$

(2.11)

Thus we obtain the condition of the condensation in its usual form,

$$\frac{1}{(2\pi)^3} \int \frac{d\vec{k}}{\exp\{k^2/2m\theta\} - 1} < n.$$

It is not hard to see that the operator $\dfrac{a_0^\dagger a_0}{V}$ is asymptotically equal to the C number:

$$\frac{a_0^\dagger a_0}{V} \sim n_0.$$

(2.12)

Let us consider the amplitudes

$$\frac{a_0^\dagger}{\sqrt{V}}, \quad \frac{a_0}{\sqrt{V}};$$

which commute with all amplitudes a_k, $a_k^\dagger$, $k \neq 0$. Since the commutator

$$\left[\frac{a_0}{\sqrt{V}}, \frac{a_0^\dagger}{\sqrt{V}}\right] = \frac{1}{V}$$

is infinitesimally small $(V \to \infty)$ we can similarly consider the amplitudes as C numbers, while in view of (2.12)

$$\frac{a_0}{\sqrt{V}} \sim \sqrt{n_0}e^{i\alpha}, \quad \frac{a_0^\dagger}{\sqrt{V}} \sim \sqrt{n_0}e^{-i\alpha}. \tag{2.13}$$

The real phase angle α is arbitrary. This is due to the gradient invariance of the 1st type, specified by the law of conservation of the number of particles, and indicates the appearance of a degeneracy.

Let us consider the regular averages

$$\left\langle \frac{a_0^\dagger}{\sqrt{V}} \right\rangle, \quad \left\langle \frac{a_0}{\sqrt{V}} \right\rangle$$

and note that because of the selection rules they are exactly equal to zero. Note also that the regular averages include an additional averaging over the angle α. In order to introduce quasi-averages and to remove degeneracy we shall include the following term in the Hamiltonian H:

$$-\nu(a_0^\dagger e^{i\varphi} + a_0 e^{-i\varphi})\sqrt{V}, \ \nu > 0.$$

Assume

$$H_{\nu,\varphi} = H - \nu(a_0^\dagger e^{i\varphi} + a_0 e^{-i\varphi})\sqrt{V} \tag{2.14}$$

where φ is some fixed angle.

To reduce (2.14) to a diagonal form we have to perform a canonical transformation on the amplitudes a_0, $a_0^\dagger$, while keeping the other amplitudes a_k, $a_k^\dagger$ fixed,

$$a_0 = -\frac{\nu}{\mu}e^{i\varphi}\sqrt{V} + a_0'$$

$$a_0^\dagger = -\frac{\nu}{\mu}e^{-i\varphi}\sqrt{V} + a_0^{\dagger\prime} \tag{2.15}$$

we find

$$H = -\mu a_0^{\dagger\prime}a_0' + \sum\left(\frac{k^2}{2m} - \mu\right)a_k^\dagger a_k + \frac{\nu^2}{\mu}V. \tag{2.16}$$

Let us now assume that:

$$\mu = -\frac{\nu}{\sqrt{n_0}}. \tag{2.17}$$

Then we have

$$\langle a_0' \rangle_{\nu, \varphi} = 0, \quad \langle a_0^{\dagger\prime} \rangle_{\nu, \varphi} = 0,$$

$$\langle a_0^{\dagger\prime} a_0' \rangle_{\nu, \varphi} = \left\{ \exp\left(\frac{\nu}{\theta\sqrt{n_0}} \right) - 1 \right\}^{-1},$$

$$\langle a_k^{\dagger\prime} a_k' \rangle_{\nu, \varphi} = \left\{ \exp\left(\frac{1}{\theta}\left(\frac{\nu}{\sqrt{n_0}} + \frac{k^2}{2m} \right) \right) - 1 \right\}^{-1},$$

where $\langle \ldots \rangle_{\nu, \varphi}$ designates the average for the Hamiltonian $H_{\nu, \varphi}$. Because of this, on the basis of (2.15) and (2.17) we have

$$N_k = \left\{ \exp\left(\frac{1}{\theta}\left(\frac{\nu}{\sqrt{n_0}} + \frac{k^2}{2m} \right) \right) - 1 \right\}^{-1},$$

$$N_0 = n_0 V + \left\{ \exp\left(\frac{\nu}{\theta\sqrt{n_0}} \right) - 1 \right\}^{-1}, \tag{2.18}$$

and

$$\frac{N}{V} = \frac{N_0}{V} + \frac{1}{V} \sum_{(k)} \left\{ \exp\left(\frac{1}{\theta}\left(\frac{\nu}{\sqrt{n_0}} + \frac{k^2}{2m} \right) \right) - 1 \right\}^{-1}. \tag{2.19}$$

Due to the presence of the "compensation" term $\dfrac{\nu}{\sqrt{n_0}}$ in the exponent, we no longer have to include the "cut-off" momentum ε. By direct transition to the thermodynamic limit in expression (2.19) we find

$$n = n_0 + \frac{1}{(2\pi)^3} \int \left\{ \exp\left(\frac{1}{\theta}\left(\frac{\nu}{\sqrt{n_0}} + \frac{k^2}{2m} \right) \right) - 1 \right\}^{-1} d\vec{k}. \tag{2.20}$$

Let us further note that

$$\left\langle \left(\frac{a_0^{\dagger}}{\sqrt{V}} - \sqrt{n_0}\, e^{-i\varphi} \right)\left(\frac{a_0}{\sqrt{V}} - \sqrt{n_0}\, e^{i\varphi} \right) \right\rangle_{\nu, \varphi}$$

$$= \frac{1}{V} \langle a_0^{\dagger\prime} a_0' \rangle = \lim_{V \to \infty} \frac{1}{V} \left\{ \exp\left(\frac{\nu}{\theta\sqrt{n_0}} \right) - 1 \right\}^{-1} = 0.$$

Consequently we have, asymptotically

$$\frac{a_0}{\sqrt{V}} \sim \sqrt{n_0}\, e^{i\varphi}, \quad \frac{a_0^{\dagger}}{\sqrt{V}} \sim \sqrt{n_0}\, e^{-i\varphi},$$

i.e. for the system with a Hamiltonian $H_{\nu,\varphi}$ the amplitudes for the condensate are asymptotically *fixed C* numbers. The results in (2.18), (2.19), (2.20) show that, by performing the limit $\nu \to 0$ (*after* the limit $V \to \infty$), we arrive at the usual result of the theory of condensation for an ideal Bose gas.

Let us introduce the quasi-averages:

$$\prec \dots \succ = \lim_{\nu \to 0} \langle \dots \rangle_{\nu,\varphi}.$$

Then we have

$$\prec \frac{a_0}{\sqrt{V}} \succ = \sqrt{n_0}\, e^{i\varphi}, \qquad \prec \frac{a_0^\dagger}{\sqrt{V}} \succ = \sqrt{n_0}\, e^{-i\varphi}.$$

As we see, the selection rules specified by the particle conservation law *are not satisfied for quasi-averages*. We also see that the quasi-averages depend upon the phase angle φ which we can arbitrary fix. Let us choose $\varphi = 0$. Then the quasi-averages become *specific* values. In other words the degeneracy is removed by adding to the Hamiltonian H the infinitesimally small term

$$-\nu(a_0 + a_0^\dagger)\sqrt{V}.$$

In the present case the quasi-averages differ from the regular averages *only* in the amplitudes of the condensate. This is due to the fact that we have an ideal gas without interaction. With the presence of interactions this difference is extended to the other amplitudes. There will appear, for example, quasi-averages different from zero of the type

$$\prec a_k a_{-k} \succ .$$

Let us now examine a more complex example. Let us consider a model system with the Hamiltonian

$$H = \sum_{(f)} T(f) a_f^\dagger a_f - \frac{1}{2V} \sum_{(f,f')} \lambda(f)\lambda(f') a_f^\dagger a_{-f}^\dagger a_{-f'} a_{f'}, \qquad (2.21)$$

which is studied in conjunction with the theory of superconductivity [1, 2]. Here we shall use the following notation:

$$f = (p, s), \quad -f = (-p, -s), \quad s = \pm 1,$$
$$T(f) = \frac{p^2}{2m} - \mu, \quad \mu > 0,$$

$$
\lambda(f) = \begin{cases} J\varepsilon(s), & \left| \dfrac{p^2}{2m} - \mu \right| < \Delta, \\[2ex] 0, & \left| \dfrac{p^2}{2m} - \mu \right| > \Delta, \end{cases}
$$

where a_f, $a_f^\dagger$ are the usual Fermi amplitudes. This example is interesting because it is not trivial. The equations of motion for the Hamiltonian (2.21) cannot be integrated exactly. However, *asymptotically exact formulas (with $V \to \infty$) can be obtained for the Green's functions of all orders.*

Let us briefly present results pertaining to the above problem which have been published [3, 4]. Let us choose an "approximate Hamiltonian"

$$
H_0 = \sum_{(f)} T(f) a_f^\dagger a_f - \frac{1}{2} \sum_{(f)} \lambda(f) \{ C^* a_f a_{-f} + C\, a_f^\dagger a_{-f}^\dagger \} + \frac{1}{2} |C|^2 V, \quad (2.22)
$$

in which C is a c-number (complex in general) defining the non-trivial solution $(C \neq 0)$ to the equation:

$$
C = \frac{1}{V} \sum_{(f)} \lambda(f) \langle a_{-f} a_f \rangle_{H_0}^{(V)}. \tag{2.23}
$$

Here:

$$
\langle \dots \rangle_{H_0}^{(V)} = \frac{\operatorname{Tr}(\dots)\, e^{-H_0/\theta}}{\operatorname{Tr}\, e^{-H_0/\theta}}.
$$

In accordance with the previous notation, the regular average $\langle \dots \rangle_H$ is defined as the limit

$$
\lim_{V \to \infty} \langle \dots \rangle_{H_0}^{(V)}
$$

Since H_0 is a quadratic form with respect to the operators a, $a^\dagger$, to within a constant term we can reduce it to "diagonal form" by means of a linear canonical transformation. We shall introduce, for this purpose new Fermi-amplitudes α and $\alpha^\dagger$, defined by

$$
\alpha_f = a_f u_f + a_{-f}^\dagger v_f, \quad \alpha_f^\dagger = a_f^\dagger u_f + a_{-f} v_f,
$$

where

$$
u_f = \frac{1}{\sqrt{2}} \left(1 + \frac{T(f)}{E(f)} \right)^{1/2}, \quad v_f = \frac{-\varepsilon(s)}{\sqrt{2}} \frac{C}{|C|} \left(1 - \frac{T(f)}{E(f)} \right)^{1/2},
$$

$$E(f) = \sqrt{\lambda^2(f)|C|^2 + T^2(f)}.$$

Then

$$H_0 = \sum_{(f)} E(f)\alpha_f^\dagger \alpha_f + K, \qquad (2.24)$$

where K is the constant

$$K = \frac{1}{2}V\left\{|C|^2 - \frac{1}{V}\sum_{(f)}[E(f) - T(f)]\right\}.$$

From this result we obtain

$$\langle \alpha_f^\dagger \alpha_f \rangle_{H_0}^{(V)} = \left\{\exp\left(\frac{E(f)}{\theta}\right) + 1\right\}^{-1},$$

and

$$\langle \alpha_{-f}\alpha_f \rangle_{H_0}^{(V)} = u_f v_f \frac{1 - \exp\left(\dfrac{E(f)}{\theta}\right)}{1 + \exp\left(\dfrac{E(f)}{\theta}\right)}$$

$$= -u_f v_v \tanh\left(\frac{E(f)}{2\theta}\right) = \frac{\lambda(f)}{2E(f)} C \cdot \tanh\left(\frac{E(f)}{2\theta}\right).$$

Using relationship (2.23) we obtain

$$\left\{1 - \frac{1}{2V}\sum_{(f)} \frac{\lambda^2(f)}{2E(f)} C \cdot \tanh\left(\frac{E(f)}{2\theta}\right)\right\} C = 0.$$

In such a way the desired non-trivial solution for C is given by the equation

$$1 = \frac{1}{V}\sum_{(f)} \frac{\lambda^2(f)}{\sqrt{\lambda^2(f)|C|^2 + T^2(f)}} \tanh\left(\frac{\sqrt{\lambda^2(f)|C|^2 + T^2(f)}}{2\theta}\right). \qquad (2.25)$$

Taking the limit $V \to \infty$, we have

$$1 = \frac{1}{(2\pi)^3} \int \frac{\lambda^2(p) d\vec{p}}{\sqrt{\lambda^2(p)|C|^2 + T^2(p)}} \tanh\left(\frac{\sqrt{\lambda^2(p)|C|^2 + T^2(p)}}{2\theta}\right). \qquad (2.26)$$

As is known this equation has a solution for θ less that a certain θ_{cr}. We shall look at only such a case ($\theta < \theta_{\text{cr}}$). Let us also note that the equation (2.25) (or (2.26)) determines only the coefficient $|C|$, and that the phase of C remains arbitrary.

Let us examine the average

$$\langle \ldots a_{f_j}^\dagger(t_j) \ldots a_{f_s}(t_s) \ldots \rangle_{H_0}^{(V)} \qquad (2.27)$$

formed form the product of any number of operators a and $a^\dagger$ (in any order). Since the Fermi-amplitudes a and $a^\dagger$ are linearly expressed through the Fermi-amplitudes α and $\alpha^\dagger$

$$a_f = u_f \alpha_f - v_f \alpha_{-f}^\dagger, \quad a_f^\dagger = u_f \alpha_f^\dagger - v_f \alpha_{-f},$$

in terms of which H_0 has the diagonal form (2.24), we see that the theorem of Wick and Bloch is applicable for the calculation of the expression (2.27). With this theorem these expressions can be written in the form of the sum of the products of "simple contractions"

$$\langle a_f^\dagger(t) a_f(\tau) \rangle_{H_0}^{(V)} = u_f^2 \frac{e^{iE(f)(t-\tau)}}{1 + e^{E(f)/\theta}} + |v|_f^2 \frac{e^{-iE(f)(t-\tau)}}{1 + e^{-E(f)/\theta}},$$

$$\langle a_f(t) a_f^\dagger(\tau) \rangle_{H_0}^{(V)} = u_f^2 \frac{e^{-iE(f)(t-\tau)}}{1 + e^{-E(f)/\theta}} + |v|_f^2 \frac{e^{iE(f)(t-\tau)}}{1 + e^{E(f)/\theta}},$$

$$\langle a_{-f}(t) a_f(\tau) \rangle_{H_0}^{(V)} = u_f v_f \left\{ \frac{e^{iE(f)(t-\tau)}}{1 + e^{E(f)/\theta}} - \frac{e^{-iE(f)(t-\tau)}}{1 + e^{-E(f)/\theta}} \right\},$$

$$\langle a_f^\dagger(t) a_{-f}^\dagger(\tau) \rangle_{H_0}^{(V)} = u_f v_f^* \left\{ \frac{e^{iE(f)(t-\tau)}}{1 + e^{E(f)/\theta}} - \frac{e^{-iE(f)(t-\tau)}}{1 + e^{-E(f)/\theta}} \right\}, \qquad (2.28)$$

The last two expressions depend not only on the magnitude of C but also on its phase. Therefore, in general, the expression (2.27) can depend upon the phase of C. It will be shown that this dependence is very simple. The Hamiltonian H_0 is invariant with respect to the substitution

$$a_f \to e^{i\varphi} a_f, \quad a_f^\dagger \to e^{-i\varphi} a_f^\dagger, \quad C \to e^{2i\varphi} C,$$

in which φ is an arbitrary (real) angle. Because of this we have

$$\langle \dots a_{f_j}^\dagger(t_j) \dots a_{f_s}(t_s) \dots \rangle_{H_0}^{(V)} \Big|_{C=e^{i\alpha}|C|}$$

$$= \langle \dots a_{f_j}^\dagger(t_j) \dots a_{f_s}(t_s) \dots \rangle_{H_0}^{(V)} \Big|_{C=|C|} e^{-in\alpha/2}, \quad (2.29)$$

where n is the difference between the number of creation operators and annihilation operators in the products considered. Clearly n here can be considered even, since with an odd n we have the identity

$$\langle \dots a_{f_j}^\dagger(t_j) \dots a_{f_s}(t_s) \dots \rangle_{H_0}^{(V)} = 0.$$

Also note that in the case when $n = 0$ it follows from (2.29) that the average (2.27) does not depend on the phase of C. As is seen, the investigation of the system with the "approximate Hamiltonian" H_0 is completely elementary. The corresponding equations of motion can be integrated exactly. We have examined a system with the Hamiltonian H_0 with the view toward proving [3, 4] the following important results:

If for the product

$$\dots a_{f_j}^\dagger(t_j) \dots a_{f_s}(t_s) \dots \quad (2.30)$$

the number $n = 0$, then[f]

---

[f]In order to clarify the basis of such a result, we present the following simple consideration. Note that the Hamiltonian H can be written in the form

$$H = \sum_{(f)} T(f)a_f^\dagger a_f - \frac{1}{2}\sum_{(f)} \lambda(f)\{\beta^\dagger a_{-f}a_f + a_f^\dagger a_{-f}^\dagger \beta\} + \frac{\beta^\dagger \beta}{2}V, \tag{1}$$

where

$$\beta = \frac{1}{V}\sum_{(f)} \lambda(f)a_{-f}a_f.$$

The equation of motion will then be:

$$i\frac{da_f}{dt} = T(f)a_f - \lambda(f)a_{-f}^\dagger \beta,$$
$$i\frac{da_f^\dagger}{dt} = -T(f)a_f^\dagger + \lambda(f)\beta^\dagger a_f. \tag{2}$$

Further we note that

$$\left|\beta a_g^\dagger - a_g^\dagger \beta\right| = \left|\frac{2}{V}\lambda(g)a_{-g}\right| < \frac{2}{V}|\lambda(g)|,$$
$$\left|\beta a_g - a_g \beta\right| = 0,$$
$$\left|\beta^\dagger \beta - \beta\beta^\dagger\right| \leq \frac{2}{V}\left|\sum_{(f)}\frac{1}{V}\lambda^2(f)\right|.$$

In such a way all the commutators involving β, $\beta^\dagger$ with themselves and with the operators a_f, $a_f^\dagger$ are infinitesimally small having magnitudes of the order of $\frac{1}{V}$. Because of this, one expects that the quantum operator nature of β and $\beta^\dagger$ disappears in the limit $V \to \infty$. Substituting in (a) and (b) the respective average values for β and $\beta^\dagger$ we arrive at the problem with the Hamiltonian $H_0(\beta = C)$. It is not hard to see that the operators β, $\beta^\dagger$ are very similar in their character to the operators $\frac{a_0}{\sqrt{V}}$, $\frac{a_0^\dagger}{\sqrt{V}}$ found in the condensation theory of a Bose gas. Both sets of operators have an arbitrary phase.

In accordance with this situation the relations (2.31) are proven only for those products (2.30) which do not depend upon this phase, i.e., which have $n = 0$. The mathematical proof is considerably simplified if we eliminate the arbitrariness of the phase, for example by including in the Hamiltonian H the term:

$$-\nu V(\beta + \beta^\dagger), \quad \nu > 0.$$

$$\langle \ldots a^\dagger_{f_j}(t_j) \ldots a_{f_s}(t_s) \ldots \rangle^{(V)}_H - \langle \ldots a^\dagger_{f_j}(t_j) \ldots a_{f_s}(t_s) \ldots \rangle^{(V)}_{H_0} \to 0, \quad V \to \infty.$$

$$(2.31)$$

Alternatively, the existence of the limit

$$\lim_{v \to \infty} \langle \ldots a^\dagger_{f_j}(t_j) \ldots a_{f_s}(t_s) \ldots \rangle_{H_0}$$

is completely determined for any product (2.30). Further, if $n \neq 0$, then due to selection rules with a Hamiltonian H which conserves the number of particles,

$$\langle \ldots a^\dagger_{f_j}(t_j) \ldots a_{f_s}(t_s) \ldots \rangle^{(V)}_H = 0,$$

and therefore

$$\langle \ldots a^\dagger_{f_j}(t_j) \ldots a_{f_s}(t_s) \ldots \rangle_H$$
$$= \begin{cases} \langle \ldots a^\dagger_{f_j}(t_j) \ldots a_{f_s}(t_s) \ldots \rangle_{H_0}, & n = 0, \\ 0, & n \neq 0. \end{cases} \quad (2.32)$$

Likewise we can calculate the regular averages (2.32) of any order, and, consequently, we can calculate the Green's functions for the model with the Hamiltonian H. In addition, with any value of the number n, it can be proved that:

$$\prec \ldots a^\dagger_{f_j}(t_j) \ldots a_{f_s}(t_s) \ldots \succ_H = \langle \ldots a^\dagger_{f_j}(t_j) \ldots a_{f_s}(t_s) \ldots \rangle_{H_0}. \quad (2.33)$$

Here, as before, the symbol $\prec \ldots \succ$ represent quasi-averages. As was noted earlier the second part of the equality (2.33) contains the factor $\exp(-in\alpha/2)$. Because of this we have,

$$\langle \ldots a^\dagger_{f_j}(t_j) \ldots a_{f_s}(t_s) \ldots \rangle_H = \frac{1}{2\pi} \int_0^{2\pi} d\alpha \prec \ldots a^\dagger_{f_j}(t_j) \ldots a_{f_s}(t_s) \ldots \succ_H,$$

i.e., the regular average is obtained from quasi-averages after the additional averaging over the arbitrary angle α is performed.

Just as in previously considered cases the quasi-averages can be introduced by adding to the Hamiltonian infinitesimally small terms, which remove the degeneracy. Let us take the Hamiltonian

$$H_\nu = H - \frac{\nu}{2} \sum_{(f)} \lambda(f)\{a_{-f}a_f + a_f^\dagger a_{-f}^\dagger\}, \quad \nu > 0, \tag{2.34}$$

containing terms which remove the degeneracy with respect to gradient invariance of the 1st type; that is terms which remove the conservation law of the total number of particles.

Let us take the approximate Hamiltonian in the form

$$H_\nu^0 = H_0 - \frac{\nu}{2} \sum_{(f)} \lambda(f)\{a_{-f}a_f + a_f^\dagger a_{-f}^\dagger\}.$$

The quantity C introduced here is defined by the equation

$$C = \frac{1}{V} \sum_{(f)} \lambda(f)\langle a_{-f}a_f \rangle_{H_\nu^0}^{(V)}$$

i.e.,

$$C = \frac{C+\nu}{2V} \sum_{(f)} \lambda^2(f) \frac{\text{th}\left\{ \dfrac{\sqrt{\lambda^2(f)(C+\nu)^2 + T^2(f)}}{2\theta} \right\}}{\sqrt{\lambda^2(f)(C+\nu)^2 + T^2(f)}}.$$

After the limit $V \to \infty$, we obtain

$$C = \frac{C+\nu}{(2\pi)^3} \int \lambda^2(p) \frac{\text{th}\left\{ \dfrac{\sqrt{\lambda^2(p)(C+\nu)^2 + T^2(p)}}{2\theta} \right\}}{\sqrt{\lambda^2(p)(C+\nu)^2 + T^2(p)}} \, d\vec{p}.$$

We shall take for C that root of this equation which approaches the positive root of equation (2.26) when $\nu \to \infty$. Then one can prove that

$$\langle \ldots a_{f_j}^\dagger(t_j) \ldots a_{f_s}(t_s) \ldots \rangle_{H_\nu} = \langle \ldots a_{f_j}^\dagger(t_j) \ldots a_{f_s}(t_s) \ldots \rangle_{H_\nu^0}.$$

Alternatively it is easy to be convinced that

$$\langle \ldots a_{f_j}^\dagger(t_j) \ldots a_{f_s}(t_s) \ldots \rangle_{H_\nu^0} \to \langle \ldots a_{f_j}^\dagger(t_j) \ldots a_{f_s}(t_s) \ldots \rangle_{H_0}, \quad \nu \to 0$$

with $C = |C|$. Therefore

$$\prec \ldots a^\dagger_{f_j}(t_j) \ldots a_{f_s}(t_s) \ldots \succ = \lim_{\substack{\nu \to 0 \\ \nu > 0}} \langle \ldots a^\dagger_{f_j}(t_j) \ldots a_{f_s}(t_s) \ldots \rangle_{H_\nu}$$

$$= \langle \ldots a^\dagger_{f_j}(t_j) \ldots a_{f_s}(t_s) \ldots \rangle_{H_0}, \quad C = |C|.$$

If we had taken the Hamiltonian $H_{\nu,\varphi}$ instead of H_ν where

$$H_{\nu,\varphi} = H - \nu \sum_{(f)} \lambda(f)\{e^{i\varphi}\, a^\dagger_f a^\dagger_{-f} + e^{-i\varphi} a_{-f} a_f\}, \quad \nu > 0,$$

then we would have obtained[g]

$$\prec \ldots a^\dagger_{f_j}(t_j) \ldots a_{f_s}(t_s) \ldots \succ = \lim_{\substack{\nu \to 0 \\ \nu > 0}} \langle \ldots a^\dagger_{f_j}(t_j) \ldots a_{f_s}(t_s) \ldots \rangle_{H_\nu}$$

$$= \langle \ldots a^\dagger_{f_j}(t_j) \ldots a_{f_s}(t_s) \ldots \rangle_{H_0}, \quad C = e^{i\varphi}|C|.$$

Thus, as one could have expected in the present case *quasi-averages depend on the arbitrary phase angle φ*. It is also essential that *for the quasi-averages here the selection rules, which are specified by the law of conservation of the number of particles, are not satisfied.* In order to have well determined values for quasi-averages we have to somehow fix this angle. Assume that $\varphi = 0$, i.e. let us agree to remove the degeneracy by including in the Hamiltonian H infinitely small terms of the type:

$$-\frac{\nu}{2} \sum_{(f)} \lambda(f)\{a_{-f} a_f + a^\dagger_f a^\dagger_{-f}\}. \tag{2.35}$$

Such a choice of the phase angle is convenient in that it make the values of all the "simultaneous" quasi-averages of the type

$$\prec \ldots a^\dagger_{f_j}(t_j) \ldots a_{f_s}(t_s) \ldots \succ$$

---

[g]As is seen the regular average

$$\langle \ldots a^\dagger_{f_j}(t_j) \ldots a_{f_s}(t_s) \ldots \rangle_H = \frac{1}{2\pi} \int \prec \ldots a^\dagger_{f_j}(t_j) \ldots a_{f_s}(t_s) \ldots \succ_H d\varphi$$

suffers a discontinuity when we add to the Hamiltonian H infinitely small terms representing pair sources:

$$-\frac{\nu}{2} \sum_{(f)} \lambda(f)\{e^{i\varphi} a_{-f} a_f + e^{-i\varphi} a^\dagger_f a^\dagger_{-f}\}$$

real. Also note that the result will not change if these additional terms (2.35) are written in more general form

$$-\nu \sum_{(f)} w(f)\{a_{-f}a_f + a_f^\dagger a_{-f}^\dagger\}, \quad \nu > 0, \tag{2.36}$$

where $w(f)$ is a real, non-trivial, and fairly regular function.

In the above we dealt with products of field functions in the momentum representation. *A similar situation arises from products of field functions*

$$\Psi(t, \vec{r}, s) = \sum_{(p)} a_{p, s}(t) e^{i(\vec{p}\vec{r})},$$

$$\Psi^\dagger(t, \vec{r}, s) = \sum_{(p)} a_{p, s}^\dagger(t) e^{-i(\vec{p}\vec{r})},$$

in the coordinate representation.

We have for example

$$\langle \Psi^\dagger(t_1, \vec{r}_1, s_1)\Psi^\dagger(t_2, \vec{r}_2, s_2)\Psi^\dagger(t_2', \vec{r}_2', s_2')\Psi(t_1', \vec{r}_1', s_1')\rangle$$
$$= \langle \Psi^\dagger(t_1, \vec{r}_1, s_1)\Psi^\dagger(t_2, \vec{r}_2, s_2)\Psi^\dagger(t_2', \vec{r}_2', s_2')\Psi(t_1', \vec{r}_1', s_1')\rangle_{H_0}$$
$$= F(t_1 - t_1', \vec{r}_1 - \vec{r}_1')F(t_2 - t_2', \vec{r}_2 - \vec{r}_2')\delta(s_1 - s_1')\delta(s_2 - s_2')$$
$$- F(t_2 - t_1', \vec{r}_2 - \vec{r}_1')F(t_1 - t_2', \vec{r}_1 - \vec{r}_2')\delta(s_2 - s_1')\delta(s_1 - s_2')$$
$$+ \Phi(t_1 - t_2, \vec{r}_1 - \vec{r}_2)\Phi(t_1' - t_2', \vec{r}_1' - \vec{r}_2')$$
$$\times \in(s_1) \in(s_1')\delta(s_1 + s_2)\delta(s_1' + s_2') \tag{2.37}$$

where

$$F(t, \vec{r}) = \frac{1}{(2\pi)^3}\int e^{-i(\vec{p}\vec{r})}\left\{\frac{u_p^2\, e^{iE(p)t}}{1 + e^{E(p)/\theta}} + \frac{v_p^2\, e^{-iE(p)t}}{1 + e^{-E(p)/\theta}}\right\}d\vec{p},$$

$$\Phi(t, \vec{r}) = \frac{1}{2(2\pi)^3}\int e^{-i(\vec{p}\vec{r})}\sqrt{1 - \frac{T^2(p)}{E^2(p)}}\left\{\frac{e^{-iE(p)t}}{1 + e^{-E(p)/\theta}} - \frac{e^{iE(p)t}}{1 + e^{E(p)/\theta}}\right\}d\vec{p}.$$

$$\tag{2.38}$$

We also have

$$F(t_1 - t'_1, \vec{r}_1 - \vec{r}'_1)\delta(s_1 - s'_1) = \langle \Psi^\dagger(t_1, \vec{r}_1, s_1)\Psi(t'_1, \vec{r}'_1, s'_1)\rangle$$
$$= \prec \Psi^\dagger(t_1, \vec{r}_1, s_1)\Psi(t'_1, \vec{r}'_1, s'_1) \succ$$
$$\Phi(t_1 - t_2, \vec{r}_1 - \vec{r}_2) \in(s_1)\delta(s_1 + s_2)$$
$$= \prec \Psi^\dagger(t_1, \vec{r}_1, s_1)\Psi^\dagger(t_2, \vec{r}_2, s_2) \succ = \prec \Psi^\dagger(t_2, \vec{r}_2, s_2)\Psi(t_1, \vec{r}_1, s_1) \succ$$

$$(2.39)$$

In the above we have investigated a number of examples of degeneracy of states of statistical equilibrium. In all of these cases such special states of statistical equilibrium were realized when the temperatures were below a certain critical temperature $(\theta < \theta_c)$. For temperatures above θ_c there appears a phase change which leads to the "normal" non-degenerate state.

In the above examples the degeneracy was dependent upon the presence of additive conservation law, or (which is the same) upon the presence of invariance with respect to corresponding transformation groups. Let us emphasize that not all the conservation laws in a given system produce degeneracy. That is, in the third and fourth examples, the degeneracy of the statistical equilibrium states depended only on the conservation law of the number of particles. In the corresponding quasi-averages only those selection rules were violated which were specified by this very law. The selection rules specified by other additive conservation laws, for example, by the law of conservation of momentum and spin (in the fourth example) were left intact.

In the second example the degeneracy depended only upon the law of conservation of momentum. The selection rules, specified, for example, by the law of conservation of the number of particles, were not violated here.

We could increase the number of such examples by investigating cases of degeneracy in connection with other groups or simultaneously with several transformation groups. However, we shall not stay to consider there points but shall turn to the general investigation, introducing the corresponding general calculations.

Let us consider a specific microscopic system with a Hamiltonian H. We now add infinitesimally small terms to H, which correspond to external fields or sources which violate the additive conservation laws. In this manner we obtain a specific Hamiltonian H_ν, $\nu \to 0$. Then, *if all the average values*

$$\langle A \rangle, \quad A = \ldots \Psi^\dagger(t_j, x_j) \ldots \Psi(t_s, x_s) \ldots \quad (2.40)$$

are changed only by an infinitesimal amount, we will say that the state of statistical equilibrium being considered is not degenerate. Alternatively, if some of the averages (2.40) are changed by a finite amount when the transition is from H to an infinitesimally changed Hamiltonian H_ν, we shall say that the state of statistical equilibrium is degenerate. It is obvious that we shall limit ourselves to observing only stable systems in as much as only they have physical meaning. Because of this the infinitesimally small variation $\delta H = H_\nu - H$ of the Hamiltonian can produce only an infinitesimally small change in those values which actually characterize the real physical property of the system.

For cases of degeneracy it is convenient to introduce instead of the regular averages the following quasi-averages

$$\prec A \succ = \lim_{\nu \to 0} \langle A \rangle_{H_\nu}.$$

As we have already seen from the series of examples for quasi-average, it is not necessary to fulfill *all* the selection rules specified by the additive conservation laws. Let us note that when determining quasi-averages we must *first* take the limit $V \to \infty$, and then let ν approach zero.

As was previously noted, the infinitesimally small terms producing the difference $H_\nu - H$ are chosen in such a way as to violate the additive conservation laws.

Generally speaking, however, it is not necessary to violate all such laws in order to obtain the Hamiltonian H, which removes the degeneracy.

For example, let infinitesimal small terms which bring about a violation of some of the laws, produce only an infinitesimally change in $\langle A \rangle_{H_\nu}$. Then it is clear that there is no need to violate these conservation laws and that H_ν, which possesses only terms which violate the rest of the conservation law, will suffice to remove the degeneracy.

In such a case for quasi-averages, just those selection rules which are specified by these last mentioned conservation law will be violated.

Let us take, in particular, the usual dynamical model for the theory of superconductivity, in which we deal with the continuum, and do not consider the *direct* presence of the crystal lattice. In this model when the external fields are absent one naturally expects total space homogeneity and that *all* the averages

$$\langle \ldots \Psi^\dagger(t_\alpha, x_\alpha) \ldots \Psi(t_\beta, x_\beta) \ldots \rangle$$

are translationally invariant.

In such a case the momentum conservation law will also hold for quasi-averages, and there is no reason to violate it in order to remove the degeneracy.

Let us also assume the presence of total spin homogeneity when the conservation law of total spin holds for quasi-averages. Then we are left only with the conservation law for the number of particles to be violated. In such case we can assume:

$$H\nu = H + \nu \sum w(f)(a_f^\dagger a_{-f}^\dagger + a_{-f}a_f), \quad w(f) = \in(\sigma)v(p) \qquad (2.41)$$

where $v(p)$ is a real function of momentum. To investigate the case when there is spin homogeneity, then we use the more general form:

$$H_\nu = H + \nu \sum \{w(p,\,\sigma,\,\sigma')a_{p\sigma}^\dagger a_{-p\sigma'}^\dagger + w^*(p,\,\sigma,\,\sigma')a_{-p\sigma'}^\dagger a_{p\sigma}^\dagger + \lambda(p,\,\sigma,\,\sigma')a_{p\sigma}^\dagger a_{p\sigma'}\}.$$

Let us now turn to the problem of applying different forms of perturbation theory (in particular, the diagram techniques) to investigate degenerate states of statistical equilibrium.

In order to remove difficulties which arise in the usual formalism discussed earlier in this section, we shall use the following general rule: *In order to use perturbation theory to investigate the degenerate states of statistical equilibrium, we must first of all remove the degeneracy, that is, we must work not with Green's functions which are constructed from regular averages satisfying all the selection rules, but instead with Green's functions which are built up from quasi-averages which do not satisfy the some of these rules.*

In such a way, the corresponding diagrams can include "anomalous" lines which are forbidden by the usual selection rules. For example, the diagrams in the theory of the crystalline state which have the "normal" lines $\overline{a_p^\dagger a_p}$ that conserve momentum, will now also include the "anomalous" lines $\overline{a_p^\dagger a_{p'}}$ $(p \neq p')$ which do not conserve momentum.

Anomalous lines $\overline{a_f a_{-f}}$, $\overline{a_f^\dagger a_{-f}^\dagger}$, etc. also appear in diagrams in the theory of superconductivity. We must keep in mind that these anomalous lines correspond to "dangerous" diagrams in that their sum gives a contribution in the limit although their very presence is formally specified by infinitesimally small complementary terms in the Hamiltonian, H_ν.

Because of this such lines must always be introduced into a calculation in a summed (even if only partially) form. One can introduce, for example, only totally summed anomalous lines, and, for the determination of their corresponding anomalous Green's functions, one can obtain an equation of the Dyson type. Actually when the calculation is carried out one can generally drop the infinitesimally small complementary terms, whose only role is to introduce quasi-averages instead of regular averages. In those cases when the Dyson equation referred to above has only a trivial solution (the anomalous Green's function are identically equal to zero), then, obviously, there is no degeneracy. Degeneracy arises if the real[h] solution is non-trivial.

As was mentioned at the end of the first section perturbation theory is usually constructed by dividing the total Hamiltonian of the system into two parts: $H = H_0 + H_1$. The Hamiltonian H_0 is selected to correspond to an "ideal gas" without interactions which possesses all those additive conservation laws which the total Hamiltonian possesses.

Such an approach to the construction of perturbation theory can be generalized for the investigation of degenerate states. In order that the anomalous (partially summed) Green's functions appear immediately in the zeroth approximation, we add to H_0 terms Δ, of the same type as the infinitesimally small additional terms in H_ν. Thus, for $H_0 + \Delta$ we remove a series of additive conservation laws which hold true for the total H and which are "responsible for degeneracy".

Let us write

$$H_0' = H_0 + \Delta; \quad H_1' = H_1 - \Delta.$$

Then, proceeding form the modified decomposition $H = H_0' + H_1'$ one can construct in the usual way degenerate perturbation theory using an expansion in powers of H_1'. By the very choice of H_0' in the zeroth approximation we obtain the corresponding anomalous Green's functions. Let us take, for example, the dynamical system which is investigated in the theory of superconductivity where the degeneracy is removed by infinitely small terms of the type (2.41).

In the normal forms of the perturbation theory which do not take into account the possibility of degeneracy the following term is included in H_0

$$\sum_{(f)} T_e(k) a_f^\dagger a_f \qquad (2.42)$$

which correspond to the "renormalization" of the kinetic energy term

$$\sum_{(f)} \left(\frac{k^2}{2m} - \mu \right) a_f^\dagger a_f. \qquad (2.43)$$

---

[h]We speak of the real situation, keeping in mind that the equations can always have a trivial solution which does not satisfy the necessary physical restrictions (for example it may have the wrong spectral structure).

For the calculations involving degeneracy we shall introduce for H'_0, instead of (2.42), a more general quadratic form in the Fermi-amplitudes:

$$\Omega = \sum_{(f)} T_e(f) a_f^\dagger a_f - \frac{1}{2} \sum_{(f)} w(f)(a_f^\dagger a_{-f}^\dagger + a_{-f} a_f), \quad w^*(f) = w(f). \quad (2.43')$$

We then must include in H'_1, in addition to the interaction terms, another compensating expression:

$$\sum_{(f)} \left(\frac{k^2}{2m} - \mu \right) a_f^\dagger a_f - \Omega.$$

The arbitrary function $w(f)$ should be chosen in such a way as to improve the degree of approximation. For example, for obtaining the fist approximation one can choose $w(f)$ on the basis that the corrections to this approximation i.e. $\langle a_{-f} a_f \rangle$, would be zero; so that this anomalous average in the zeroth approximation would already be "summed" from the point of view of the usual first approximation. Let us note in conjunction with this, that in the specific case of the model system, considered previously, with the Hamiltonian (2.21) we can thus obtain an asymptotically exact solution. For this, it is only necessary to take for $T_e(k)$ its non-renormalized value from (2.43) and assume:

$$w(f) = \lambda(f) \sum_{(f')} \lambda(f') \langle a_{-f'} a_{f'} \rangle.$$

Then, in fact, the "zeroth approximation", determined by the Hamiltonian H'_0, will give an asymptotically exact solution, and corrections of *any* order will be asymptotically equal to zero.

Let us note that Ω reduces to the diagonal form:

$$\sum_{(f)} E(f) \alpha_f^\dagger \alpha_f + \text{const}, \quad E(f) = \sqrt{T_e^2(k) + w^2(f)}$$

by means of the canonical $u - v$ transformation:

$$\begin{aligned} \alpha_f &= a_f u_f + a_{-f}^\dagger v_f, \\ \alpha_f^\dagger &= a_f^\dagger u_f + \alpha_{-f} v_f^*. \end{aligned} \quad (2.44)$$

Thus, for the construction of degenerate perturbation theory it is absolutely equivalent to modify the expression H_0, by the substitution, $H_0 \rightarrow H_0'$, or to use as the Hamiltonian of zeroth approximation the Hamiltonian of the ideal gas:

$$\sum_{(\nu)} E(\nu) a_\nu^\dagger a_\nu$$

in which the "new Fermi-amplitudes", α, are coupled with the "old" by $u - v$ transformation.

In our first papers [2,5] on the theory of superconductivity we made muse the $u - v$ transformation to obtain the correct modification of perturbation theory. The last observation is general in character and does not apply only to the case of the quadratic form Ω (2.43) considered above. Actually, if we take an arbitrary quadratic form

$$\Omega = \sum A(f, f') a_f^\dagger a_{f'} + \sum C(f, f') a_f a_{f'} + \sum C^*(f, f') a_{f'}^\dagger a_f^\dagger, \tag{2.45}$$
$$A^*(f, f') = A(f', f)$$

requiring only that it be positive definite then by means of the general $u - v$ transformation:

$$a_f = \sum_{(\nu)} u_{f\nu} \alpha_\nu + \sum_{(\nu)} v_{f\nu} \alpha_\nu^\dagger$$

(2.45) can be reduced to a diagonal form

$$\sum_{(f)} E(f) a_f^\dagger a_f + \text{const.}$$

In conclusion, we note that if one works with completely summed Green's functions (with "thick lines" in diagrams of the Feynman type), then the final equations are invariant with respect to the special form H_0', and it only necessary to introduce into the diagrams the corresponding anomalous lines. The method of Green's functions is especially convenient if we must take damping into account, if we have to deal with higher approximations.

3.   Principle of Correlation Weakening

In this paragraph we will try to formulate the intuitive concept, generally accepted in statistical mechanics, that the correlations between space distant parts of a macroscopic system in vanishingly small.

Consider the average:

$$F(t_1, x_1, \ldots, t_n, x_n) = \langle \ldots \Psi^\dagger(t_j, x_j) \ldots \Psi^\dagger(t_s, x_s) \ldots \rangle, \quad x = (\vec{r}, \sigma) \quad (3.1)$$

and arbitrarily divide the set of arguments $t_1, x_1; \ldots, t_n, x_n$ into a series of groups:

$$\{\ldots, t_\alpha, x_\alpha, \ldots\}, \quad \{\ldots, t_\beta, x_\beta, \ldots\}, \quad \ldots .$$

The asymptotic form of F will be considered with the time points, $t_1, \ldots, t_n$ fixed and the distances between the points $\vec{r}$, from *different* groups, tending toward infinity. First of all we postulate that under the average, the field functions,

$$\varphi(t_1, \boldsymbol{r}_1, \sigma_1), \quad \varphi(t_2, \boldsymbol{r}_2, \sigma_2), \quad (\varphi = \Psi^\dagger \text{ or } \Psi),$$

with t_1 and t_2 fixed and $|\vec{r}_1 - \vec{r}_2| \to \infty$, will exactly commute or anticommute among themselves in the limit.

Then, in order to find the asymptotic form F, we can reorder the field functions $\varphi(t_i, x_i)$ in expression (3.1) and thus, obtain the field functions for each given group of arguments together in one set. We will thus have

$$F(t_1, x_1, \ldots, t_n, x_n) -$$
$$\eta \langle \mathscr{U}_1(\ldots, t_\alpha, x_\alpha, \ldots) \mathscr{U}_2(\ldots, t_\beta, x_\beta, \ldots) \ldots \rangle \to 0, \quad \eta = \pm 1, \quad (3.2)$$

where $\mathscr{U}_1(\ldots, t_\alpha, x_\alpha, \ldots)$ represents the product of field functions with arguments from only the 1st group, and $\mathscr{U}_2(\ldots, t_\beta, x_\beta, \ldots)$ represents the corresponding product with arguments from only the 2nd group, etc. The statement made about the asymptotic commutation expresses, in our opinion, a universal law for real dynamical systems of statistical mechanics.

As is known in quantum field theory, all the field functions $\varphi(t_1, x_1)$, $\varphi(t_2, x_2)$ must *exactly* commute or anti-commute, if the four dimension vector $t_1 - t_2$, $\vec{r}_1 - \vec{r}_2$ is space-like. In problems of statistical mechanics, where we deal with interactions which are formally non-local this characteristic of commutation rules must be satisfied, at least approximately, and more exactly as $|\vec{r}_1 - \vec{r}_2|$ increases with fixed t_1, t_2.

Let us now turn to the investigation of the asymptotic structure of the expression

$$\langle \mathscr{U}_1(\ldots, t_\alpha, x_\alpha, \ldots) \mathscr{U}_2(\ldots, t_\beta, x_\beta, \ldots) \ldots \rangle \quad (3.3)$$

in the limit of infinite spatial separation between the points $\vec{r}$ from different groups (the temporal arguments $t_1, \ldots, t_n$ being fixed. Since the correlation between dynamical quantities $\mathscr{U}_1$, $\mathscr{U}_2$, ... must become weaker and practically disappear for large enough distances, the corresponding asymptotic form for (3.3) breaks up into products of the form:

$$\langle \mathscr{U}_1(\ldots, t_\alpha, x_\alpha, \ldots)\rangle\langle \mathscr{U}_2(\ldots, t_\beta, x_\beta, \ldots)\rangle \ldots . \qquad (3.4)$$

Here it is necessary to specify the type of "averages" we are dealing with in our formulation of the principle of correlation weakening. In the nondegenerate case the expressions $\langle \ldots \rangle$, are obviously *regular averages*. However, one should note that in the case where the state of statistical equilibrium is degenerate, the expressions $\langle \ldots \rangle$, entering into our formulation, must be understood as *quasi-averages*. The formulation of the principle of correlation weakening presented above is incorrect if one considers $\langle \ldots \rangle$ as regular averages.

Let us now investigate the crystalline state once again. In this case when we refer to the correlation weakening between dynamical quantities $\mathscr{U}_1$, $\mathscr{U}_2$, ... we intuitively mean that the crystal lattice, as a whole, is fixed in space. Even though the crystal position is arbitrary fixed the calculation of the averages of $\mathscr{U}_1$ and $\mathscr{U}_2$, etc. involve just this one fixed position. In other words we now assume that all the averages considered here depend upon the same fixed position of the crystal lattice. Thus we are dealing with quasi-averages, and not with regular averages which are obtained from quasi-averages by an additional average over all possible positions and orientations of the crystal lattice.

In other cases of degeneracy of the statistical equilibrium state similar situations arise with parameters which are fixed in the same way for all parts of the system. As further examples we have either the magnetic moment (ferromagnetism) or the phase angle (superfluidity or superconductivity), etc.

Thus, in our formulation of the principle of correlation weakening it follows that we should consider the expressions $\langle \ldots \rangle$ as quasi-averages.[i] Note that we cannot prove exactly the principle of correlation weakening for macroscopic dynamical systems considered in statistical mechanics. We can develop an exact proof only in a number of simple models such as in the

---

[i]Since in the degenerate cases we will always deal with quasi-averages and in non-degenerate cases the quasi-averages and regular averages coincide, we will no longer use the special symbol $\prec \ldots \succ$ for denoting quasi-averages, but will use the symbol $\langle \ldots \rangle$ everywhere since this will no longer lead to misunderstanding.

models mentioned in the previous paragraph. For the general can we can use either intuitive ideas or arguments borrowed from perturbation theory. In this respect the principle of correlation weakening is no different from other generally accepted important assumptions in statistical mechanics.

Thus, for example, the problem of the proof of a considerably simpler assumption; namely, the existence of the limit

$$- \lim_{V \to \infty} \frac{\theta \ln \mathrm{Tr}\, e^{-H/\theta}}{V}$$

which represent the free energy per unit volume is in almost the same situation. Thus, we will not investigate here the difficult mathematical problem of the formulation of the correlation weakening principle but we will restrict ourselves to its physical implementation. Let us first examine the application of this principle in the construction of a somewhat different, generally more "physical" definition, of the meaning of quasi-averages.

Consider, as an example, the case investigated in the theory of superconductivity with a statistical equilibrium state where the degeneracy depends only upon the law of conservation of the number of particles. Let us examine the expression

$$\langle \Psi^\dagger(t_1,\, x_1) \Psi^\dagger(t_2,\, x_2) \Psi(t'_2,\, x'_2) \Psi(t'_1,\, x'_1) \rangle. \tag{3.5}$$

Since the operator

$$\Psi^\dagger(t_1,\, x_1) \Psi^\dagger(t_2,\, x_2) \Psi(t'_2,\, x'_2) \Psi(t'_1,\, x'_1)$$

conserves the number of particles, expression (3.5) will be a regular average.

Let us increase without limit the distance between two groups of spatial points $(\vec{r}_1,\, \vec{r}_2)$ and $(\vec{r}'_1,\, \vec{r}'_2)$ with time variable fixed. The on the basis of the correlation weakening principle the expression (3.5) will approach the product

$$\langle \Psi^\dagger(t_1,\, x_1) \Psi^\dagger(t_2,\, x_2) \rangle \langle \Psi(t'_2,\, x'_2) \Psi(t'_1,\, x'_1) \rangle.$$

Proceeding from such an asymptotic decomposition of the regular average (3.5), we can now define the quasi-averages

$$\langle \Psi^\dagger(t_1,\, x_1) \Psi^\dagger(t_2,\, x_2) \rangle, \quad \langle \Psi(t'_2,\, x'_2) \Psi(t'_1,\, x'_1) \rangle.$$

By using the analogous procedure one can introduce quasi-averages of higher order products of field functions. Previously we introduced quasi-averages by adding infinitesimal terms to the Hamiltonian, without necessarily having a clear physical meaning. Now, with the principle of correlation weakening we are able to introduce quasi-averages by examining the asymptotic forms of the regular averages with the given and unaltered Hamiltonian which corresponds to the dynamical system under consideration.

However, we must point out that the method involving infinitesimally small additions to the Hamiltonian is more convenient for a formal derivation of the generalized diagram technique, (using anomalous lines), in as much as it automatically reduces this problem to the previously solved one.

Let us examine a system of spinless Bose particles, in a spatially homogeneous statistical equilibrium state, and consider the expression:

$$F(\vec{r}_1 - \vec{r}_2) = \langle \Psi^\dagger(t, \vec{r}_1)\Psi(t, \vec{r}_2)\rangle = \langle \Psi^\dagger(\vec{r}_1)\Psi(\vec{r}_2)\rangle,$$
$$\Psi(\vec{r}) = \Psi(0, \vec{r}). \tag{3.6}$$

Here, transformation to the momentum representation gives:

$$F(\vec{r}_1 - \vec{r}_2) = \frac{1}{V}\sum_k \langle a_k^\dagger a_k\rangle e^{-\vec{k}(\vec{r}_1 - \vec{r}_2)}. \tag{3.7}$$

Therefore, in the Fourier integral

$$F(\vec{r}) = \int w(k)e^{-\vec{k}\cdot\vec{r}}d\vec{k}, \tag{3.8}$$

the product $w(k)d\vec{k}$ expresses the number density of particles with momenta in the infinitesimal momentum volume $\vec{k}$. From this it follows that $\rho = \dfrac{N}{V}$ represents the particle number density

$$w(k) \geq 0, \quad \int w(k)d\vec{k} = \rho.$$

Let us further consider the case when a quiescent condensate is present in the system. Then

$$w(k) = \rho_0\delta(\vec{k}) + w_1(k)$$

where $w_1(k)$ is a regular function characterizing the continuous momentum distribution of the particles not located in the condensate and ρ_0 is the number density of particle in the condensate. However, in as much as $w_1(k)$ is well-behaved we have

$$\int w_1(k)e^{-\vec{k}\cdot\vec{r}}d\vec{k} \to 0, \quad |\vec{r}| \to \infty,$$

and thus

$$\langle \Psi^\dagger(r_1)\Psi(r_2)\rangle = F(\vec{r}_1 - \vec{r}_2) = \rho_0 + \int w_1(k)e^{-\vec{k}(\vec{r}_1-\vec{r}_2)}\,d\vec{k} \to \rho_0 \neq 0,$$

as $|\vec{r}_1 - \vec{r}_2| \to \infty$. Therefore $\langle \Psi(\vec{r}_1)\rangle \neq 0$.

On the other hand if the statistical equilibrium state was not degenerate with respect to the law of conservation of the number of particles, then on the strength of the selection rules corresponding to this law we would have had the identity, $\langle \Psi(\vec{r}_1)\rangle = 0$. Thus, for the systems with a condensate, the selection rules specified by the law of conservation of particle number are not satisfied and this statistical equilibrium state will be degenerate.

One can show that an analogous situation also arises for fermi systems when a condensate of coupled pairs appears. It is now necessary to define the meaning of "coupled pair". We proceed to do this in the following paragraph.

4.   Particle Pair States

We will investigate here the spatially homogeneous statistical equilibrium states for macroscopic systems composed of identical Fermi particles. For these states let us try to clarify such ideas as "wave function of a pair of particles", [7] "state of a pair of particles", and in particular "coupled state of a pair", etc. These ideas have a clear meaning in the case where the dynamical system consists of two particles. We wish to generalize these ideas to systems of macroscopically large number of particles which interact one with another. With this goal in mind let us look at a pair correlation function (corresponding to one instant of time):

$$F(x_1, x_2; x_1', x_2') = \langle \Psi^\dagger(x_1)\Psi^\dagger(x_2)\Psi(x_2')\Psi^\dagger(x_1')\rangle. \tag{4.1}$$

Using the Hermitian property

$$F(x_1', x_2'; x_1, x_2) = F(x_1, x_2; x_1', x_2') \tag{4.2}$$

we can expand F in the orthonormal system of eigenfunctions Ψ_ν:

$$F(x_1, x_2; x_1', x_2') = \sum_{(\nu)} N_\nu \Psi_\nu^*(x_1, x_2)\Psi_\nu(x_1', x_2') \tag{4.3}$$

with the normalization

$$\iint_V \left| \Psi_\nu(x_1,\, x_2) \right|^2 dx_1\, dx_2 = 1 \tag{4.4}$$

where, generally.

$$\int_V \dots dx = \sum_\sigma \int \dots d\vec{r}.$$

In the case of a low density gas, to the first approximation $\Psi_\nu(x_1,\, x_2)$ will be the usual wave function of the two body problem (which is very natural since to the first approximation the action of the other particles upon the *given pair* of particles can be neglected).

Because of this analogy, *we will call the eigenfunctions $\Psi_\nu(x_1,\, x_2)$ the wave functions of pairs of particles. We will interpret the coefficients N_ν as the average number of pairs of particles in the state with wave function Ψ_ν.* From (4.1), (4.3), and (4.4) it follows that[j]

$$\langle N^2 - N \rangle = \sum_\nu N_\nu$$

i.e. the sum of all the N_ν *represents the total number of pairs.*

Let also note that due to (4.1):

$$F(x_2, x_1;\, x_1', x_2') = -F(x_1, x_2;\, x_1', x_2')$$
$$F(x_1, x_2;\, x_2', x_1') = -F(x_1, x_2;\, x_1', x_2')$$

and thus

$$\Psi_\nu(x_2,\, x_1) + \Psi_\nu(x_1,\, x_2) = 0.$$

As is seen, the function Ψ_ν must be antisymmetric just like the usual wave functions of two Fermi particles. Now, let us write the expansion (4.3)

---

[j]Actually, we really have from (4.1), (4.3), (4.4):

$$\sum_\nu N_\nu = \iint_V \langle \Psi^\dagger(x_1)\Psi^\dagger(x_2)\Psi(x_2)\Psi^\dagger(x_1) \rangle dx_1\, dx_2$$

$$= \iint_V \langle \Psi^\dagger(x_1)\Psi^\dagger(x_1)\Psi^\dagger(x_2)\Psi(x_2) \rangle dx_1\, dx_2 - \int_V \langle \Psi^\dagger(x)\Psi(x) \rangle dx$$

while $\int_V \Psi^\dagger(x)\Psi(x)dx = N.$

in a more detailed form.   We make use of the law of conservation of
momentum and separate out from the wave function $\Psi_\nu(x_1, x_2)$ a factor
which corresponds to the motion of the center of mass with momentum $\vec{q}$

$$\Psi_\nu(x_1,\ x_2) = \Psi_{\omega,q}(\vec{r}_1 - \vec{r}_2, \sigma_1, \sigma_2) \exp\left\{i\vec{q}\left(\frac{\vec{r}_1 + \vec{r}_2}{2}\right)\right\}.$$

Assume the index $\nu = (\omega, \vec{q})$ includes the momentum $\vec{q}$ and, possibly, some
other indices ω. Then the relation (4.3) takes the form:

$$F(x_1, x_2;\ x_1', x_2') = 2\sum_{(\omega, q)} N_{\omega, q}\Psi^*_{\omega, q}(\vec{r}_1 - \vec{r}_2;\ \sigma_1, \sigma_2)\Psi_{\omega, q}(\vec{r}_1' - \vec{r}_2';\ \sigma_1', \sigma_2')$$

$$\times \exp\left\{i\vec{q}\left(\frac{\vec{r}_1' + \vec{r}_2' - \vec{r}_1 - \vec{r}_2}{2}\right)\right\}. \quad (4.5)$$

Here, $N_{\omega,q}$ represents the average number of particle pais in the state $\Psi_{\omega, q}$,
where each pair is counted once (and not twice as before). From (4.4) the
following normalization occurs in (4.5)

$$\sum_{\sigma_1, \sigma_2} \int \left|\Psi_{\omega, q}(\vec{r};\ \sigma_1,\ \sigma_2)\right|^2 d\vec{r} = \frac{1}{V}. \quad (4.6)$$

Let us now write the expansion (4.5) in integral form. We will switch
to a more convenient normalization.   Consider the wave function of a
pair, $\Psi_{\omega,q}(\vec{r};\ \sigma_1, \sigma_2)$ for a given *fixed* momentum $\vec{q}$. Since the correlation
between particles in the pair must disappear for large enough distances r,
the asymptotic form $(r \to \infty)$ of the considered functions is either equal
to zero or becomes a plane wave corresponding to relative free motion with
relative momentum $\vec{p}$. Let us look at the first possibility and assume in this
case:

$$\Psi_{\omega,q}(\vec{r};\ \sigma_1,\ \sigma_2) = \frac{1}{\sqrt{V}}\varphi_{\omega,q}(\vec{r};\ \sigma_1,\ \sigma_2)$$

so that

$$\sum_{\sigma_1, \sigma_2} \int \left|\varphi_{\omega, q}(\vec{r};\ \sigma_1,\ \sigma_2)\right|^2 d\vec{r} = 1. \quad (4.7)$$

Let us then say that $\varphi_{\omega, q}$ represents a *bound state of a particle pair*, with
total momentum $\vec{q}$. The discrete index ω indicates, so to speak, the number
of the bound state.

Now let us examine the other possibility where the asymptotic form of $\Psi_{\omega,q}$ is a plane wave representing the relative motion of the particles in the pair with momentum $\vec{p}$. Assume

$$\omega = (\vec{p},\, j), \quad \Psi_{\omega,q}(\vec{r};\, \sigma_1,\, \sigma_2) = \frac{1}{V}\, \varphi_{p,q,j}(\vec{r};\, \sigma_1,\, \sigma_2).$$

In this case we will say that $\varphi_{p,q,j}$ represents the wave function of an unbound or "dissociated" state of a particle pair. For $\varphi_{p,q,j}$ we have the normalization in this situation:

$$\sum_{\sigma_1,\sigma_2} \frac{1}{V} \int \left|\varphi_{p,q,j}(\vec{r};\, \sigma_1,\, \sigma_2)\right|^2 d\vec{r} = 1. \tag{4.8}$$

We can write the expansion (4.5) in the following form

$$
\begin{aligned}
F(x_1, x_2;\, x_1', x_2') = & 2 \sum_{(\omega, q)} \frac{N_{\omega, q}}{V} \varphi_{\omega, q}^*(\vec{r}_1 - \vec{r}_2;\, \sigma_1, \sigma_2) \varphi_{\omega, q}(\vec{r}_1' - \vec{r}_2';\, \sigma_1', \sigma_2') \\
& \times \exp\left\{ i\vec{q}\left(\frac{\vec{r}_1' + \vec{r}_2' - \vec{r}_1 - \vec{r}_2}{2} \right) \right\} \\
& + 2 \sum_{(p, q, j)} \frac{N_{p,q,j}}{V^2} \varphi_{p,q,j}^*(\vec{r}_1 - \vec{r}_2;\, \sigma_1, \sigma_2) \varphi_{p,q,j}(\vec{r}_1' - \vec{r}_2';\, \sigma_1', \sigma_2') \\
& \times \exp\left\{ i\vec{q}\left(\frac{\vec{r}_1' + \vec{r}_2' - \vec{r}_1 - \vec{r}_2}{2} \right) \right\}.
\end{aligned}
$$

In the limit $V \to \infty$, we go from the momentum sums to the corresponding integrals and obtain

$$
\begin{aligned}
F(x_1, x_2;\, x_1', x_2') = & 2 \sum_{(\omega)} \int d\vec{q}\, w(\omega, q) \varphi_{\omega, q}^*(\vec{r}_1 - \vec{r}_2;\, \sigma_1, \sigma_2) \varphi_{\omega, q}(\vec{r}_1' - \vec{r}_2';\, \sigma_1', \sigma_2') \\
& \times \exp\left\{ i\vec{q}\left(\frac{\vec{r}_1' + \vec{r}_2' - \vec{r}_1 - \vec{r}_2}{2} \right) \right\} \\
& + 2 \sum_{(j)} \int d\vec{p}\, d\vec{q}\, w_j(p, q) \varphi_{p,q,j}^*(\vec{r}_1 - \vec{r}_2;\, \sigma_1, \sigma_2) \varphi_{p,q,j}(\vec{r}_1' - \vec{r}_2';\, \sigma_1', \sigma_2') \\
& \times \exp\left\{ i\vec{q}\left(\frac{\vec{r}_1' + \vec{r}_2' - \vec{r}_1 - \vec{r}_2}{2} \right) \right\}. \tag{4.9}
\end{aligned}
$$

As seen

$$w(\omega, q)d\vec{q}$$

in this formula represents the number density of bound state pairs with momentum $\vec{q}$ in an infinitesimally small momentum volume $d\vec{q}$;

$$w_j(p, q)d\vec{q}\,d\vec{q}$$

represents the number density of unbound pairs with relative momentum $\vec{p}$ and center of mass momentum $\vec{q}$ in the infinitesimally small volumes $d\vec{p}$ and $d\vec{q}$. Let us take any wave function of the bound state:

$$\varphi_{\omega,q}(\vec{r},\, \sigma_1,\, \sigma_2).$$

If the linear dimension, l, of that space region in which $\varphi_{\omega,q}$ is essentially localized is considerably smaller than the average distance, $\vec{r}$, between particles (from different pairs) in the macroscopic system, then it is natural to say that *the given $\varphi_{\omega,q}$ corresponds to a molecule* composed of two particles which is in the state ω and moves with the momentum $\vec{q}$. In the case where l is of the same order of magnitude or larger than $\vec{r}$, then we can add prefixes of "quasi" or "pseudo" to the word "molecule".

Let us compare the integral representation (4.9) with the representation of the simple average,

$$F(x,\, x') = \langle\Psi^\dagger(x)\Psi(x') = \Delta(\sigma - \sigma')\int d\vec{q}\,w(q)\,e^{i\vec{r}(\vec{r}\,' - \vec{r})}. \qquad (4.10)$$

We see that although (4.10) describes the distribution of the particles by "single particle states", i.e. the plane waves, the formula (4.9) characterizes the distribution of the particles by "pair states". With the above correlation function we can introduce in similar manner the concept of wave functions for a group of three or more particles. [7]

We recall at this point that Schafroth in his early investigations proposed the hypothesis, which was later completely verified, that the phenomenon of superconductivity depends upon the formation of a condensate consisting of quasi-molecules, formed from pairs of electrons in the system of conduction electrons. In this connection consider the case of a fermion system (with the usual spin 1/2), with a condensate of quasi-molecule pairs, which are for example, in S states. In other words, we will consider the case where in formula (4.9) we put[k]

$$w(\omega, q) = \rho_0 \Delta(\omega - \omega_0)\delta(\vec{q}) + w_1(\omega, q)$$

$$\varphi_{\omega_0,0}(\vec{r}, \sigma_1, \sigma_2) = \in(\sigma_1)\Delta(\sigma_1 + \sigma_2)\frac{1}{\sqrt{2}}\varphi(r) \qquad (4.11)$$

where,

1. $w_1(\omega, q)$ and $w_j(p, q)$ correspond to the usual continuous particle pair state momentum distribution function.

2. $\varphi(r)$ is a real, radially symmetric function, and due to (4.7) its normalization is:

$$\int \varphi^2(r)\, d\vec{r} = 1.$$

We now write formula (4.9) in the following form

$$F(x_1, x_2; x_1', x_2')$$
$$= \rho_0 \in(\sigma_1)\in(\sigma_1')\Delta(\sigma_1 + \sigma_2)\Delta(\sigma_1' + \sigma_2')\varphi(\vec{r}_1 - \vec{r}_2)\varphi(\vec{r}_1' - \vec{r}_2')$$
$$+ 2\sum_{(\omega)} \int d\vec{q}\, w(\omega, q)\varphi_{\omega,q}^*(\vec{r}_1 - \vec{r}_2; \sigma_1, \sigma_2)\varphi_{\omega,q}(\vec{r}_1' - \vec{r}_2'; \sigma_1', \sigma_2')$$
$$\times \exp\left\{ i\vec{q}\left(\frac{\vec{r}_1' + \vec{r}_2' - \vec{r}_1 - \vec{r}_2}{2}\right)\right\}$$
$$+ 2\sum_{(j)} \int d\vec{p}\, d\vec{q}\, w_j(p, q)\varphi_{p,q,j}^*(\vec{r}_1 - \vec{r}_2; \sigma_1, \sigma_2)\varphi_{p,q,j}(\vec{r}_1' - \vec{r}_2'; \sigma_1', \sigma_2')$$
$$\times \exp\left\{ i\vec{q}\left(\frac{\vec{r}_1' + \vec{r}_2' - \vec{r}_1 - \vec{r}_2}{2}\right)\right\}. \qquad (4.12)$$

Here ρ_0 represents the bound pair number density in the condensate. Note that we neglect the problem of the existence of bound states *which are not in the condensate.* If such bound states do not exist we would then put $w_1(\omega, q) = 0$ in (4.12)

---

[k]Here $\Delta(S)$ is the discrete S-function:

$$\Delta(S) = \begin{cases} 1, & S = 0; \\ 0, & S \neq 0. \end{cases}$$

Let us consider, for example, the model dynamical system, studied in the section 2 and make the formula (2.37). For this system we obtain in the present notation:

$$
\begin{aligned}
F(x_1, x_2; &x_1', x_2') \\
&= \phi(\vec{r}_1 - \vec{r}_2)\phi(\vec{r}_1' - \vec{r}_2')\in(\sigma_1)\in(\sigma_1')\Delta(\sigma_1 + \sigma_2)\Delta(\sigma_1' + \sigma_2') \\
&\quad + F(\vec{r}_1 - \vec{r}_1')F(\vec{r}_2 - \vec{r}_2')\Delta(\sigma_1 - \sigma_1')\Delta(\sigma_2' - \sigma_2') \\
&\quad - F(\vec{r}_2 - \vec{r}_1')F(\vec{r}_1 - \vec{r}_2')\Delta(\sigma_2 - \sigma_1')\Delta(\sigma_1' - \sigma_2') \qquad (4.13)
\end{aligned}
$$

where

$$
\phi(\vec{r}) = \phi(0, \vec{r}); \quad F(\vec{r}) = F(0, \vec{r}).
$$

Substituting the integral representation,

$$
F(\vec{r}) = \int w(k)\, e^{-i(\vec{k}\cdot\vec{r})}\, d\vec{k}
$$

we bring (4.13) into the form (4.12).

Note that in this case $w_1(\omega, q) = 0$, and the pair states $\varphi_{p,q,j}$ are regular plane waves. Consequently we have only one bound state with total momentum zero, and the rest of the pair states which have total momentum $q \neq 0$ will be the same as those for non-interacting particles. Such a result is completely natural since in our model system interactions are possible only between particle pairs having total momentum equal to zero.

Let us now go back to the "general case of Schafroth" and apply the principle of correlation weakening. We break up the arguments of the function (4.1) into two groups

$$
(x_1, x_2); \quad (x_2', x_2')
$$

and increase without limit the distance, $\vec{r}$, between points from different groups. Then, due the principle of correlation weakening the corresponding asymptotic form for F will be:

$$
\langle \Psi^\dagger(x_1)\Psi^\dagger(x_2)\rangle\langle \Psi(x_1')\Psi(x_2')\rangle.
$$

Alternatively, from (4.12) we obtain for this asymptotic form the product,

$$
\rho_0\in(\sigma_1)\in(\sigma_1')\Delta(\sigma_1 + \sigma_2)\Delta(\sigma_1' + \sigma_2')\varphi(\vec{r}_1 - \vec{r}_2)\varphi(\vec{r}_1' - \vec{r}_2').
$$

Thus we can write

$$\langle\Psi^\dagger(x_1)\Psi^\dagger(x_2)\rangle = \sqrt{\rho_0}\,\epsilon(\sigma_1)\Delta(\sigma_1+\sigma_2)\varphi(\vec{r}_1-\vec{r}_2)$$
$$\langle\Psi(x_1')\Psi(x_2')\rangle = \sqrt{\rho_0}\,\epsilon(\sigma_1')\Delta(\sigma_1'+\sigma_2')\varphi(\vec{r}_1'-\vec{r}_2'). \qquad (4.14)$$

Hence, we see that these quasi-averages are not zero and they do not satisfy the selection rules which are specified by the law of conservation of the number of particles. Thus, *if a condensate of quasi-molecules of pairs exists for this state of statistical equilibrium, then this statistical equilibrium state will be degenerate. The degeneracy here is dependent upon the law of conservation of the number of particles.*

In conclusion note that the formulas (4.14) give a simple interpretation of the "anomalous quasi-averages" $\langle\Psi^\dagger\Psi^\dagger\rangle$, $\langle\Psi\Psi\rangle$. That is, these quasi-averages are proportional to the wave function of quasi-molecule in the condensate. The normalization

$$\sum_{(\sigma_2)}\int\left|\langle\Psi(x_1)\Psi(x_2)\rangle\right|^2 d\vec{r}_2 = \rho_0 \qquad (4.15)$$

gives the number density of such quasi-molecules.

5.   Certain Inequalities

We now investigate averages of the product of two operators; $\langle AB\rangle$, as bilinear forms A and B (linear with respect to each of these operators). If symbol $\langle\ldots\rangle$ represents a regular average then one can easily see that

$$\langle AB\rangle^* = \langle B^\dagger A^\dagger\rangle$$
$$\langle AA^\dagger\rangle \geq 0. \qquad (5.1)$$

In as much as the quasi-averages can be investigated can be investigated as regular averages taken for the system with an infinitely small variation Hamiltonian, then the same relations (5.1) hold for quasi-averages. Further if $A(t)$ and $B(t)$ are operators in the Heisenberg picture, then in the case of regular averages the following spectral formulas can be proven

$$\langle B(\tau)A(t)\rangle = \int\limits_{-\infty}^{+\infty} J_{A,B}(\omega)\,e^{-i\omega(t-\tau)}\,d\omega,$$

$$\langle A(t)B(\tau)\rangle = \int\limits_{-\infty}^{+\infty} J_{A,B}(\omega)\, e^{\omega/\theta}\, e^{-i\omega(t-\tau)}\, d\omega, \tag{5.2}$$

where the spectral density $J_{A,B}(\omega)$ is a bilinear form with respect to the operators A and B. Due to the argument just presented, the same formulas remain correct for quasi-averages.

Using the properties (5.1) and (5.2), we shall now establish certain inequalities which will be needed in the next chapter. Here the symbol $\langle\ldots\rangle$ can represent a quasi-averages as well as a regular average.

First of all let us prove that,

$$J_{A,A^{\dagger}}(\omega) \geq 0. \tag{5.3}$$

For this assume an arbitrary function, $f(\omega)$, which is sufficiently regular enough and which goes to zero at infinity. If we are able to prove that for every such function

$$\int\limits_{-\infty}^{+\infty} J_{A,A^{\dagger}}(\omega)|f(\omega)|^2 d\omega \geq 0 \tag{5.4}$$

holds, then (5.3) will thereby established in as much as we can always localize $|f(\omega)|^2$ in as narrow a vicinity as needed, of any point ω_0.

In order to prove the inequality (5.4) construct the function

$$h(t) = \frac{1}{2\pi} \int\limits_{-\infty}^{+\infty} f(\omega)\, e^{i\omega t}\, d\omega$$

and note that

$$f(\omega) = \int\limits_{-\infty}^{+\infty} h(t)\, e^{-i\omega t}\, dt.$$

Thus we have

$$\int\limits_{-\infty}^{+\infty} J_{A,A^{\dagger}}(\omega)|f(\omega)|^2 d\omega =$$

$$\int\limits_{-\infty}^{+\infty} dt \int\limits_{-\infty}^{+\infty} d\tau \int\limits_{-\infty}^{+\infty} d\omega\, J_{A,A^{\dagger}}(\omega)h(t)h^*(\tau)\, e^{-i\omega(t-\tau)}$$

$$= \int\limits_{-\infty}^{+\infty} dt \int\limits_{-\infty}^{+\infty} d\tau \langle A^\dagger(\tau) A(t) \rangle h(t) h^*(\tau) = \langle \mathscr{U} \mathscr{U}^\dagger \rangle,$$

where

$$\mathscr{U} = \int\limits_{-\infty}^{+\infty} A^\dagger(\tau) h^*(\tau)\, d\tau, \quad \mathscr{U}^\dagger = \int\limits_{-\infty}^{+\infty} A(t) h(t)\, dt.$$

The inequality (5.4) follows from this using (5.1). Now we prove that

$$J^*_{A,B}(\omega) = J_{B^\dagger, A^\dagger}(\omega). \tag{5.5}$$

We have

$$\langle A^\dagger(\tau) B^\dagger(t) \rangle = \int\limits_{-\infty}^{+\infty} J_{B^\dagger, A^\dagger}(\omega) e^{-i\omega(t-\tau)}\, d\omega$$

and thus

$$\langle A^\dagger(t) B^\dagger(\tau) \rangle = \int\limits_{-\infty}^{+\infty} J_{B^\dagger, A^\dagger}(\omega) e^{i\omega(t-\tau)}\, d\omega. \tag{5.6}$$

Alternatively,

$$\langle A^\dagger(t) B^\dagger(\tau) \rangle = \langle B(\tau) A^{(t)} \rangle^*$$

$$= \left\{ \int\limits_{-\infty}^{+\infty} J_{A,B}(\omega) e^{-i\omega(t-\tau)}\, d\omega \right\}^* = \int\limits_{-\infty}^{+\infty} J^*_{A,B}(\omega) e^{i\omega(t-\tau)}\, d\omega. \tag{5.7}$$

By comparing (5.6) and (5.7) we obtain (5.5). Now let $Z(A, B)$ be an arbitrary bilinear form of A, B possessing the properties,

$$Z(A, A^\dagger) \geq 0,$$
$$\{Z(A, B)\}^* = Z(B^\dagger, A^\dagger). \tag{5.8}$$

We shall demonstrate that the following inequality always exists,

$$|Z(A, B)|^2 \leq Z(A, A^\dagger) Z(B, B^\dagger). \tag{5.9}$$

For the prove, not that on the basis (5.8)

$$Z(xA + y^* B^\dagger, x^* A^\dagger + yB) \geq 0,$$

where x, y are arbitrary c-numbers. From this, by expansion we obtain,

$$xx^* Z(A, A^*) + xy Z(A, B) + y^* x^* Z(B^\dagger, A^\dagger) + y^* y Z(B^\dagger, B) \geq 0. \qquad (5.10)$$

If we take

$$x^* = -Z(A, B), \quad x = -\{Z(A, B)\}^* = -Z(B^\dagger, A^\dagger), \quad y = y^* = Z(A, A^\dagger)$$

then we obtain,

$$-|Z(A, B)|^2 Z(A, A^*) + \{Z(A, A^\dagger)\}^2 Z(B^\dagger, B) \geq 0.$$

Form this, if $Z(A, A^\dagger) \neq 0$, we obtain the inequality (5.9). It remains for us to show that if

$$Z(A, A^\dagger) = 0 \qquad (5.11)$$

then we also have

$$Z(A, B) = 0 \qquad (5.12)$$

For this purpose let us substitute (5.11) into (5.10). We then set

$$x^* = -Z(A, B)R, \quad x = -Z(B^\dagger, A^\dagger)R, \quad y = y^* = 1,$$

where R is an arbitrary positive number. We obtain

$$-2R|Z(A, B)|^2 + Z(B^\dagger, B) \geq 0. \qquad (5.13)$$

Let R approach infinity. Then, if (5.12) does not hold, the left side of (5.13) must approach $-\infty$, which is not possible. This completes the proof of the inequality (5.9).

Now, note that the choice,

$$Z(A, B) = J_{A,B}(\omega)$$

satisfies the condition (5.8), since the relationships (5.3) and (5.5) hold for $J_{A,B}$. Thus, in this case, we can make use of the inequality (5.9) and write:

$$|J_{A,B}|^2 \leq J_{A,A^\dagger}(\omega) J_{B^\dagger,B}(\omega). \qquad (5.14)$$

We can also take

$$Z(A, B) = \frac{1}{2\pi} \int\limits_{-\infty}^{+\infty} J_{A,B}(\omega) \frac{e^{\omega/\theta} - 1}{\omega} d\omega \qquad (5.15)$$

since the conditions (5.8) are again satisfies because of (5.3), (5.5), and the positive nature of the function

$$\frac{e^{\omega/\theta} - 1}{\omega}.$$

Let us relate the function (5.15) to Green's functions. We shall investigate the following [8] retarded and advanced Green's functions

$$\ll A(t), B(\tau) \gg_r = -i\theta(t - \tau)\langle A(t)B(\tau) - B(\tau)A(t)\rangle,$$
$$\ll A(t), B(\tau) \gg_a = i\theta(\tau - t)\langle A(t)B(\tau) - B(\tau)A(t)\rangle. \qquad (5.16)$$

On the basis of the spectral representation, (5.2), it is clear that their Fourier transforms, due to the nature of the step function, $\theta(t)$, will be respectively

$$\ll A, B \gg_{E+i\varepsilon}, \quad \ll A, B \gg_{E-i\varepsilon}.$$

We see that,

$$\ll A, B \gg_E$$

is given by the following formula as a function of the complex variable E

$$\ll A, B \gg_E = \frac{1}{2\pi} \int_{-\infty}^{+\infty} J_{A,B}(\omega) \frac{e^{\omega/\theta} - 1}{E - \omega} d\omega. \qquad (5.17)$$

We can see from this that the expression (5.15) may be written as

$$- \ll A, B \gg_{E=0} . \qquad (5.18)$$

Thus the inequality, (5.9), takes the form

$$| \ll A, B \gg_{E=0} |^2 \leq \ll A, A^\dagger \gg_{E=0} \ll B^\dagger, B \gg_{E=0} \qquad (5.19)$$

in the present case. We will apply this result later.

In conclusion, let us consider one important application of the Green's function of the type (5.18). Give the Hamiltonian H an infinitesimal increment, δH (independent of time). The corresponding variation of the average of an operator $A(t)$ will be given by, [8]

$$\delta\langle A\rangle = \langle A\rangle_{H+\delta H} - \langle A\rangle_H = 2\pi \ll A, \delta H \gg_{E=0} . \qquad (5.20)$$

**Part B.   CHARACTERISTIC THEOREMS ABOUT THE
$1/q^2$ TYPE INTERACTION IN THE THEORY
OF SUPERCONDUCTIVITY OF BOSE AND
FERMI SYSTEMS**

6.   Symmetry Properties of Basic Green's Functions for Bose Systems in the Presence of a Condensate

Consider a dynamical system of identical spinless Bose particles with a Hamiltonian of the form,

$$H = -\frac{1}{2m}\int_V \Psi^\dagger \Delta\Psi \, d\vec{r} - \mu \int_V \Psi^\dagger \Psi \, d\vec{r} + U(\Psi^\dagger, \Psi)$$

$$= \sum \left(\frac{k^2}{2m} - \mu\right) a_k^\dagger a_k + U(\Psi^\dagger, \Psi), \tag{6.1}$$

$$U(\Psi^\dagger, \Psi) = \frac{1}{2}\iint_V \phi(\vec{r}_1 - \vec{r}_2)\Psi^\dagger(\vec{r}_1)\Psi^\dagger(\vec{r}_2)\Psi(\vec{r}_2)\Psi(\vec{r}_2) \, d\vec{r}_1 \, d\vec{r}_2. \tag{6.2}$$

Here $\psi(r)$ is a real function of distance and represents the interaction energy of a pair of particle. In addition, we limit ourselves to a system at a given temperature θ, with a Bose condensate.

As previously noted in section 3, the corresponding statistical equilibrium state must be degenerate, in such a case. The degeneracy here depends upon the law of conservation of the number of particles. In order to remove the degeneracy consider the Hamiltonian

$$H_\nu = H - \nu\sqrt{V}(a_0 + a_0^\dagger), \tag{6.3}$$

which contains additional infinitesimal terms of the form,

$$-\nu\sqrt{V}(a_0 + a_0^\dagger), \quad \nu > 0. \tag{6.4}$$

We are assuming that other types of degeneracy do not exist[1] and, thus, the introduction of the term (6.4) is sufficient for the removal of the degeneracy.

In this way for quasi-averages $\langle A(t)B(\tau)\rangle$, where A, $B = a_{\pm k}$, $a_{\pm k}^\dagger$ ($\vec{k} \neq 0$), the selection rules resulting from the law of conservation of momentum

---

[1]Actually we are here assuming that our system is in a spatially homogeneous phase with no molecules of two or more particles.

must be satisfied; but, the selection rules specified by the law of conservation of the number of particles can be violated.

We introduce the following Green's Functions

$$\ll A(t), B(\tau) \gg^{\mathrm{ret}} = -i\theta(t - \tau)\langle A(t)B(\tau) - B(\tau)A(t)\rangle,$$
$$\ll A(t), B(\tau) \gg^{\mathrm{adv}} = i\theta(\tau - t)\langle A(t)B(\tau) - B(\tau)A(t)\rangle,$$
$$\ll A(t), B(\tau) \gg^{\mathrm{c}} = -\langle T(A(t)B(\tau))\rangle$$
$$= -i\{\theta(t - \tau)\langle A(t)B(\tau)\rangle + \theta(\tau - t)\langle B(\tau)A(t)\rangle. \quad (6.5)$$

We determine their "energy representation" with the Fourier integrals:

$$\ll A(t), B(\tau) \gg = \int_{-\infty}^{+\infty} \ll A, B \gg \mathrm{e}^{-iE(t-\tau)}dE. \quad (6.6)$$

Using the spectral formulas,

$$\langle B(\tau)A(t)\rangle = \int_{-\infty}^{+\infty} J_{A,B}\,\mathrm{e}^{-i\omega(t-\tau)}d\omega,$$

$$\langle A(t)B(\tau)\rangle = \int_{-\infty}^{+\infty} J_{A,B}\,\mathrm{e}^{\omega/\theta}\mathrm{e}^{-i\omega(t-\tau)}d\omega,$$

we obtain

$$\ll A, B \gg_E^{\mathrm{ret}} = \frac{1}{2\pi} \int_{-\infty}^{+\infty} (\mathrm{e}^{\omega/\theta} - 1)J_{A,B}(\omega)\frac{d\omega}{E - \omega + i\varepsilon}$$

$$\ll A, B \gg_E^{\mathrm{adv}} = \frac{1}{2\pi} \int_{-\infty}^{+\infty} (\mathrm{e}^{\omega/\theta} - 1)J_{A,B}(\omega)\frac{d\omega}{E - \omega - i\varepsilon}$$

$$\ll A, B \gg_E^{\mathrm{c}} = \frac{1}{2\pi} \int_{-\infty}^{+\infty} J_{A,B}(\omega)\left\{\frac{\mathrm{e}^{\omega/\theta}}{E - \omega + i\varepsilon} - \frac{1}{E - \omega + i\varepsilon}\right\} d\omega.$$

For the special case of zero temperatures the spectral formulas for these averages can be written in the form,

$$\langle A(t)B(\tau)\rangle = \int_{0}^{+\infty} I_{A,B}(\omega)\mathrm{e}^{-i\omega(t-\tau)}d\omega$$

$$\langle B(\tau)A(t), \rangle = \int_{-\infty}^{0} I_{A,B}(\omega)e^{-i\omega(t-\tau)}d\omega.$$

We then have for the energy representation of the Green's functions

$$\ll A, B \gg_E^{\mathrm{ret}} = \frac{1}{2\pi} \int_{-\infty}^{+\infty} \frac{\in(\omega)I_{A,B}(\omega)}{E - \omega + i\varepsilon} d\omega$$

$$\ll A, B \gg_E^{\mathrm{adv}} = \frac{1}{2\pi} \int_{-\infty}^{+\infty} \frac{\in(\omega)I_{A,B}(\omega)}{E - \omega - i\varepsilon} d\omega$$

$$\ll A, B \gg_E^{\mathrm{c}} = \frac{1}{2\pi} \int_{-\infty}^{+\infty} \frac{\in(\omega)I_{A,B}(\omega)}{E - \omega + i\varepsilon\in(\omega)} d\omega,$$

where

$$\in(\omega) = \begin{cases} +1, & \omega > 0, \\ -1, & \omega < 0. \end{cases}$$

Obviously, the retarded and advanced Green's functions,

$$\ll A, B \gg_E^{\mathrm{ret}}, \quad \ll A, B \gg_E^{\mathrm{adv}},$$

are boundary values of the function of the complex variable E,

$$\ll A, B \gg = \frac{1}{2\pi} \int_{-\infty}^{+\infty} J_{A,B}(\omega) \frac{e^{\omega/\theta} - 1}{E - \omega} d\omega. \tag{6.7}$$

In general the causal Green's functions, $\ll A, B \gg_E^{\mathrm{c}}$, possesses this property, only in the limit of zero temperature. Now, note that from the definition (6.5) the following holds,

$$\ll A(t), B(\tau) \gg^{\mathrm{c}} = \ll B(\tau), A(t) \gg^{\mathrm{c}}$$
$$\ll A(t), B(\tau) \gg^{\mathrm{ret}} = \ll B(\tau), A(t) \gg^{\mathrm{adv}}.$$

From (6.6) we obtain,

$$\ll A, B \gg_E^{\mathrm{c}} = \ll B, A \gg_{-E}^{\mathrm{c}} \tag{6.8}$$

and also,

$$\ll A, B \gg_E^{\text{ret}} = \ll B, A \gg_{-E}^{\text{adv}} . \tag{6.9}$$

By continuing relation (6.9) into the complex E plane, we can convince ourselves that for the complex function, (6.7), the following equality holds,

$$\ll A, B \gg_E = \ll B, A \gg_{-E} . \tag{6.10}$$

Let us investigate the matrix Green's function:

$$G(E, k) = \begin{vmatrix} G_{11}(E, K); & G_{21}(E, K) \\ G_{12}(E, K); & G_{22}(E, K) \end{vmatrix} \tag{6.11}$$

where

$$G_{11}(E, k) = \ll a_k, a_k^\dagger \gg_E, \quad G_{21}(E, k) = \ll a_{-k}^\dagger, a_k^\dagger \gg_E,$$
$$G_{12}(E, k) = \ll a_k, a_{-k} \gg_E, \quad G_{22}(E, k) = \ll a_{-k}^\dagger, a_{-k} \gg_E . \tag{6.12}$$

For $\ll A, B \gg_E$ we mean either the function of the complex variable, (6.7), or causal Green's function for real E. In both cases, due to (6.8) and (6.10), we have,

$$G_{22}(E, k) = G_{11}(-E, -k);$$
$$G_{\alpha\beta}(E, k) = G_{\alpha\beta}(-E, -k); \quad \text{if } \alpha \neq \beta. \tag{6.13}$$

Now note that the Hamiltonian H_ν is invariant with respect to the canonical transformation,

$$a_k \rightarrow a_{-k}, \quad a_k^\dagger \rightarrow a_{-k}^\dagger.$$

Because of this, the averages

$$\langle a_k^\dagger(t) a_k(\tau) \rangle, \quad \langle a_k(t) a_k^\dagger(\tau) \rangle$$
$$\langle a_k(t) a_{-k}(\tau) \rangle, \quad \langle a_{-k}^\dagger(t) a_k^\dagger(\tau) \rangle$$

can not change under the transformation $\vec{k} \rightarrow -\vec{k}$. Consequently, for the Green's functions we will also have

$$G_{\alpha\beta}(E, k) = G_{\alpha\beta}(E, -k). \tag{6.14}$$

Note further, that since all the coefficients are real in the expression for the Hamiltonian H_ν, the corresponding equations of motion must be invariant with respect to time inversion ($t \to -t$, accompanied by the substitution i for $-i$).

Thus, the average

$$\langle a_{-k}(\tau)a_k(t)\rangle = \int_{-\infty}^{+\infty} J_k(\omega)\,e^{-i\omega(t-\tau)}\,d\omega \qquad (6.15)$$

does not change under the transformation $t \to -t$, $\tau \to -\tau$, $i \to -i$. Because of this:

$$\int_{-\infty}^{+\infty} J_k(\omega)\,e^{-i\omega(t-\tau)}\,d\omega = \int_{-\infty}^{+\infty} J_k^*(\omega)\,e^{-i\omega(t-\tau)}\,d\omega.$$

From this it follows that the spectral intensity, $J_k(\omega)$, is a real function, i.e.

$$J_k^*(\omega) = J_k(\omega). \qquad (6.16)$$

Thus from (6.15) we have

$$\langle a_k^\dagger(t)a_{-k}^\dagger(\tau)\rangle = \langle a_{-k}(\tau)a_k(t)\rangle^* = \int_{-\infty}^{+\infty} J_k(\omega)\,e^{i\omega(t-\tau)}\,d\omega$$

and

$$\langle a_k^\dagger(\tau)a_{-k}^\dagger(t)\rangle = \langle a_{-k}(\tau)a_k(t)\rangle.$$

The corresponding relation for Green's functions is

$$\ll a_k^\dagger, a_{-k}^\dagger \gg_E = \ll a_{-k}, a_k \gg_E .$$

Or, in our notation we have,

$$G_{21}(E,k) = G_{12}(E,k). \qquad (6.17)$$

Let us now introduce the matrix, $\Sigma(E,k)$,

$$\Sigma(E,k) = \frac{1}{2\pi}G^{-1}(E,k) \qquad (6.18)$$

or,

$$2\pi\Sigma(E,k)G(E,k) = 1 \qquad (6.19)$$

where 1 is the unit matrix. We can interpret the matrix $\Sigma(E,k)$ as the total "mass operator". In the particular case of zero temperature, when the Feynman diagram technique is applicable, $\Sigma(E,k)$ represents the usual "self-energy" part.

It is also clear from the definition (6.18) that the elements $\Sigma_{\alpha\beta}(E,k)$ always satisfy the same symmetry relations (6.13), (6.14), (6.17) as $G_{\alpha\beta}(E,k)$. Explicitly writing out the matrix equality, (6.19), we obtain:

$$\Sigma_{11}(E,k)G_{11}(E,k) + \Sigma_{12}(E,k)G_{21}(E,k) = \frac{1}{2\pi},$$

$$\Sigma_{21}(E,k)G_{11}(E,k) + \Sigma_{22}(E,k)G_{21}(E,k) = 0.$$

However, in view of the above, we have

$$\Sigma_{21}(E,k) = \Sigma_{12}(E,k), \quad \Sigma_{22}(E,k) = \Sigma_{11}(-E,k).$$

Thus we can write

$$\Sigma_{11}(E,k) \ll a_k, a_k^\dagger \gg_E + \Sigma_{12}(E,k) \ll a_{-k}^\dagger, a_k^\dagger \gg_E = \frac{1}{2\pi},$$

$$\Sigma_{21}(E,k) \ll a_k, a_k^\dagger \gg_E + \Sigma_{11}(-E,k) \ll a_{-k}^\dagger, a_k^\dagger \gg_E = 0.$$

(6.20)

These functions $\Sigma_{\alpha\beta}$ possess the following symmetry properties,

$$\Sigma_{11}(E,k) = \Sigma_{11}(E,-k), \quad \Sigma_{12}(E,k) = \Sigma_{12}(E,-k);$$

$$\Sigma_{12}(E,k) = \Sigma_{12}(-E,k).$$

(6.21)

In view of this, we can obtain from (6.20) the following formulas which express the Green's functions in terms of Σ_{11} and Σ_{12}

$$\ll a_k, a_k^\dagger \gg_E = \frac{1}{2\pi} \frac{\Sigma_{11}(-E,k)}{\Sigma_{11}(E,k)\Sigma_{11}(-E,k) - \Sigma_{12}^2(E,k)}$$

$$\ll a_{-k}^\dagger, a_k^\dagger \gg_E = -\frac{1}{2\pi} \frac{\Sigma_{12}(E,k)}{\Sigma_{11}(E,k)\Sigma_{11}(-E,k) - \Sigma_{12}^2(E,k)}.$$

$$k \neq 0$$

(6.22)

7. Model with a Condensate

We observe that since the coefficients in the hamiltonian H_ν are real, the expression $\langle a_0 \rangle$ is also real and thus

$$\left\langle \frac{a_0}{\sqrt{V}} \right\rangle = \left\langle \frac{a_0^\dagger}{\sqrt{V}} \right\rangle.$$

(7.1)

Let us consider the average,

$$\left\langle \frac{a_0^\dagger a_0}{V} \right\rangle = \rho_0$$

and note that

$$\rho_0 = \frac{1}{V^2} \iint_V \langle \Psi^\dagger(\vec{r}_1)\Psi(\vec{r}_2)\rangle \, d\vec{r}_1 \, d\vec{r}_2. \tag{7.2}$$

Since $V \to \infty$ here, it is clear that all contribution to the integral in (7.2) comes the region where the points $\vec{r}_1$ and $\vec{r}_2$ are infinitely separated. Thus, by applying the principle of correlation weakening, we obtain asymptotically,

$$\rho_0 = \frac{1}{V} \int_V \langle \Psi^\dagger(\vec{r}_1), d\vec{r}_1 \, \frac{1}{V} \int_V \langle \Psi(\vec{r}_2)\rangle \, d\vec{r}_2 = \left\langle \frac{a_0^\dagger}{\sqrt{V}} \right\rangle \left\langle \frac{a_0}{\sqrt{V}} \right\rangle,$$

where from (7.1), we have

$$\left\langle \frac{a_0}{\sqrt{V}} \right\rangle = \left\langle \frac{a_0^\dagger}{\sqrt{V}} \right\rangle = \sqrt{\rho_0}. \tag{7.3}$$

Let us now look at expressions of the type,

$$\langle \ldots A(t_\alpha) \ldots \varphi(t_\beta, \vec{r}_\beta) \ldots \rangle$$

in which

$$A = \frac{a_0}{\sqrt{V}}, \ \frac{a_0^\dagger}{\sqrt{V}}; \quad \varphi = \Psi, \ \Psi^\dagger.$$

Let us apply the principle of correlation weakening to them as in (7.2). We will obtain the equality,[m]

$$\langle \ldots A(t_\alpha) \ldots \varphi(t_\beta, \vec{r}_\beta) \ldots \rangle = \left(\sqrt{\rho_0}\right)^l \langle \ldots \varphi(t_\beta, \vec{r}_\beta) \ldots \rangle$$

where l is the number of A's involved in the average. We introduce the relations,

$$\tilde{\varphi} = \eta \frac{a_0}{\sqrt{V}} + \Psi_1, \quad \eta \frac{a_0^\dagger}{\sqrt{V}} + \Psi_1^\dagger$$

---

[m]Of course, equalities of this type are asymptotic in character and become exact only in the limit $V \to \infty$. However, since we are always dealing here with limit relations, we will not mention this explicitly.

$$\eta = 0,\ 1; \quad \Psi_1 = \frac{1}{\sqrt{V}} \sum_{k \neq 0} a_k\, e^{i(\vec{k}\,\vec{r})}.$$

Then on the basis of above result we find that in the *calculation of averages of the form:*

$$\langle \ldots \tilde{\varphi}(t_\alpha, \vec{r}_\alpha) \ldots \rangle$$

we can substitute the c-number $\sqrt{N_0}$, *(where* $N_0 = \rho_0 V$*) for the operators* a_0 *and* $a_0^\dagger$, *involved in* $\tilde{\varphi}$. Taking this property into account, we will show that the problem with the Hamiltonian H_ν can be reduced to the problem with the $H_\nu(N_0)$, which is obtained from the expression (6.3) for H_ν with the substitution,

$$\Psi(\vec{r}) \to \sqrt{\rho_0} + \Psi_1(\vec{r}); \quad \Psi^\dagger(\vec{r}) \to \sqrt{\rho_0} + \Psi_1^\dagger(\vec{r})$$

i.e., by the substitution of the *c-number* $\sqrt{N_0}$ for the operators a_0 and $a_0^\dagger$ in H_ν. Let us now examine the system of "double time" Green's functions of the type

$$\ll \varphi_1(t, \vec{r}_1) \ldots \varphi(t, \vec{r}_s); \varphi(\tau, \vec{x}_1) \ldots \varphi(\tau, \vec{x}_m) \gg_{\text{adv}}^{\text{ret}}$$
$$= \frac{-\theta(t - \tau)}{\theta(t - \tau)} \langle [\varphi_1(t, \vec{r}_1) \ldots \varphi(t, \vec{r}_s); \varphi(\tau, \vec{x}_1) \ldots \varphi(\tau, \vec{x}_m)] \rangle \quad (7.4)$$

where $\varphi_1 = \Psi_1,\ \Psi_1^\dagger$.

In order to obtain a chain of equations connecting these functions, let us express the derivative

$$i\frac{\partial}{\partial t} \ll \varphi_1(t, \vec{r}_1) \ldots \varphi(t, \vec{r}_s); \varphi(\tau, \vec{x}_1) \ldots \varphi(\tau, \vec{x}_m) \gg \quad (7.5)$$

using the equations of motion. We have for the Hamiltonian H

$$i\frac{\partial \Psi(t, \vec{r})}{\partial t} = D(t; \vec{r}; \Psi, \Psi^\dagger) \equiv$$
$$\equiv \left(-\frac{1}{2m}\Delta - \mu \right) \Psi(t, \vec{r}) - \nu + \int \phi(\vec{r} - \vec{r}')\Psi^\dagger(t, \vec{r}')\Psi(t, \vec{r}')\, d\vec{r}'\, \Psi(t, \vec{r})$$

and, thus,

$$i\frac{\partial \Psi_1(t, \vec{r})}{\partial t} = D(t; \vec{r}; \Psi, \Psi^\dagger) - \frac{1}{V}\int_V D(t; \vec{r}; \Psi, \Psi^\dagger)\, d\vec{r},$$

$$-i\frac{\partial \Psi_1^\dagger(t, \vec{r})}{\partial t} = D^\dagger(t; \vec{r}; \Psi, \Psi^\dagger) - \frac{1}{V}\int_V D^\dagger(t; \vec{r}; \Psi, \Psi^\dagger)\, d\vec{r}.$$

Since D and $D^\dagger$ will enter into expression for the derivatives, (7.5), only as averages of the form

$$\langle \ldots \varphi_1 \ldots D \ldots \varphi_1 \rangle, \quad \langle \ldots \varphi_1 \ldots D^\dagger \ldots \varphi_1 \rangle$$

thus, we can perform in the expressions

$$D(t, \vec{r}; \Psi, \Psi^\dagger), \quad D^\dagger(t, \vec{r}; \Psi, \Psi^\dagger),$$

the following substitution,

$$\Psi \to \sqrt{\rho_0} + \Psi_1, \quad \Psi^\dagger \to \sqrt{\rho_0} + \Psi_1^\dagger.$$

Similarly, to deal with terms of the type

$$\ll \varphi_1(t, \vec{r}_1) \ldots \frac{\partial \varphi_1(t, \vec{r}_1)}{\partial t} \ldots, \ldots \varphi_1(\tau, \vec{x}_m) \gg \tag{7.6}$$

we will make use of equations of the type

$$i\frac{\partial \Psi_1(t, \vec{r})}{\partial t} = D(t, \vec{r}; \sqrt{\rho_0} + \Psi_1, \sqrt{\rho_0} + \Psi_1^\dagger)$$
$$- \frac{1}{V}\int_V D(t, \vec{r}; \sqrt{\rho_0} + \Psi_1, \sqrt{\rho_0} + \Psi_1^\dagger)\, d\vec{r},$$

$$-i\frac{\partial \Psi_1^\dagger(t, \vec{r})}{\partial t} = D^\dagger(t, \vec{r}; \sqrt{\rho_0} + \Psi_1, \sqrt{\rho_0} + \Psi_1^\dagger)$$
$$- \frac{1}{V}\int_V D^\dagger(t, \vec{r}; \sqrt{\rho_0} + \Psi_1, \sqrt{\rho_0} + \Psi_1^\dagger)\, d\vec{r}. \tag{7.7}$$

With the help of these equations the terms (7.6) will be introduced through differentiating Green's functions of the type under consideration.

In expression (7.5), in addition to the sum of terms of type (7.6), there will be present an additional "inhomogeneous member",

$$\delta(t - \tau)\langle[\varphi_1(t, \vec{r}_1) \ldots, \ldots \varphi_1(t, \vec{x}_m)]\rangle.$$

We will obtain these averages from averages of products containing no more than $s + m - z$ simultaneous field functions φ. These averages can again be expressed through Green's functions (of lower order) with the help of spectral representations.

For their calculation it will be convenient to use a momentum energy representation. Let us designate the Green's functions in this representation by

$$\mathscr{G}_{H_\nu}(E; p_1, \ldots p_s; q_1, \ldots q_m)$$

Then we can write the hierarchy of equations in the form

$$E\mathscr{G}_{H_\nu}(E; p_1, \ldots p_s; q_1, \ldots q_m) = \mathscr{L}(E; p_1, \ldots p_s; q_1, \ldots q_m; \mathscr{G}_{H_\nu})$$

where the $\mathscr{L}$ are expressions which depend upon the functions $\mathscr{G}_{H_\nu}$ of different orders.

Since the spectral representations are "universal", then it is clear from the above that only those terms in $\mathscr{L}$ which result from terms of the type (7.6) will depend upon the specific form of the Hamiltonian. In their evaluation we have made use of equations (7.7). It is not hard to see, however, *that these equations are the exact equations of motion using the Hamiltonian $H_\nu(N_0)$. In such a way $\mathscr{G}_{H_\nu}$ satisfy the same hierarchy of equations as the corresponding Green's functions $\mathscr{G}_{H_\nu(N_0)}$ with the Hamiltonian $H_\nu(N_0)$.*

Alternatively, the "boundary conditions" for the functions of the complex variable E

$$\mathscr{G}_{H_\nu}, \quad \mathscr{G}_{H_\nu(N_0)}$$

on the real E axis (a type of dispersion relation), which are defined by the spectral representations, are also identical.

From this we conclude that[n]

$$\mathscr{G}_{H_\nu} = \mathscr{G}_{H_\nu(N_0)} \tag{7.8}$$

or

$$\begin{aligned}
&\ll \varphi_1(t, \vec{r}_1) \ldots \varphi(t, \vec{r}_s); \varphi(\tau, \vec{x}_1) \ldots \varphi(\tau, \vec{x}_m) \gg_{H_\nu} \\
&= \ll \varphi_1(t, \vec{r}_1) \ldots \varphi(t, \vec{r}_s); \varphi(\tau, \vec{x}_1) \ldots \varphi(\tau, \vec{x}_m) \gg_{H_\nu(N_0)} .
\end{aligned} \tag{7.9}$$

Further, since averages of the type $\ll \varphi_1(t, \vec{r}_1) \ldots \varphi(t, \vec{r}_n) \gg$ can be expressed in terms of our Green's functions, then we will also have

$$\begin{aligned}
\langle \ldots \Psi^\dagger(\vec{r}_\alpha) \ldots \Psi(\vec{r}_\beta) \ldots \rangle_{H_\nu} &= \langle \ldots (\sqrt{\rho_0} + \Psi_1^\dagger(\vec{r}_\alpha)) \ldots (\sqrt{\rho_0} + \Psi_1(\vec{r}_\beta)) \ldots \rangle_{H_\nu} \\
&= \langle \ldots (\sqrt{\rho_0} + \Psi_1^\dagger(\vec{r}_\alpha)) \ldots (\sqrt{\rho_0} + \Psi_1(\vec{r}_\beta)) \ldots \rangle_{H_\nu(N_0)}.
\end{aligned} \tag{7.10}$$

From this follows the equality of the corresponding average energies, and thus, also the free energies for both dynamical systems.

Thus, we arrive at the conclusion that *the investigation of a dynamical system with the Hamiltonian H_ν can be reduced to the investigation of a "model system with a condensate", characterized by the Hamiltonian $H_\nu(N_0)$.* Note here that in the model with a condensate the value N_0 can be formally considered as an "arbitrary" external parameter.

In order to obtain an equation for N_0 we will again turn to the original system with the Hamiltonian H_ν and evaluate the expression,

$$0 = i \frac{d}{dt} \left\langle \frac{a_0}{\sqrt{V}} \right\rangle = \left\langle iV^{-1/2} \frac{da_0}{dt} \right\rangle \tag{7.11}$$

---

[n]The same conclusion would have been arrived at with the same method, if we had investigated, the Schwinger "multi-time" Green's functions of the type,

$$\langle T(\varphi_1(t_1, \vec{r}_1) \ldots \varphi_1(t_n, \vec{r}_n)) \rangle$$

instead of the "double time" Green's functions, (7.4). The corresponding Schwinger hierarchy of equations would then have been obtained. The Green's functions

$$\langle T(\ldots \varphi_1(t_\alpha, \vec{r}_\alpha) \ldots) \rangle_{H_\nu}$$

would again satisfy the same hierarchy of equations as

$$\langle T(\ldots \varphi_1(t_\alpha, \vec{r}_\alpha) \ldots) \rangle_{H_\nu(N_0)}.$$

The corresponding spectral properties defined by the structure of averages from the T-product would also be identical.

using the equation of motion (which "is missing" in the model system)

$$i\frac{da_0}{dt} = \frac{1}{2V}\sum_{(k_1',k_2',k_2)} \{\tilde{\phi}(k_1') + \tilde{\phi}(k_2')\}\Delta(\vec{k}_2 - \vec{k}_1' - \vec{k}_2')a^\dagger_{k_2}a_{k_2'}a_{k_1'}$$

$$- \nu\sqrt{V} - \mu a_0$$

where

$$\tilde{\phi}(k) = \int \phi(r)e^{i(\vec{k}\cdot\vec{r})}\,d\vec{r}.$$

Now, separating amplitudes with zero momentum, we find

$$i\frac{da_0}{dt} = -\mu a_0 + \frac{a^\dagger_0 a_0 a_0}{V}\tilde{\phi}(0) - \nu\sqrt{V}$$

$$+ \frac{1}{V}\sum_{(p\neq 0)}\{\tilde{\phi}(p) + \tilde{\phi}(0)\}a^\dagger_p a_p a_0 + \frac{1}{V}\sum_{(p\neq 0)}\tilde{\phi}(p)a^\dagger_0 a_p a_{-p}$$

$$+ \frac{1}{2V}\sum_{(p\neq 0, p_1\neq 0, p_2\neq 0)}\{\tilde{\phi}(p_1) + \tilde{\phi}(p_2)\}\Delta(p - p_1 - p_2)a^\dagger_p a_{p_1} a_{p_2}.$$

Substituting this equation into the right hand side of equation (7.11), we obtain

$$S = 0 \tag{7.12}$$

where

$$S = \left\langle\frac{a^\dagger_0 a_0 a_0}{V^{3/2}}\right\rangle_{H_\nu} - \nu - \mu\left\langle\frac{a_0}{\sqrt{V}}\right\rangle_{H_\nu}$$

$$+ \frac{1}{V^{3/2}}\sum_{(p\neq 0)}\{\tilde{\phi}(p) + \tilde{\phi}(0)\}\langle a^\dagger_p a_p a_0\rangle_{H_\nu} + \frac{1}{V^{3/2}}\sum_{(p\neq 0)}\tilde{\phi}(p)\langle a^\dagger_0 a_p a_{-p}\rangle_{H_\nu}$$

$$+ \frac{1}{2V^{3/2}}\sum_{(p\neq 0, p_1\neq 0, p_2\neq 0)}\{\tilde{\phi}(p_1) + \tilde{\phi}(p_2)\}\Delta(p - p_1 - p_2)\langle a^\dagger_p a_{p_1} a_{p_2}\rangle_{H_\nu}.$$

In accordance with the above discussion we can substitute $\sqrt{N_0}$ for a_0 and $a^\dagger_0$, and the remaining averages,

$$\langle a^\dagger_p, a_p\rangle_{H_\nu}, \quad \langle a_p, a_{-p}\rangle_{H_\nu}, \quad \langle a^\dagger_p, a_{p_1}a_{p_2}\rangle_{H_\nu}$$

can be replaced by the averages

$$\langle a^\dagger_p, a_p\rangle_{H_\nu(N_0)}, \quad \langle a_p, a_{-p}\rangle_{H_\nu(N_0)}, \quad \langle a^\dagger_p, a_{p_1}a_{p_2}\rangle_{H_\nu(N_0)}.$$

We then obtain

$$
\begin{aligned}
S \;=\;& \rho_0^{3/2}\tilde{\phi}(0) - \nu - \mu\rho_0^{1/2} \\
&+ \frac{\rho_0^{1/2}}{V} \sum_{(p\neq 0)} \{\tilde{\phi}(p) + \tilde{\phi}(0)\}\langle a_p^\dagger a_p\rangle_{H_\nu(N_0)} + \frac{\rho_0^{1/2}}{V} \sum_{(p\neq 0)} \tilde{\phi}(p)\langle a_p a_{-p}\rangle_{H_\nu(N_0)} \\
&+ \frac{1}{2V^{3/2}} \sum_{(p\neq 0, p_1\neq 0, p_2\neq 0)} \{\tilde{\phi}(p_1) + \tilde{\phi}(p_2)\}\Delta(p - p_1 - p_2)\langle a_p^\dagger a_{p_1} a_{p_2}\rangle_{H_\nu(N_0)}.
\end{aligned}
$$

$$(7.13)$$

On the other hand we have,

$$
H_\nu(N_0) = H(N_0) - 2\nu\sqrt{N_0}V^{1/2} \tag{7.14}
$$

where $H(N_0)$ is obtained from H by the substitution of $\sqrt{N_0}$ for the operators a_0 and $a_0^\dagger$, i.e.

$$
\begin{aligned}
H(N_0) \;=\;& -\mu N_0 + \frac{N_0^2}{2V}\tilde{\phi}(0) + \sum_{(k\neq 0)}\left(\frac{k^2}{2m} - \mu\right)a_k^\dagger a_k \\
&+ \frac{N_0}{V}\sum_{(k\neq 0)}\{\tilde{\phi}(k) + \tilde{\phi}(0)\}a_k^\dagger a_k + \frac{N_0}{2V}\sum_{(k\neq 0)}\tilde{\phi}(k)(a_k a_{-k} + a_{-k}^\dagger a_k^\dagger) \\
&+ \frac{\sqrt{N_0}}{2V}\sum_{(k\neq 0, k_1\neq 0, k_2\neq 0)}\{\tilde{\phi}(k_1) + \tilde{\phi}(k_2)\}\Delta(k - k_1 - k_2)a_{k_1}^\dagger a_{k_2}^\dagger a_k \\
&+ \frac{\sqrt{N_0}}{2V}\sum_{(k\neq 0, k_1\neq 0, k_2\neq 0)}\{\tilde{\phi}(k_1) + \tilde{\phi}(k_2)\}\Delta(k - k_1 - k_2)a_k^\dagger a_{k_1} a_{k_2} \\
&+ \frac{1}{2V}\sum_{(k_1\neq 0, k_2\neq 0, k_1'\neq 0, k_2'\neq 0)}\tilde{\phi}(k_1 - k_1')\Delta(k_1 + k_2 - k_1' - k_2') \\
&\times a_{k_1}^\dagger a_{k_2}^\dagger a_{k_2'} a_{k_1'}.
\end{aligned}
$$

$$(7.15)$$

From this, using (7.12), we can write

$$
\left\langle \frac{\partial H_\nu(N_0)}{\partial N_0}\right\rangle_{N_\nu(N_0)} = \frac{S + S^*}{2\sqrt{\rho_0}} = \frac{S}{\sqrt{\rho_0}} = 0. \tag{7.16}
$$

Let us construct an expression for the free energy with the Hamiltonian $N_\nu(N_0)$,

$$
F_\nu(N_0, \mu, \theta) = -\theta \ln \mathrm{Tr}\, e^{-\frac{H_\nu(N_0)}{\theta}}.
$$

In this expression N_0 is considered to be an arbitrary macroscopic parameter. We have,

$$\left\langle \frac{\partial H_\nu(N_0)}{\partial N_0} \right\rangle_{N_\nu(N_0)} = \frac{\partial F_\nu(N_0, \mu, \theta)}{\partial N_0}$$

and thus equation (7.16) for the determination of N_0 becomes,

$$\frac{\partial F_\nu(N_0, \mu, \theta)}{\partial N_0} = 0. \tag{7.17}$$

This equation obviously agrees with thermodynamic consideration. In reality, since in the model system N_0 is an outside parameter, its value for a given μ and θ must be such as to minimize the free energy. Equation (7.17), in this interpretation, expresses the required condition for a minimum.

Let us now consider how the auxiliary parameter ν (which must approach zero only after the limit $V \to 0$) is introduced into the calculation of our model system. Note, first that since H_ν differs from $H(N_0)$ only by the last term, $-2V\sqrt{N_0 V}$, then all the Green's functions and averages of field operator products will not, obviously, depend on ν *for the given* N_0. They will be the same as for the system with the Hamiltonian $H(N_0)$. We have

$$F_\nu = F - 2\nu\sqrt{N_0}\,V^{1/2} \tag{7.18}$$

where $F = F(N_0, \mu, \theta)$ represents the free energy of the system with the Hamiltonian $H(N_0)$. Because of this equation (7.17) can be written in the form

$$\frac{\partial F(N_0, \mu, \theta)}{\partial N_0} = \frac{\nu}{\sqrt{\rho}} \tag{7.19}$$

and we see that the parameter ν will enter the calculation only through N_0. Since we must take the limit $\nu \to 0$, we finally obtain,

$$\frac{\partial F(N_0, \mu, \theta)}{\partial N_0} = 0. \tag{7.20}$$

The model with a condensate was first introduced in 1947 [9]. The approximate Hamiltonian investigated there was diagonalized by means of $u-v$ transformation. Green's functions of the type (6.11) and the diagram technique were applied to it in 1958 [10]. This model has since been carefully considered [11], more recently.

8. The $1/q^2$ Theorem and its Application

Let us return to the investigation of those Green's functions which were considered in section 6. We shall show that

$$
| \ll a_q, a_q^\dagger \gg_{E=0} | \geq \frac{\text{const}}{q^2},
$$
$$
|\Sigma_{11}(0, q) - \Sigma_{12}(0, q)| \leq \text{const } q^2.
\tag{8.1}
$$

Gradient transformation will play important part in this proof. Our quasi-averages were not introduced by the gradient-invariant method. Thus, for completeness, we shall use the Hamiltonian H_ν and first obtain a series of inequalities for the corresponding expressions based on regular averages with H. We will then obtain formulas of the type (8.1) by taking the limit $\nu \to 0$.

Let us now investigate the gradient transformation,

$$
\Psi(\vec{r}) \to \Psi'(\vec{r}) = e^{-i\chi(\vec{r})} \Psi(\vec{r}),
$$
$$
\Psi^\dagger(\vec{r}) \to \Psi'^\dagger(\vec{r}) = e^{i\chi(\vec{r})} \Psi^\dagger(\vec{r})
\tag{8.2}
$$

and construct the "transformed" Hamiltonian,

$$
H_\nu'(\Psi^\dagger, \Psi) = H_\nu'(\Psi'^\dagger, \Psi').
\tag{8.3}
$$

Note that,

$$
\langle \ldots \Psi^\dagger(\vec{r}_\alpha) \ldots \Psi(\vec{r}_\beta) \ldots \rangle_{H_\nu} = \frac{\text{Tr}\left\{ (\ldots \Psi^\dagger(\vec{r}_\alpha) \ldots \Psi(\vec{r}_\beta) \ldots) e^{-\frac{H_\nu(\Psi^\dagger, \Psi)}{\theta}} \right\}}{\text{Tr } e^{-\frac{H_\nu(\Psi^\dagger, \Psi)}{\theta}}}
$$

holds. From which, using (8.3) we have

$$
\langle \ldots \Psi^\dagger(\vec{r}_\alpha) \ldots \Psi(\vec{r}_\beta) \ldots \rangle_{H_\nu} = \langle \ldots \Psi'^\dagger(\vec{r}_\alpha) \ldots \Psi'(\vec{r}_\beta) \ldots \rangle_{H_\nu'}.
$$

However, due to (8.3), we have directly,

$$
\langle \ldots \Psi^\dagger(\vec{r}_\alpha) \ldots \Psi(\vec{r}_\beta) \ldots \rangle_{H_\nu'}
$$
$$
= \langle \ldots \Psi^\dagger(\vec{r}_\alpha) \ldots \Psi(\vec{r}_\beta) \ldots \rangle_{H_\nu'} \exp i \left(\sum_\alpha \chi(\vec{r}_\alpha) - \sum_\beta \chi(\vec{r}_\beta) \right).
$$

Hence we find

$$\langle \ldots \Psi^\dagger(\vec{r}_\alpha) \ldots \Psi(\vec{r}_\beta) \ldots \rangle_{H'_\nu}$$

$$= \langle \ldots \Psi^\dagger(\vec{r}_\alpha) \ldots \Psi(\vec{r}_\beta) \ldots \rangle_{H_\nu} \exp i\left(\sum_\alpha \chi(\vec{r}_\beta) - \sum_\beta \chi(\vec{r}_\alpha)\right). \qquad (8.4)$$

In this manner we obtain the following rule: in order to calculate the average of products of field functions for the Hamiltonian H'_ν we must take these averages for the Hamiltonian H_ν and perform the gradient transformations inverse to (8.2)

$$\begin{aligned} \Psi(\vec{r}) &\to e^{i\chi(\vec{r})}\Psi(\vec{r}), \\ \Psi^\dagger(\vec{r}) &\to e^{-i\chi(\vec{r})}\Psi^\dagger(\vec{r}). \end{aligned} \qquad (8.5)$$

For our purpose it is sufficient to consider only the infinitesimal gradient transformations, in which

$$\chi(\vec{r}) = \delta\chi(\vec{r}) = \left(e^{i\vec{q}\cdot\vec{r}} + e^{-i\vec{q}\cdot\vec{r}}\right)\delta\xi$$

where $\delta\xi$ is a real, infinitesimal quantity. In this case the transformations (8.2) will take the form

$$\begin{aligned} \Psi'(\vec{r}) &= \Psi(\vec{r}) - i\Psi(\vec{r})\delta\chi(\vec{r}) \\ \Psi'^\dagger(\vec{r}) &= \Psi^\dagger(\vec{r}) + i\Psi^\dagger(\vec{r})\delta\chi(\vec{r}). \end{aligned}$$

In the momentum representation this becomes,

$$\begin{aligned} a'_k &= a_k - i(a_{k+q} + a_{k-q})\delta\xi \\ a'^\dagger_k &= a^\dagger_k + i(a^\dagger_{k+q} + a^\dagger_{k-q})\delta\xi. \end{aligned} \qquad (8.6)$$

Let us now examine

$$\delta H_\nu = H_\nu(\Psi'^\dagger, \Psi') - H_\nu(\Psi^\dagger, \Psi).$$

Due to (6.2) we have,

$$U(\Psi'^\dagger, \Psi') = U(\Psi^\dagger, \Psi). \qquad (8.7)$$

Thus, we obtain

$$\delta H_\nu = -\frac{1}{2m}\int_V \Psi^\dagger(\vec{r})\left(\vec{p}\,\frac{\partial\delta\chi}{\partial\vec{r}} + \frac{\partial\delta\chi}{\partial\vec{r}}\,\vec{p}\right)d\vec{r} - \nu V^{1/2}(\delta a_0 + \delta a_0^\dagger).$$

From this it follows that

$$\delta H_\nu = \mathscr{U}_a \delta\xi \tag{8.8}$$

where

$$\mathscr{U}_q = \frac{q^2}{2\mu} S_q + i\nu(a_q + a_{-q} - a_q^\dagger - a_{-q}^\dagger)$$

$$S_q = -i \sum_{(k)} \frac{(2\vec{k} + \vec{q}) \cdot \vec{q}}{q^2}(a_{k+q}^\dagger a_k - a_k^\dagger a_{k+q}). \tag{8.9}$$

Let us now obtain the increment,

$$\langle \mathscr{U}_q \rangle_{H_\nu + \delta H_\nu} - \langle \mathscr{U}_q \rangle_{H_\nu}.$$

For this purpose we need the quantities,

$$\langle a_k \rangle_{H_\nu + \delta H_\nu}, \quad \langle a_k^\dagger \rangle_{H_\nu + \delta H_\nu}, \quad \langle a_{k_1}^\dagger a_{k_2} \rangle_{H_\nu + \delta H_\nu}.$$

Using the rule mentioned above we can calculate these quantities by substituting $\langle \ldots \rangle_{H_\nu}$ for $\langle \ldots \rangle_{H_\nu + \delta H_\nu}$ and simultaneously subjecting the amplitudes a and $a^\dagger$ to the gradient transformation inverse to (8.6), namely,

$$a_k \to a_k + i(a_{k+q} + a_{k-q})\delta\xi$$
$$a_k^{\prime\dagger} \to a_k^\dagger - i(a_{k+q}^\dagger + a_{k-q}^\dagger)\delta\xi.$$

In this way we obtain,

$$\langle a_k \rangle_{H_\nu + \delta H_\nu} - \langle a_k \rangle_{H_\nu} = i\{\langle a_{k+q} + a_{k-q} \rangle_{H_\nu}\}\delta\xi.$$

However we have

$$\langle a_k \rangle_{H_\nu} = \begin{cases} \sqrt{N_0}, & k = 0 \\ 0, & k \neq 0 \end{cases}$$

and thus we obtain

$$\langle a_k \rangle_{H_\nu + \delta H_\nu} - \langle a_k \rangle_{H_\nu} = i\sqrt{N_0}\{\Delta(k + q) + \Delta(k - q)\}\delta\xi. \tag{8.10}$$

Similarly we obtain,

$$\langle a_k^\dagger \rangle_{H_\nu + \delta H_\nu} - \langle a_k^\dagger \rangle_{H_\nu} = -i\sqrt{N_0}\{\Delta(k - q) + \Delta(k + q)\}\delta\xi. \tag{8.11}$$

We have, further

$$\langle a^\dagger_{k_1} a_{k_2}\rangle_{H_\nu + \delta H_\nu} - \langle a^\dagger_{k_1} a_{k_2}\rangle_{H_\nu} = -i\langle (a^\dagger_{k_1+q} + a^\dagger_{k_1-q})a_{k_2}\rangle_{H_\nu}\delta\xi$$
$$+ i\langle a^\dagger_{k_1}(a_{k_2+q} + a_{k_2-q})\rangle_{H_\nu}\delta\xi.$$

Now, note that $\langle a^\dagger_{p_1} a_{p_2}\rangle = \Delta(p_1 - p_2)N_{p_1}$ where $N_{p_1} = \langle a^\dagger_{p_1} a_{p_2}\rangle_{H_\nu}$. We therefore obtain,

$$\langle a^\dagger_{k_1} a_{k_2}\rangle_{H_\nu + \delta H_\nu} - \langle a^\dagger_{k_1} a_{k_2}\rangle_{H_\nu} = i\{\Delta(k_1 - k_2 + q) + \Delta(k_1 - k_2 - q)\}$$
$$\times (N_{k_1} - N_{k_2})\delta\xi.$$

Making use of this equality, we find using (8.9),

$$\langle S_q\rangle_{H_\nu + \delta H_\nu} - \langle S_q\rangle_{H_\nu} = -4N\delta\xi$$

where

$$N = \sum_{(k)} N_k = \sum_{(k)} \langle a^\dagger_k a_k\rangle_{H_\nu}$$

represents the total number of particles in our system. Taking into account (8.10) and (8.11) we obtain,

$$\langle \mathscr{U}_q\rangle_{H_\nu + \delta H_\nu} - \langle \mathscr{U}_q\rangle_{H_\nu} = -4\left(N\frac{q^2}{2m} + \nu\sqrt{N_0}\cdot v^{1/2}\right)\delta\xi,$$
$$\langle a_q\rangle_{H_\nu + \delta H_\nu} - \langle a_q\rangle_{H_\nu} = i\sqrt{N_0}\,\delta\xi, \qquad (8.12)$$
$$\langle a_q - a^\dagger_{-q}\rangle_{H_\nu + \delta H_\nu} - \langle a_q - a^\dagger_{-q}\rangle_{H_\nu} = 2i\sqrt{N_0}\,\delta\xi.$$

Alternately, we can compute the increment $\langle A\rangle_{H_\nu + \delta H_\nu} - \langle A\rangle_{H_\nu}$ where $A = \mathscr{U}_q, a_q, a_q - a^\dagger_{-q}$ using formula (5.20).

Taking into account (8.8), we obtain

$$\langle \mathscr{U}_q\rangle_{H_\nu + \delta H_\nu} - \langle \mathscr{U}_q\rangle_{H_\nu} = 2\pi \ll \mathscr{U}_q, \mathscr{U}_q \gg_{E=0} \delta\xi,$$
$$\langle a_q\rangle_{H_\nu + \delta H_\nu} - \langle a_q\rangle_{H_\nu} = 2\pi \ll a_q, \mathscr{U}_q \gg_{E=0} \delta\xi, \qquad (8.13)$$
$$\langle a_q - a^\dagger_{-q}\rangle_{H_\nu + \delta H_\nu} - \langle a_q - a^\dagger_{-q}\rangle_{H_\nu} = 2\pi \ll (a_q - a^\dagger_{-q}), \mathscr{U}_q \gg_{E=0} \delta\xi.$$

Comparing these formulas with (8.12) we see that,

$$\ll \mathscr{U}_q, \mathscr{U}_q \gg_{E=0} = -\frac{2}{\pi}\left(N\frac{q^2}{2m} + \nu\sqrt{N_0}\cdot v^{1/2}\right) \qquad (8.14)$$

and

$$| \ll a_q, \mathscr{U}_q \gg_{E=0} |^2 = \frac{N_0}{4\pi^2},$$

$$| \ll (a_q - a^\dagger_{-q}), \mathscr{U}_q \gg_{E=0} |^2 = \frac{N_0}{\pi^2}.$$

(8.15)

Let us now make use of inequality (5.19), in which we assume,

$$A = a_q, \ a_q - a^\dagger_{-q},; \quad B = \mathscr{U}_q.$$

Note on the basis of (8.15), that $\mathscr{U}_q = \mathscr{U}_q^\dagger$. We the obtain from (8.15),

$$\frac{N_0}{4\pi^2} \leq | \ll a_q, a^\dagger_q \gg_{E=0} \ll \mathscr{U}_q, \mathscr{U}_q \gg_{E=0} |,$$

$$\frac{N_0}{4\pi^2} \leq | \ll (a_q - a^\dagger_{-q}), (a^\dagger_q - a_{-q}) \gg_{E=0} \ll \mathscr{U}_q, \mathscr{U}_q \gg_{E=0} |.$$

From which, using (8.14), we will have

$$| \ll a_q, a^\dagger_q \gg_{E=0} | \geq \frac{\rho_0 m}{4\pi(q^2\rho + 2\nu m\sqrt{\rho_0})},$$

$$| \ll (a_q - a^\dagger_{-q}), (a^\dagger_q - a_{-q}) \gg_{E=0} | \geq \frac{\rho_0 m}{(q^2\rho + 2\nu m\sqrt{\rho_0})}.$$

(8.16)

Let us now take the limit $\nu \to 0$. As is seen, *these expressions, in the neighborhood of $q \sim 0$, have the characteristic behavior*, constant q^{-2}, as is seen from,

$$| \ll a_q, a^\dagger_q \gg_{E=0} | \geq \left(\frac{\rho_0 m}{4\pi\rho}\right) \frac{1}{q^2},$$

(8.17)

$$| \ll (a_q - a^\dagger_{-q}), (a^\dagger_q - a_{-q}) \gg_{E=0} | \geq \left(\frac{\rho_0 m}{\pi\rho}\right) \frac{1}{q^2}.$$

(8.18)

We note that

$$\ll (a_q - a^\dagger_{-q}), (a^\dagger_q - a_{-q}) \gg_{E=0} = \ll a_q, a^\dagger_q \gg_{E=0} - \ll a^\dagger_{-q}, a^\dagger_q \gg_{E=0}$$
$$- \ll a_q, a_{-q} \gg_{E=0} + \ll a^\dagger_{-q}, a_{-q} \gg_{E=0}.$$

Using the symmetry properties (6.13) and (6.17) gives

$$\ll a_q, a_{-q} \gg_{E=0} = \ll a^\dagger_{-q}, a^\dagger_q \gg_{E=0}, \quad \ll a^\dagger_{-q}, a_{-q} \gg_{E=0} = \ll a_q, a^\dagger_q \gg_{E=0}$$

and thus we have,

$$\ll (a_q - a_{-q}^\dagger), (a_q^\dagger - a_{-q}) \gg_{E=0} = 2 \ll a_q, a_q^\dagger \gg_{E=0} -2 \ll a_{-q}^\dagger, a_q^\dagger \gg_{E=0}.$$
(8.19)

Similarly, due to (6.22) we can write

$$\ll a_q, a_q^\dagger \gg_{E=0} - \ll a_{-q}^\dagger, a_q^\dagger \gg_{E=0}$$
$$= \frac{1}{2\pi} \frac{\Sigma_{11}(0, q) + \Sigma_{12}(0, q)}{\Sigma_{11}^2(0, q) - \Sigma_{12}^2(0, q)} = \frac{1}{2\pi} \frac{1}{\Sigma_{11}(0, q) - \Sigma_{12}(0, q)}.$$

Thus, taking into account (8.18) and (8.19), we obtain,

$$\left| \frac{1}{\Sigma_{11}(0, q) - \Sigma_{12}(0, q)} \right| \geq \left(\frac{\rho_0 m}{\rho} \right) \frac{1}{q^2}$$

or,

$$|\Sigma_{11}(0, q) - \Sigma_{12}(0, q)| \leq \frac{\rho}{\rho_0 m} q^2.$$
(8.20)

The inequalities (8.17) and (8.20) are just the inequalities (8.1) which were to be proved.

Let us now consider some the applications. We write the spectral formulas,

$$\langle a_q(\tau) a_q^\dagger(t) \rangle = \int_{-\infty}^{+\infty} J_q(\omega) e^{-i\omega(t-\tau)} d\omega,$$

$$\langle a_q^\dagger(\tau) a_q(t) \rangle = \int_{-\infty}^{+\infty} J_q(\omega) e^{\omega/\theta} e^{-i\omega(t-\tau)} d\omega,$$

$$\langle a_q a_q^\dagger \rangle = -\frac{1}{2\pi} \int_{-\infty}^{+\infty} J_q(\omega) \frac{e^{\omega/\theta} - 1}{\omega} d\omega;$$

$$J_q(\omega) \geq 0.$$
(8.21)

and note that,

$$\frac{e^{\omega/\theta} - 1}{\omega} = \frac{(1 + e^{\omega/\theta})}{\omega} \operatorname{th} \frac{\omega}{2\theta} \leq \frac{1}{2\omega} (1 + e^{\omega/\theta}).$$

Thus,

$$| \ll a_q, a_q^\dagger \gg_{E=0} | \leq \frac{1}{4\pi\theta} \int_{-\infty}^{+\infty} J_q(\omega)(1 + e^{\omega/\theta}) \, d\omega$$

$$= \frac{1 + 2\langle a_q^\dagger a_q \rangle}{4\pi\theta}$$

holds. From this we have, on the basis of (8.17)

$$1 + 2\langle a_q^\dagger a_q \rangle \geq \frac{m\theta}{q^2} \frac{\rho_0}{\rho}. \tag{8.22}$$

Referring, for example, to (3.7) and (3.8) we see that

$$\langle a_q^\dagger a_q \rangle = (2\pi)^3 w(q) = (2\pi)^3 W_1(q); \quad q \neq 0.$$

We can also write

$$W_1(q) \geq \frac{1}{2(2\pi)^3} \left\{ \frac{m\theta}{q^2} \frac{\rho_0}{\rho} - 1 \right\}. \tag{8.23}$$

From this it follows that *the density of the continuous particle momentum distribution approaches infinity as* q^{-2} *as* $q \to 0$.° This statement holds only for the case $\theta > 0$. In order to obtain information about the situation when $\theta = 0$ let us make use of the spectral formulas for this particular case (refer to section 6),

$$- \ll a_q, a_q^\dagger \gg_{E=0} = \frac{1}{2\pi} \int_{-\infty}^{+\infty} I_q(\omega) \frac{\in(\omega)}{\omega} \, d\omega = \frac{1}{2\pi} \int_{-\infty}^{+\infty} I_q(\omega) \frac{d\omega}{|\omega|}$$

$$2\langle a_q^\dagger a_q \rangle + 1 = \langle a_q^\dagger a_q + a_q a_q^\dagger \rangle = \int_{-\infty}^{+\infty} I_q(\omega) \, d\omega,$$

$$I_q(\omega) \geq 0.$$

For small enough $|q|$, when one can speak of "elementary excitations" possessing a known energy, it is natural to assume that the spectral intensity

---

°If we had investigated the auxiliary system with fixed $\nu > 0$ and did note take into the limit $\nu \to 0$, then this characteristic behavior would not have appeared.

$I_q(\omega)$ is almost equal to zero for $|\omega| < E(q)$. Here $E(q)$ is the minimum energy of the elementary excitation for the momentum q. Then we have

$$\frac{1}{2\pi} \int\limits_{-\infty}^{+\infty} I_q(\omega) \frac{d\omega}{|\omega|} \leq \frac{1}{2\pi E(q)} \int\limits_{-\infty}^{+\infty} I_q(\omega)\, d\omega = \frac{2\langle a_q^\dagger a_q \rangle + 1}{2\pi E(q)}$$

and thus, on the basis of (8.17), we obtain

$$\frac{2\langle a_q^\dagger a_q \rangle + 1}{2\pi E(q)} \geq \frac{m}{4\pi q^2} \frac{\rho_0}{\rho}$$

or

$$2(2\pi)^3 W_1(q) \geq \frac{E(q)}{2q^2} \frac{m\rho_0}{\rho} - 1.$$

If the elementary excitation spectra possesses a phonon character, $E(q) = c|q|$, then $W_1(q)$ approaches infinity as $q \to 0$ not slower than $\text{const} \cdot |q|^{-1}$.

We will now show that by using our inequalities one can determine the character of the excitation spectra. For this, let us turn to the relation (8.20), from which it follows that,

$$\Sigma_{11}(0,0) - \Sigma_{12}(0,0) = 0. \tag{8.24}$$

Note that the equality for the case of zero temperature was first derived by Hegenholtz and Pines using perturbation theory. In their work [11] a model system with a condensate was considered and the diagram technique was used for its investigation. They were able to show that the equalities (8.24) hold in any order of perturbation theory. The importance of this relationship lies in its connection with the structure of the energy spectra of the perturbed system.

Let us consider the "secular" equation,

$$\Sigma_{11}(E,k)\Sigma_{11}(-E,k) - \Sigma_{12}^2(E,k) = 0 \tag{8.25}$$

and assume that the mass operator, $\Sigma(E,k)$ is regular in the neighborhood of the point $E = 0$, $k = 0$. We write (8.25) in the form

$$\left[\frac{\Sigma_{11}(E,k) + \Sigma_{11}(-E,k)}{2}\right]^2 - \Sigma_{12}^2(E,k) = \left[\frac{\Sigma_{11}(E,k) - \Sigma_{11}(-E,k)}{2}\right]^2. \tag{8.26}$$

Note that due to the radial symmetry in our problem the function $\Sigma_{\alpha\beta}(E, k^2)$ will depend on k only the scalar k^2.

Further, the left hand side of the equation is an even function of E. Due to (8.24) it becomes zero when $E = 0$, $k = 0$. Thus for sufficiently small E and k we can write

$$\left[\frac{\Sigma_{11}(E, k) + \Sigma_{11}(-E, k)}{2}\right]^2 - \Sigma_{12}^2(E, k) = \beta k^2 + \gamma E^2$$

$$\beta, \; \gamma = \text{const.}$$

We also note, that the expression

$$\frac{\Sigma_{11}(E, k) + \Sigma_{11}(-E, k)}{2}$$

will be an odd function of E. If we retain only the first order term, we will obtain, for sufficiently small E and k,

$$\frac{\Sigma_{11}(E, k) + \Sigma_{11}(-E, k)}{2} = \alpha E, \quad \alpha = \text{const.}$$

In this way equations (8.25) and (8.26) yield,

$$\alpha^2 E^2 = \beta k^2 + \gamma E^2.$$

let us exclude from this investigation the special cases when $\alpha^2 - \gamma$ and β become zero. Then, we have

$$E^2 = sk^2, \quad s \neq 0, \quad s = \frac{\beta}{\alpha^2 - \gamma}.$$

We see that the magnitude, s, must be positive, since the pole of the Green's function must lie on the real axis of the complex E-plane. Thus, *for the excitation energy, we obtain an "acoustical" dependency without gap,*

$$E = \sqrt{s}\,|k|. \tag{8.27}$$

From previous consideration, it is seen that equation (8.24) is related to the gradient invariance of the "potential energy" U. Thus, it is not surprising that if we violate this invariance property then we also will violate equation (8.24) and will obtain formulas for $E(k)$ which contain an energy gap. This occurs, for example, when the investigating model system in which, in the expression for U, only the interaction of pairs with opposite momenta is kept. The same situation arises if we include a term which is not gradient invariant, $-\nu\sqrt{V}(a_0 + a_0^\dagger)$, with a fixed $\nu > 0$, in the Hamiltonian.

Let us illustrate this fact by a simple example where we keep only the lowest terms in the interaction, ϕ, for the case $\theta = 0$. For the actual construction of such an approximation we will deal with a model system with a Hamiltonian $H_\nu(N_0)$. From the forms (7.14) and (7.15) of this Hamiltonian it is not difficult to note, that by taking into account only the first order terms, we have,

$$\Sigma_{11}(E, k) = E - \frac{k^2}{2m} + \mu - \rho\{\tilde{\phi}(k) + \tilde{\phi}(0)\},$$
$$\Sigma_{12}(E, k) = -\rho_0\tilde{\phi}(k),$$
$$F_\nu = -\mu N_0 + \frac{N_0^2}{2V}\tilde{\phi}(0) - 2\nu\sqrt{N_0 V}. \tag{8.28}$$

Because of this equation (7.17) yields,

$$0 = \frac{\partial F_\nu}{\partial N_0} = -\mu + \rho_0\tilde{\phi}(0) - \frac{\nu}{\sqrt{\rho_0}}.$$

Thus we have

$$\Sigma_{11}(0, 0) - \Sigma_{12}(0, 0) = -\frac{\nu}{\sqrt{\rho_0}} < 0.$$

Let us further consider the secular equation (8.25). In the present approximation we obtain from (8.28),

$$\left(\frac{k^2}{2m} + \rho_0\tilde{\phi}(k) + \frac{\nu}{\sqrt{\rho_0}}\right)^2 - \left(\rho_0\tilde{\phi}(k)\right)^2 = E^2.$$

From this we the formula for the energy of the elementary excitation

$$E(k) = \sqrt{\frac{\nu^2}{\rho_0} + 2\frac{\nu}{\sqrt{\rho_0}}\left(\frac{k^2}{2m} + \rho_0\tilde{\phi}(k)\right) + \rho_0\tilde{\phi}(k)\frac{k^2}{m} + \left(\frac{k^2}{2m}\right)^2}.$$

which contains an energy gap.

The gap disappears after taking the limit $\nu \to 0$, whereupon we arrive at the usual expression,

$$E(k) = \sqrt{\rho_0\tilde{\phi}(k)\frac{k^2}{m} + \left(\frac{k^2}{2m}\right)^2}.$$

This spectrum has a quasi-acoustic character when k is small. Let us finally note that the formulas "with a gap" in $E(k)$ can be also obtained with $\nu = 0$, if one "mismatches" the approximations. For example, we would obtain a gap in $E(k)$ if we used for $\Sigma_{\alpha\beta}(E, k)$ the formulas of the first approximation, and substituted in the equation

$$\frac{\partial F_\nu}{\partial N_0} = 0$$

for E_ν, the formulas of the second approximation.

9. The $1/q^2$ Theorem for Fermi Systems

Let us now turn to the derivation of the "$1/q^2$ theorem" for the Fermi systems which were investigated in section 4. We consider the case when the system has a condensate of pair quasi-molecules, in S-state, (4.11), and when the Hamiltonian has the usual form,

$$H = \sum_{(\sigma)} \int \Psi^\dagger(\vec{r}, \sigma) \left(\frac{p^2}{2m} - \mu\right) \Psi(\vec{r}, \sigma) \, d\vec{r} + U(\Psi^\dagger, \Psi). \tag{9.1}$$

Here the expression U is gradient invariant. Note that the model of Fröhlich, in which the electrons interact with the phonon field, is of the type which was considered. Actually, we can include the energy of the phonons and the energy of their interaction with the electrons in U. Note that U must be invariant with respect to gradient transformation which act *only* on the Fermi functions Ψ, $\Psi^\dagger$.

Let us return to the arguments of the previous paragraph and consider them in detail. For brevity we will not introduce the infinitesimal terms which remove the degeneracy into the expression for the Hamiltonian, and we will agree to deal directly with the corresponding quasi-averages.

Thus, let us consider the infinitesimal gradient transformations,

$$\Psi(x) \rightarrow \Psi'(x) = \Psi(x) - i\Psi(x)\delta\chi$$
$$\Psi^\dagger(x) \rightarrow \Psi'^\dagger(x) = \Psi^\dagger(x) + i\Psi^\dagger(x)\delta\chi$$
$$\delta\chi = \left(e^{i\vec{q}\vec{r}} + e^{-i\vec{q}\vec{r}}\right)\delta'\xi.$$

We construct the variation,

$$\delta H = H(\Psi'^\dagger, \Psi') - H(\Psi^\dagger, \Psi).$$

We then have,

$$\delta H = -\frac{1}{2m} \int \sum_{\sigma} \Psi^{\dagger}(\vec{r}, \sigma)\left(\vec{p}\,\frac{\partial \delta \xi}{\partial \vec{r}} + \frac{\partial \delta \xi}{\partial \vec{r}}\,\vec{p}\right)\Psi(\vec{r}, \sigma)\, d\vec{r}.$$

From this follows

$$\delta H = \frac{q^2}{2m} S_q \delta \chi \tag{9.2}$$

where q is a given non-zero momentum and

$$S_q = -i \sum_{(k,\sigma)} \frac{(2\vec{k} + \vec{q}) \cdot \vec{q}}{q^2}\left(a^{\dagger}_{k+q,\sigma} a_{k,\sigma} - a^{\dagger}_{k,\sigma} a_{k+q,\sigma}\right). \tag{9.3}$$

Now note that the increment $\langle \mathscr{U} \rangle_{H+\delta H} - \langle \mathscr{U} \rangle_H$ can be computed by two methods.

The direct calculations gives

$$\langle \mathscr{U} \rangle_{H+\delta H} - \langle \mathscr{U} \rangle_H = \langle \mathscr{U}' - \mathscr{U} \rangle_H \tag{9.4}$$

where $\mathscr{U}'$ is obtained from $\mathscr{U}$ with the inverse gradient transformation,

$$\Psi \to \Psi + i\Psi \delta \chi, \quad \Psi^{\dagger} \to \Psi^{\dagger} - i\Psi^{\dagger} \delta \chi.$$

In the momentum representation this becomes,

$$a_{k,\sigma} \to a_{k,\sigma} + i\{a_{k+q,\sigma} + a_{k-q,\sigma}\}\delta \xi,$$
$$a^{\dagger}_{k,\sigma} \to a^{\dagger}_{k,\sigma} - i\{a^{\dagger}_{k+q,\sigma} + a^{\dagger}_{k-q,\sigma}\}\delta \xi. \tag{9.5}$$

Alternatively, we can make use of the formula (5.20) and write

$$\langle \mathscr{U} \rangle_{H+\delta H} - \langle \mathscr{U} \rangle_H = \frac{\pi q^2}{m} \ll \mathscr{U}, S_q \gg_{E=0} d\xi.$$

We therefore have,

$$\langle \mathscr{U}' - \mathscr{U} \rangle_H = \frac{\pi q^2}{m} \ll \mathscr{U}, S_q \gg_{E=0} d\xi. \tag{9.6}$$

If we take $\mathscr{U} = V^{-1}S_q$, we then find

$$\ll V^{-1/2}S_q, V^{-1/2}S_q \gg_{E=0} = -\frac{4\rho m}{\pi q^2}; \quad \rho = \frac{N}{V}. \tag{9.7}$$

Let us introduce the operators,

$$\beta_k = \frac{1}{\sqrt{V}} \sum_{(\sigma_1,\sigma_2)} \in(\sigma_1)\Delta(\sigma_1 + \sigma_2) \iint_V \Psi(x_2)\Psi(x_1)\vartheta(\vec{r}_1 - \vec{r}_2) \, e^{-i\vec{k}\frac{\vec{r}_1+\vec{r}_2}{2}} \, d\vec{r}_1 \, d\vec{r}_2$$

$$(9.8)$$

where $\vartheta(r)$ is a radially symmetric real function of r, which decreases rapidly enough as $r \to \infty$ and which satisfies the condition,

$$\gamma \equiv \int \varphi(r)\vartheta(r) \, d\vec{r} \neq 0. \tag{9.9}$$

We then have

$$\langle V^{-1/2}(\beta'_q - \beta - q)\rangle$$

$$= i \sum_{(\sigma_1,\sigma_2)} \in(\sigma_1)\Delta(\sigma_1 + \sigma_2)\frac{1}{V} \iint_V \left\{ e^{i\vec{q}\vec{r}_2} + e^{i\vec{q}\vec{r}_1} + e^{-i\vec{q}\vec{r}_2} + e^{-i\vec{q}\vec{r}_1} \right\}$$

$$\times \langle \Psi(x_2)\Psi(x_1)\rangle \, e^{-i\vec{k}\frac{\vec{r}_1+\vec{r}_2}{2}} \, d\vec{r}_1 \, d\vec{r}_2\delta\xi$$

$$= i \sum_{(\sigma_1,\sigma_2)} \in(\sigma_1)\Delta(\sigma_1 + \sigma_2)\frac{1}{V} \iint_V 2\cos\left(\frac{\vec{q}(\vec{r}_1 - \vec{r}_2)}{2}\right)$$

$$\times \left\{ 1 + e^{-i\vec{q}(\vec{r}_1+\vec{r}_2)} \right\} \langle \Psi(x_2)\Psi(x_1)\rangle \, d\vec{r}_1 \, d\vec{r}_2\delta\xi$$

and therefore, using (4.14) we obtain,

$$\langle V^{-1/2}(\beta'_q - \beta_q)\rangle = 4i\sqrt{\rho_0} \int \varphi(r)\vartheta(r) \cos\left(\frac{\vec{q}\vec{r}}{2}\right) d\vec{r}\,\delta\xi. \tag{9.10}$$

From this, because the functions φ and ϑ are real we have,

$$\langle V^{-1/2}(\beta'^{\dagger}_{-q} - \beta^{\dagger}_{-q})\rangle = -4i\sqrt{\rho_0} \int \varphi(r)\vartheta(r) \cos\left(\frac{\vec{q}\vec{r}}{2}\right) d\vec{r}\,\delta\xi.$$

Consequently,

$$\langle V^{-1/2}(\beta'_q - \beta'^{\dagger}_{-q}) - (\beta_q - \beta^{\dagger}_{-q})\rangle = 8i\sqrt{\rho_0} \int \varphi(r)\vartheta(r) \cos\left(\frac{\vec{q}\vec{r}}{2}\right) d\vec{r}\,\delta\xi. \tag{9.11}$$

In the relations (9.6) we substitute,

$$\mathscr{U} = V^{-1/2}\beta_q, \quad V^{-1/2}(\beta_q + \beta^{\dagger}_{-q}).$$

Then, due to (9.10) and (9.11) we will have,

$$| \ll \beta_q, V^{-1/2}S_q \gg_{E=0} |^2 = \left(\frac{4m\gamma(q)}{\pi q^2}\right)^2 \rho_0$$

$$| \ll (\beta_q - \beta^\dagger_{-q}), V^{-1/2}S_q \gg_{E=0}|^2 = \left(\frac{8m\gamma(q)}{\pi q^2}\right)^2 \rho_0 \qquad (9.12)$$

where

$$\gamma(q) = \int \varphi(r)\vartheta(r)\cos\left(\frac{\vec{q}\vec{r}}{2}\right) d\vec{r}. \qquad (9.13)$$

Let us maximize the left hand side of equation (9.12) using the inequalities (5.19), in which we assume,

$$A = \beta_q, \ (\beta_q - \beta^\dagger_{-q}); \quad B = V^{-1/2}S_q.$$

Since on the basis of (9.3), $S_q = S^\dagger_q$, we arrive at inequalities of the form,

$$\left(\frac{4m\gamma(q)}{\pi q^2}\right)^2 \rho_0 \leq | \ll \beta_q, \beta^\dagger_q \gg_{E=0} \ll V^{-1/2}S_q, V^{-1/2}S_q \gg_{E=0} |$$

$$\left(\frac{8m\gamma(q)}{\pi q^2}\right)^2 \rho_0 \leq | \ll (\beta_q - \beta^\dagger_{-q}), (\beta^\dagger_q - \beta_{-q}) \gg_{E=0}$$

$$\times \ll V^{-1/2}S_q, V^{-1/2}S_q \gg_{E=0} |.$$

From this, taking into account (9.7), we obtain

$$| \ll \beta_q, \beta^\dagger_q \gg_{E=0} | \geq \frac{4m\gamma^2(q)}{\pi q^2} \frac{\rho_0}{\rho}$$

$$\ll (\beta_q - \beta^\dagger_{-q}), (\beta^\dagger_q - \beta_{-q}) \gg_{E=0} \geq \frac{16m\gamma^2(q)}{\pi q^2} \frac{\rho_0}{\rho}. \qquad (9.14)$$

Due to (9.9) and (9.13) we have

$$\gamma^2(0) = \gamma^2 > 0.$$

And thus the "$1/q^2$ theorem" is proven for the present case. Apparently it is associated with the property of the energy spectrum of "collective excitations". Here we will not consider this problem, but will limit ourselves to the application of the proven theorem for estimating the number of pairs with momentum $q \neq 0$ in the case when $\theta > 0$.

Let us consider the spectral function

$$- \ll \beta_q, \beta_q^\dagger \gg_{E=0} = \frac{1}{2\pi} \int\limits_{-\infty}^{+\infty} \frac{e^{\omega/\theta} - 1}{\omega} J(\omega)\, d\omega, \quad J(\omega) > 0$$

$$\langle \beta_q \beta_q^\dagger + \beta_q^\dagger \beta_q \rangle = \int\limits_{-\infty}^{+\infty} \left(1 + e^{\omega/\theta}\right) J(\omega)\, d\omega. \tag{9.15}$$

Taking into account that

$$\frac{e^{\omega/\theta} - 1}{\omega} < \frac{e^{\omega/\theta} + 1}{2\theta}$$

we can write

$$\langle \beta_q \beta_q^\dagger + \beta_q^\dagger \beta_q \rangle \geq 4\pi\theta \ll \beta_q, \beta_q^\dagger \gg_{E=0}.$$

Because of this, using (9.14)

$$\langle \beta_q \beta_q^\dagger + \beta_q^\dagger \beta_q \rangle \geq \frac{16m\theta\gamma^2(q)}{q^2} \frac{\rho_0}{\rho}. \tag{9.16}$$

Let us now estimate $\langle \beta_q \beta_q^\dagger + \beta_q^\dagger \beta_q \rangle$. For this purpose we introduce the quantity (9.8) in a somewhat more abstract form

$$\beta_q = \int \mathscr{K}(x_1, x_2) \Psi(x_2) \Psi(x_1)\, dx_1\, dx_2 \tag{9.17}$$

where $\mathscr{K}(x_2, x_1) = -\mathscr{K}(x_1, x_2)$. Let us make use of the commutation relationships,

$$\Psi(x)\Psi^\dagger(x') + \Psi^\dagger(x')\Psi(x) = \delta(x - x').$$

We obtain

$$\beta_q \beta_q^\dagger - \beta_q^\dagger \beta_q = 2 \int \left| \mathscr{K}(x_1, x_2) \right|^2 dx_1\, dx_2$$

$$- 4 \int \left\{ \int \mathscr{K}(x_1, x_2) \mathscr{K}^*(x_1', x_2)\, dx_2 \right\} \Psi^\dagger(x_1') \Psi(x_1)\, dx_1\, dx_1'.$$

By specifying the form of the function $\mathscr{K}$ so that (9.7) agrees with (9.8), we obtain

$$\langle \beta_q \beta_q^\dagger - \beta_q^\dagger \beta_q \rangle = 4 \int \vartheta^2(r)\, d\vec{r}$$

$$-\frac{4}{V}\sum_{(p,\sigma)}\langle a_{p\sigma}^\dagger a_{p\sigma}\rangle \int\{\int \vartheta(\vec{r}+\vec{r}')\vartheta(\vec{r}')\,d\vec{r}'\}\,e^{i\left(\vec{p}-\frac{\vec{q}}{2}\right)\vec{r}}\,d\vec{r}.$$

Thus, we have

$$|\langle \beta_q \beta_q^\dagger - \beta_q^\dagger \beta_q\rangle| \le C$$

where

$$C = 4\int \vartheta^2(r)\,d\vec{r} + 4\rho\left(\int |\vartheta(r)|\,d\vec{r}\right)^2.$$

Consequently, we obtain from (9.16)

$$\langle \beta_q^\dagger \beta_q\rangle \ge \frac{8m\theta\gamma^2(q)}{q^2}\frac{\rho_0}{\rho} - \frac{C}{2}. \tag{9.18}$$

Let us now note that

$$\langle \beta_q^\dagger \beta_q\rangle = \sum_{(\sigma_1,\sigma_2,\sigma_1',\sigma_2')} \in(\sigma_1)\in(\sigma_1')\Delta(\sigma_1+\sigma_2)\Delta(\sigma_1'+\sigma_2')$$

$$\times \frac{1}{V}\int \langle \Psi^\dagger(x_1)\Psi^\dagger(x_2)\Psi(x_2')\Psi(x_1')\rangle$$

$$\times \exp i\vec{q}\left(\frac{\vec{r}_1+\vec{r}_2-\vec{r}_1'-\vec{r}_2'}{2}\right)\,d\vec{r}_1,d\vec{r}_2\,d\vec{r}_1'\,d\vec{r}_2'.$$

Utilizing expression (4.5), we find

$$\langle \beta_q^\dagger \beta_q\rangle = 2\sum_{(\omega)} N_{\omega q}V\left|\sum_{(\sigma_1,\sigma_2)}\in(\sigma_1)\Delta(\sigma_1+\sigma_2)\int \Psi_{\omega,q}(\vec{r},\sigma_1,\sigma_2)\vartheta(r)\,d\vec{r}\right|^2. \tag{9.19}$$

However, the functions $\Psi_{\omega,q}$ in formula (4.5), possess the following orthonormalization properties for a given $\vec{q}$,

$$\sum_{(\sigma_1,\sigma_2)}V\int \Psi_{\omega_1,q}(\vec{r},\sigma_1,\sigma_2)\Psi_{\omega_2,q}(\vec{r},\sigma_1,\sigma_2)\,d\vec{r} = \delta(\omega_1-\omega_2).$$

Thus, by applying the corresponding Bessel inequality we obtain

$$\sum_{(\omega)}V\left|\sum_{(\sigma_1,\sigma_2)}\in(\sigma_1)\Delta(\sigma_1+\sigma_2)\int \Psi_{\omega,q}(\vec{r},\sigma_1,\sigma_2)\vartheta(r)\,d\vec{r}\right|^2$$

$$\le \sum_{(\sigma_1,\sigma_2)}\Delta(\sigma_1+\sigma_2)\int \theta^2(r)\,d\vec{r} = 2\int \theta^2(r)\,d\vec{r}.$$

From this, using (9.19),

$$\langle \beta_q^\dagger \beta_q \rangle \le 4 \Big(\max_{(\omega)} N_{\omega,q} \Big) \int \theta^2(r)\, d\vec{r}$$

and thus, due to (9.18),

$$\max_{(\omega)} N_{\omega,q} \ge \left(\frac{2m\theta\gamma^2(q)}{\int \theta^2(r)\, d\vec{r}} \frac{\rho_0}{\rho} \right) \frac{1}{q^2} - \frac{C}{8 \int \theta^2(r)\, d\vec{r}}. \tag{9.20}$$

Let $\theta > 0$ and let ζ^2 be a positive quantity satisfying the inequality,

$$\zeta^2 \le \frac{2m\theta\gamma^2(q)}{\int \theta^2(r)\, d\vec{r}} \frac{\rho_0}{\rho}.$$

Then, *for small enough q* we obtain

$$\max_{(\omega)} N_{\omega,q} \ge \frac{\zeta^2}{q^2}.$$

Thus, for small enough momentum, q, there is a pair state $(\omega, \vec{q})$ such that the average number of particle existing in this state,

$$n_q = N_{\omega,\vec{q}} \tag{9.21}$$

will satisfy the inequality

$$n_q \ge \frac{\zeta^2}{q^2}. \tag{9.22}$$

As can be seen, this is analogous tom the inequality (8.22), which was determined for Bose system in the presence of a condensate.

In conclusion, let us emphasize that the inequality (9.22) is proven *only for the case when* $U(\psi^\dagger, \psi)$, which enters into expression (9.1) of the Hamiltonian H, *is gradient invariant.* For the model system considered in section 2, U is not gradient invariant and, thus, it is not surprising that in this case the inequality (9.22) is not satisfied.

References

1. J. Bardeen, L. Cooper and J. Schrieffer, *Phys. Rev.*, **106**, 162 (1957); *Phys. Rev.*, **108**, 1175 (1957).

2. N. N. Bogoliubov, D. N. Zubarev, and Yu. A. Tserkovnikov, *Docl. Akad. Nauk SSSR*, **117**, 788 (1957); *Sov. Phys. "Doklady", English Transl.*, **2**, 535, (1957).

3. N. N. Bogoliubov, D. N. Zubarev, and In. Tserkovnikov, *Zh. Exper. i Teor. Fiz.*, **39**, 120 (1960); *Sov. Phys. JETP, English Transl.*, **12**, 88 (1961).

4. N. N. Bogoliubov, "On the model Hamiltonian in superconductivity theory" preprint, (1960), see p. 167 this volume.

5. N. N. Bogoliubov, *Zh. Exper. i Teor. Fiz.*, **34**, 58 (1958); *Sov. Phys. JETF, English Transl.*, **7**, 41 (1958).

6. N. N. Bogoliubov, *Docl. Acad. Nauk SSSR*, **119**, 52 (1958); *Sov. Phys. "Doklady", English Transl.*, **3**, 279 (1958); *Usp. Fiz. Nauk* **67**, 549 (1959); *Sov. Phys. – Usp., English Transl.*, **2**, 236 (1959).

7. N. N. Bogoliubov, *Lectures on quantum statistics*, monograph, ed. "Radiansika Shkola", Kiev, 1949; *Lectures on Quantum Statistics*, Vol. 1, Gordon and Breach, New York, (1967).

8. D. N. Zubarev, *Usp. Fiz. Nauk* **71**, 71 (1960); *Sov. Phys. – Usp., English Transl.*, **3**, 320 (1960); V. L. Bonch-Bruevich and S. V. Tyablikov, *The Green Function Method in Statistical Mechanics* (English transl.) North Holland Pub. Co., Amsterdam (1962).

9. N. N. Bogoliubov, *Izv. Akad. Nauk SSSR Ser. Fiz.* 11, 77 (1947); *Journal* **7**, 43 (1947).

10. S. T. Beliaev, *Zh. Exper. i Theor. Fiz.*, **34**, 417 (1958); *Sov. Phys. JETP, English Transl.*, **7**, 289 (1958).

11. H. Hugenholtz and D. Pines, *Phys. Rev.* **116**, 489 (1959).

CHAPTER 3

HYDRODYNAMICS EQUATIONS IN STATISTICAL MECHANICS

1. In the present paper the equations of hydrodynamics are derived on the basis of classical mechanics of a system of molecules. As usual, we restrict ourselves to the simplest scheme and consider a system of a very large number N of identical monoatomic molecules enclosed in a macroscopic volume V. We suppose also that the interaction between the molecules is due to central forces described by the potential energy of a molecule pair $\Phi(r)$, which depends only on the distance between them.

We shall use the results and the notations of our preceding works [1, 2]. We introduce the distribution function of dynamical variables of the molecule set

$$F_s = F_s(t, \, \boldsymbol{q}_1, \ldots, \boldsymbol{q}_s, \, \boldsymbol{p}_1, \ldots, \boldsymbol{p}_s), \quad s = 1, \, 2, \, 3, \, \ldots,$$

in such a way that the expressions

$$\frac{1}{V^s} F_s \, d\boldsymbol{q}_1 \ldots d\boldsymbol{q}_s d\boldsymbol{p}_1 \ldots d\boldsymbol{p}_s$$

give the probabilities of finding the coordinates and the momenta of the 1-st, ..., s-th molecules from some arbitrary set of molecules in the infinitesimal volumes $d\boldsymbol{q}_1 \ldots d\boldsymbol{q}_s$, $d\boldsymbol{p}_1 \ldots d\boldsymbol{p}_s$ of the coordinate and momentum spaces at time t.

Since our aim is to study only volume properties of the system, we restrict ourselves to considering the leading asymptotic terms in the equation of motion for the function F_s that has been derived in the preceding works

$$\frac{\partial F_s}{\partial t} = [H_s, F_s] + \frac{1}{\nu} \int \Big[\sum_{(i \le i \le s)} \Phi_{i,s+1}, F_{s+1} \Big] \, d\boldsymbol{q}_{s+1} d\boldsymbol{p}_{s+1}, \tag{1}$$

100

where
$$\Phi_{i,s+1} = \Phi(|\boldsymbol{q}_i - \boldsymbol{q}_{s+1}|), \quad \nu \frac{V}{N}$$

and H_s denotes the Hamiltonian of a system of s isolated particles

$$H_s = \sum_{(i \leq i \leq s)}) \frac{|\boldsymbol{p}_i|^2}{2m} + \sum_{(i \leq i < j \leq s)} \Phi_{i,j}. \tag{2}$$

Equation (1) must be augmented by the normalized conditions

$$F_s = \lim_{V \to \infty} \frac{1}{V} \int_V d\boldsymbol{q}_{s+1} \int dp_{s+1} F_{s+1},$$

$$\lim_{V \to \infty} \frac{1}{V} \int_V d\boldsymbol{q}_1 \int dp_1 F_1 = 1 \tag{3}$$

and by conditions of correlation weakening, which can be represented, for instance, in the following form

$$S_{-\tau}^{(s)} \left\{ F_s - \prod_{(i \leq i \leq s)} F_1(t, \boldsymbol{q}_i, \boldsymbol{p}_i) \right\} \to 0, \quad \tau \to +\infty, \tag{4}$$

where $S_{-\tau}^{(s)}$ denotes the operator replacing the coordinates

$$\boldsymbol{q}_1, \ldots, \boldsymbol{q}_s$$

by

$$\boldsymbol{q}_1 - \frac{\boldsymbol{p}_1}{m} \tau, \ldots, \boldsymbol{q}_s - \frac{\boldsymbol{p}_s}{m} \tau,$$

respectively, with the values of the momenta remaining unaltered. Finally, it is clear that the functions F_s must be symmetric with regard to permutation of the molecule's dynamic variables

$$P_{ij} F_s = F_s \tag{5}$$

where P_{ij} is the operator replacing the variables $(\boldsymbol{q}_i, \boldsymbol{p}_i)$ by $(\boldsymbol{q}_j, \boldsymbol{p}_j)$ and vice versa.

2.   Such dynamical variables as the particle number density $\sigma(t, \boldsymbol{q})$, the particle number flux density $J(t, \boldsymbol{q})$, and the internal energy flux

density $E(t, \boldsymbol{q})$ are of special importance for the problems of hydrodynamics. These quantities can be represented as follows

$$\sigma(t, \boldsymbol{q}) = \sum_{(i \leq i \leq s)} \delta(\boldsymbol{q}_i - \boldsymbol{q}),$$

$$J(t, \boldsymbol{q}) = \sum_{(i \leq i \leq s)} P_i \, \delta(\boldsymbol{q}_i - \boldsymbol{q}),$$

$$E(t, \boldsymbol{q}) = \sum_{(i \leq i \leq s)} \frac{\boldsymbol{p}_i^2}{2m} \delta(\boldsymbol{q}_i - \boldsymbol{q})$$

$$+ \frac{1}{2} \sum_{(i \leq i < j \leq s)} \Phi(|\boldsymbol{q}_i - \boldsymbol{q}_j|) \{ \delta(\boldsymbol{q}_i - \boldsymbol{q}) + \delta(\boldsymbol{q}_j - \boldsymbol{q}) \}. \qquad (6)$$

However, it should be emphasized that such a definition of quantities σ, J and E is necessary only under microscopic consideration of the system's motion. As far as the usual macroscopic hydrodynamics is concerned, it is sufficient to know their average values, which can be determined with the aid of the function F_1 and F_2

$$\bar{\sigma} = \rho = \frac{1}{\nu} \int F_1(t, \boldsymbol{q}, \boldsymbol{p}) \, d\boldsymbol{p},$$

$$\bar{J} = m \rho \bar{u} = \frac{1}{\nu} \int \boldsymbol{q} \, F_1(t, \boldsymbol{q}, \boldsymbol{p}) \, d\boldsymbol{p},$$

$$\bar{E} = \frac{1}{\nu} \int \frac{|\boldsymbol{p}|^2}{2m} F_1(t, \boldsymbol{q}, \boldsymbol{p}) \, d\boldsymbol{p}$$

$$+ \frac{1}{2\nu^2} \int \Phi(|\boldsymbol{q} - \boldsymbol{q}'|) F_2(t, \boldsymbol{q}, \boldsymbol{q}', \boldsymbol{p}, \boldsymbol{p}') \, d\boldsymbol{q}' d\boldsymbol{p} d\boldsymbol{p}'. \qquad (7)$$

Since, due to above definition, the quantity $\rho(t, \boldsymbol{q}) d\boldsymbol{q}$ provides the average number of molecules whose coordinates at time t are in the infinitesimal volume $d\boldsymbol{q}$, we can consider the quantity ρ as a distribution function for the molecules' coordinates. The vector $\boldsymbol{u}$ is, obviously, the average velocity, the ratio E/ρ is the average energy per molecule. If we subtract the kinetic energy of ordered motion from E/ρ, we obtain the average internal energy

$$\varepsilon = E/\rho - \frac{1}{2} m u^2.$$

From Equation (2) it is not difficult to obtain

$$\rho \varepsilon = \frac{1}{\nu} \int \frac{|\boldsymbol{p} - m \boldsymbol{u}|^2}{2m} F_1(t, \boldsymbol{q}, \boldsymbol{p}) \, d\boldsymbol{p}$$

$$+ \frac{1}{2\nu^2} \int \Phi(|\boldsymbol{q} - \boldsymbol{q}'|) F_2(t, \boldsymbol{q}, \boldsymbol{q}', \boldsymbol{p}, \boldsymbol{p}') \, d\boldsymbol{q}' d\boldsymbol{p} d\boldsymbol{p}'. \tag{8}$$

Our principle task is to derive the equations which describes evolution of the main hydrodynamics functions ρ, u^α, and ε.[P]

3. In the theory of kinetic equations, in studying the distribution functions approaching their stationary form, which corresponds to the state of statistical equilibrium, particular attention is paid to the so called spatially homogeneous case, when the functions $F_1(t, \boldsymbol{q}, \boldsymbol{p})$ do not depend on $\boldsymbol{q}$ and the higher correlation functions are invariant under the spatial transformations

$$\boldsymbol{q}_1 \to \boldsymbol{q}_1 + \boldsymbol{q}_0, \quad \ldots, \quad \boldsymbol{q}_s \to \boldsymbol{q}_s + \boldsymbol{q}_0$$

(where $\boldsymbol{q}_0$ is arbitrary). It should be stressed that the equations considered always have a spatially homogeneous solution. This can be seen by taking into account Equation (1) and noting the fact that, if there is spatial inhomogeneity at the initial moment, it still exists at later time t. It can also be established that in the spatially homogeneous case the hydrodynamic functions ρ, $\boldsymbol{u}$, and ε are independent of t and do not vary with time. Thus, from the physical point of view, this case corresponds to complete homogeneity of the macroscopic spatial distribution of molecules.

In hydrodynamics, when introducing these functions we actually average them over a region the linear dimensions of which are large compared with the radius of intermolecular forces, and over an interval of time large compared with the so called molecular unit of time $r_0 m/|\boldsymbol{p}|_{\mathrm{av}}$, where $|\boldsymbol{p}|_{\mathrm{av}}$ is the average value of the modules of the molecule's momentum. Thus, to obtain the usual macroscopic hydrodynamics we should consider distribution functions sufficiently close to the spatially homogeneous ones in order to ensure that the quantities ρ, $\boldsymbol{u}$, and ε determined by Equations (7) and (8) being considered as functions of $\boldsymbol{q}$ and t vary sufficiently smoothly with respect to the molecular scale r_0, $r_0 m/|\boldsymbol{p}|_{\mathrm{av}}$.

In this connection it is necessary to introduce some small parameter μ and to determine its physical significance. First of all, we consider the expression

$$F_s(t, \boldsymbol{q}, \boldsymbol{q}_2 - \boldsymbol{q}_1 + \boldsymbol{q}, \ldots, \boldsymbol{q}_s - \boldsymbol{q}_1 + \boldsymbol{q}, \boldsymbol{p}_1, \ldots, \boldsymbol{p}_s)$$
$$= \tilde{F}_s(t, \boldsymbol{q}, \boldsymbol{q}_1, \ldots, \boldsymbol{q}_s, \boldsymbol{p}_1, \ldots, \boldsymbol{p}_s) \tag{9}$$

[P]The superscripts α, β, γ appearing after the vectors denote a Descartes components.

and note that

$$F_s(t, \boldsymbol{q}, \boldsymbol{q}_1, \ldots, \boldsymbol{q}_s, \boldsymbol{p}_1, \ldots, \boldsymbol{p}_s) = \widetilde{F}_s(t, \boldsymbol{q}, \boldsymbol{q}_1, \ldots, \boldsymbol{q}_s, \boldsymbol{p}_1, \ldots, \boldsymbol{p}_s),$$

since expression (9) is spatially homogeneous with respect to $\boldsymbol{q}_1, \ldots, \boldsymbol{q}_s$

$$F_s(t, \boldsymbol{q}, \boldsymbol{q}_1 + \boldsymbol{q}_0, \ldots, \boldsymbol{q}_s + \boldsymbol{q}_0, \boldsymbol{p}_1, \ldots, \boldsymbol{p}_s)$$
$$= \widetilde{F}_s(t, \boldsymbol{q}_1 + \boldsymbol{q}_0, \boldsymbol{q}_1, \ldots, \boldsymbol{q}_s, \boldsymbol{p}_1, \ldots, \boldsymbol{p}_s)$$

Thus, the slower the variation of $\widetilde{F}_s(t, \boldsymbol{q}, \boldsymbol{q}_1, \ldots, \boldsymbol{q}_s, \boldsymbol{p}_1, \ldots, \boldsymbol{p}_s)$ with varying $\boldsymbol{q}$, the smaller is the variation of the functions F_s under the translations

$$\boldsymbol{q}_1 \rightarrow \boldsymbol{q}_1 + \boldsymbol{q}_0, \quad \ldots, \quad \boldsymbol{q}_s \rightarrow \boldsymbol{q}_s + \boldsymbol{q}_0.$$

Therefore, restricting ourselves to distributions close to the spatially homogeneous, we seek a solution for F_s in the form

$$F_s = f_s(t, \mu \boldsymbol{q}_1, \boldsymbol{q}_1, \ldots, \boldsymbol{q}_s, \boldsymbol{p}_1, \ldots, \boldsymbol{p}_s, \mu),$$
$$F_1 = f_1(t, \mu \boldsymbol{q}_1, \boldsymbol{p}_1, \ldots, \mu), \tag{10}$$

where μ is a small parameter and the functions

$$f_s(t, \xi, \boldsymbol{q}_1, \ldots, \boldsymbol{q}_s, \boldsymbol{p}_1, \ldots, \boldsymbol{p}_s, \mu), \quad s = 1, \ 2, \ 3 \ldots \tag{11}$$

are asymptotically regular in the vicinity of $\mu = 0$. After taking the limit $\mu \rightarrow 0$ expressions (10) yield distributions which are more closer to the spatially homogeneous.

We next substitute expressions (10) into Equations (7) and (8) and take into account the fact that the functions ρ, $\boldsymbol{u}$, and ε depend on $\boldsymbol{q}$ only through the product $\mu \boldsymbol{q} = \xi$, and, therefore, are the smooth functions of $\boldsymbol{q}$. Further, we note that the derivatives $\partial \rho / \partial t$, $\partial \boldsymbol{u} / \partial t$, $\partial \varepsilon / \partial t$ are proportional to μ, so that the functions ρ, $\boldsymbol{u}$, and ε are smooth functions of t as well. In this connection the parameter μ introduced above can be physically interpreted as a number of the order of the ratio r_0/l of the radius of intermolecular forces r_0 to the length l characterizing the average size of macroscopic inhomogeneity. The equations of hydrodynamics to be derived below are of obvious asymptotic character of an expansion in powers of the parameter μ. It should be stressed immediately that we need the expressions for the hydrodynamic parameters ρ, $\boldsymbol{u}$, and ε to be represented as functions of t and ξ. We substitute Equation (10) into Equations (7) and (8) to obtain

$$\rho(t,\xi) = \frac{1}{\nu} \int f_1(t,\xi,\boldsymbol{p}_1)\,d\boldsymbol{p}_1, \tag{12}$$

$$m\rho u^{\alpha}(t,\xi) = \frac{1}{\nu} \int p_1^{\alpha}\, f_1(t,\xi,\boldsymbol{p}_1)\,d\boldsymbol{p}_1, \tag{13}$$

$$\rho\varepsilon(t,\xi) = \frac{1}{\nu} \int \frac{|\boldsymbol{p}_1 - m\boldsymbol{u}|^2}{2m}\, f_1(t,\xi,\boldsymbol{p}_1)\,d\boldsymbol{p}_1$$
$$+ \frac{1}{2\nu^2} \int \Phi(|q_1 - q_2|) f_2(t,\xi,\boldsymbol{q}_1,\boldsymbol{q}_2,\boldsymbol{p}_1,\boldsymbol{p}_2)\,d\boldsymbol{q}_2 d\boldsymbol{p}_1 d\boldsymbol{p}_2. \tag{14}$$

Since the function f_2 is spatially homogeneous with respect to $\boldsymbol{q}_1$ and $\boldsymbol{q}_2$, it depends on their difference $\boldsymbol{q}_1 - \boldsymbol{q}_2$

$$f_2(t,\xi,\boldsymbol{q}_1,\boldsymbol{q}_2,\boldsymbol{p}_1,\boldsymbol{p}_2) = \varphi(t,\xi,\boldsymbol{q}_1 - \boldsymbol{q}_2,\boldsymbol{p}_1,\boldsymbol{p}_2), \tag{15}$$

and hence the expression on the right-hand side of Equation (14) is independent of $\boldsymbol{q}_1$ although the integration over this variable has not been performed.

Now on the basis of Equation (1) we obtain the equations determining the time evolution of the functions f_s. Note that according to Equations (10) we can write

$$\frac{\partial F_s}{\partial q_j} = \frac{\partial f_s}{\partial q_j}, \quad j = 2,\ 3,\ \ldots,\ s,$$

$$\frac{\partial F_s}{\partial p_j} = \frac{\partial f_s}{\partial p_j}, \quad j = 1,\ 2,\ 3,\ \ldots,\ s,$$

and

$$\frac{\partial F_s}{\partial q_1} = \frac{\partial f_s}{\partial q_1} + \mu\frac{\partial f_s}{\partial \xi}.$$

Consequently, Equations (1) give the main equations determining the evolution of the functions f_s in the form

$$\frac{\partial f_s}{\partial t} = -\mu \sum_{(1\leq\alpha\leq3)} \frac{p_1^{\alpha}}{m}\frac{\partial f_s}{\partial \xi^{\alpha}} + [H_s,\ f_s]$$

$$+ \frac{1}{\nu} \int \Big[\sum_{(1 \leq \alpha \leq 3)} \Phi_{i,s+1}, \, f_{s+1} \Big] \, d\boldsymbol{q}_{s+1} d\boldsymbol{p}_{s+1}, \tag{16}$$

where the Poisson bracket acts only upon the canonical variables $\boldsymbol{q}_1, \, \ldots, \, \boldsymbol{q}_s,$ $\boldsymbol{p}_1, \, \ldots, \, \boldsymbol{p}_s$ and not upon the auxiliary variable ξ. If we take into account the identity

$$\int \frac{\partial \Phi_{1,2}}{\partial q_2^\alpha} \frac{\partial f_2}{\partial p_2^\alpha} \, d\boldsymbol{p}_2 = 0,$$

we can write the equation

$$\frac{\partial f_1}{\partial t} = -\mu \sum_{(1 \leq \alpha \leq 3)} \frac{p_1^\alpha}{m} \frac{\partial f_s}{\partial \xi^\alpha} + \frac{1}{\nu} \int \sum_{(1 \leq \alpha \leq 3)} \frac{\partial \Phi_{1,2}}{\partial q_1^\alpha} \frac{\partial f_2}{\partial p_1^\alpha} \, d\boldsymbol{q}_2 d\boldsymbol{p}_2 \tag{17}$$

as a particular case of Equation (16).

Now we turn to the symmetry conditions (5). In view of expressions (10) it is obvious that

$$P_{ij} f_s = f_s, \quad i = 2, \, 3, \, \ldots, \, s, \quad j = 2, \, 3, \, \ldots, \, s. \tag{18}$$

Then we have

$$f_s(t, \mu \boldsymbol{q}_1, \boldsymbol{q}_1, \ldots, \boldsymbol{q}_s, \boldsymbol{p}_1, \ldots, \boldsymbol{p}_s)$$
$$= f_s(t, \mu \boldsymbol{q}_j, \boldsymbol{q}_j, \ldots \boldsymbol{q}_1, \ldots, \boldsymbol{q}_s, \boldsymbol{p}_j, \ldots, \boldsymbol{p}_1, \ldots, \boldsymbol{p}_s)$$
$$= f_s(t, \mu \boldsymbol{q}_1 + \mu(\boldsymbol{q}_j - \boldsymbol{q}_1), \boldsymbol{q}_j, \ldots \boldsymbol{q}_1, \ldots, \boldsymbol{q}_s, \boldsymbol{p}_j, \ldots, \boldsymbol{p}_1, \ldots, \boldsymbol{p}_s)$$

Let us agree to consider P_{1j} as an operator that acts only on the canonical variables but not on the variable ξ. Then we can write

$$f_s(t, \xi, \boldsymbol{q}_1, \ldots, \boldsymbol{q}_s, \boldsymbol{p}_1, \ldots, \boldsymbol{p}_s)$$
$$= P_{1j} f_s(t, \xi - \mu(\boldsymbol{q}_j - \boldsymbol{q}_1), \boldsymbol{q}_1, \ldots, \boldsymbol{q}_s, \boldsymbol{p}_1, \ldots, \boldsymbol{p}_s)$$

or

$$P_{1j} f_s(t, \xi, \boldsymbol{q}_1, \ldots, \boldsymbol{q}_s, \boldsymbol{p}_1, \ldots, \boldsymbol{p}_s)$$
$$= f_s(t, \xi - \mu(\boldsymbol{q}_j - \boldsymbol{q}_1), \boldsymbol{q}_1, \ldots, \boldsymbol{q}_s, \boldsymbol{p}_1, \ldots, \boldsymbol{p}_s). \tag{19}$$

Hence we find

$$P_{1j} f_s = f_s - \mu \sum_\alpha (\boldsymbol{q}_j - \boldsymbol{q}_1)^\alpha \frac{\partial f_s}{\partial \xi^\alpha}$$

$$+ \frac{\mu^2}{2} \sum_{\alpha\beta} (\boldsymbol{q}_j - \boldsymbol{q}_1)^\alpha (\boldsymbol{q}_j - \boldsymbol{q}_1)^\beta \frac{\partial^2 f_s}{\partial \xi^\alpha \partial \xi^\beta} + \mu^3 \dots \qquad (20)$$

Note now that the normalization conditions (3) and the correlation weakening conditions (4) for functions f_s can be represented in the form

$$f_s = \lim_{V \to \infty} \frac{1}{V} \int_V d\boldsymbol{q}_{s+1} \int d\boldsymbol{p}_{s+1}\, f_{s+1}, \qquad (21)$$

$$S_{-\tau}^{(s)} \Big\{ f_s(t, \xi, \boldsymbol{q}_1, \dots, \boldsymbol{q}_s, \boldsymbol{p}_1, \dots, \boldsymbol{p}_s)$$
$$- \prod_{1 \le i \le s} f_1(t, \xi + \mu(\boldsymbol{q}_i - \boldsymbol{q}_1), \boldsymbol{p}_1) \Big\} \to 0, \quad \tau \to +\infty, \qquad (22)$$

where the operator $S_{-\tau}^{(s)}$ acts only on the canonical variables. The differential equations (16) together with the symmetry (18), (20), normalization (21), and the correlation weakening condition (22) determine the distribution functions f_s.

4. Now we proceed to calculate the derivatives

$$\frac{\partial \rho}{\partial t}, \quad \frac{\partial(\rho u^\alpha)}{\partial t}, \quad \frac{\partial(\rho \varepsilon)}{\partial t}.$$

We differentiate Equation (12) and take into account Equation (17) to obtain

$$\frac{\partial \rho}{\partial t} = \frac{1}{\nu} \int \frac{\partial f_1(t, \xi, \boldsymbol{p}_1)}{\partial t} d\boldsymbol{p}_1$$
$$= -\mu \sum_\alpha \frac{1}{m\nu} \int p_1^\alpha \frac{\partial f_1}{\partial \xi^\alpha} d\boldsymbol{p}_1 + \frac{1}{\nu^2} \sum_\alpha \int \frac{\partial \Phi_{1,2}}{\partial q_1^\alpha} \frac{\partial f_2}{\partial p_1^\alpha} d\boldsymbol{p}_1 d\boldsymbol{q}_2 d\boldsymbol{p}_2.$$

Since

$$\int \frac{\partial \Phi_{1,2}}{\partial q_1^\alpha} \frac{\partial f_2}{\partial p_1^\alpha}, d\boldsymbol{p}_1 = 0$$

and due to Equation (13)

$$\frac{1}{m\nu} \int p_1^\alpha \frac{\partial f_1}{\partial \xi^\alpha} d\boldsymbol{p}_1 = \frac{\partial}{\partial \xi^\alpha} \frac{1}{m\nu} \int p_1^\alpha f_1\, d\boldsymbol{p}_1 = \frac{\partial(\rho u^\alpha)}{\partial \xi^\alpha},$$

we arrive at the usual continuity equation

$$\frac{\partial \rho}{\partial t} = -\mu \sum_\alpha \frac{\partial(\rho u^\alpha)}{\partial \xi^\alpha}. \tag{23}$$

If we differentiate expression (13) and take into account Equation (17), we obtain

$$\frac{\partial(\rho u^\alpha)}{\partial t} = \frac{1}{m\nu} \int p_1^\alpha \frac{\partial f_1}{\partial t} d\boldsymbol{p}_1 = -\mu \sum_\beta \frac{1}{m^2 \nu} \frac{\partial}{\partial \xi^\beta} \int p_1^\alpha p_1^\beta f_1 \, d\boldsymbol{p}_1$$

$$+ \frac{1}{m\nu^2} \sum_\beta \int \frac{\partial \Phi_{1,2}}{\partial q_1^\beta} \frac{\partial f_2}{\partial p_1^\beta} p_1^\alpha \, d\boldsymbol{p}_1 d\boldsymbol{q}_2 d\boldsymbol{p}_2,$$

whence after integrating the second term by parts, we find

$$\frac{\partial(\rho u^\alpha)}{\partial t} = - \mu \sum_\beta \frac{1}{m^2 \nu} \frac{\partial}{\partial \xi^\beta} \int p_1^\alpha p_1^\beta f_1 \, d\boldsymbol{p}_1$$

$$+ \frac{1}{m\nu^2} \int \frac{\partial \Phi_{1,2}}{\partial q_1^\alpha} f_2 \, d\boldsymbol{p}_1 d\boldsymbol{q}_2 d\boldsymbol{p}_2. \tag{24}$$

Now we note that according to Equation (15) we have

$$\int \frac{\partial \Phi_{1,2}}{\partial q_1^\alpha} f_2 \, d\boldsymbol{p}_1 d\boldsymbol{q}_2 d\boldsymbol{p}_2$$

$$= \int \frac{\partial \Phi(|\boldsymbol{q}_1 - \boldsymbol{q}_2|)}{\partial q_1^\alpha} \varphi(t, \xi, \boldsymbol{q}_1 - \boldsymbol{q}_2, \boldsymbol{p}_1, \boldsymbol{p}_2) \, d\boldsymbol{q}_2 d\boldsymbol{p}_1 d\boldsymbol{p}_2$$

$$= \int \frac{\partial \Phi(|\boldsymbol{q}|)}{\partial q_1^\alpha} \varphi(t, \xi, \boldsymbol{q}, \boldsymbol{p}_1, \boldsymbol{p}_2) \, d\boldsymbol{q} d\boldsymbol{p}_1 d\boldsymbol{p}_2,$$

and, consequently, changing the indices of the integration variables, we arrive at

$$\int \frac{\partial \Phi_{1,2}}{\partial q_1^\alpha} f_2 \, d\boldsymbol{p}_1 d\boldsymbol{p}_2 d\boldsymbol{q}_2$$

$$= \int \frac{\partial \Phi(|\boldsymbol{q}|)}{\partial q_1^\alpha} \varphi(t, \xi, \boldsymbol{q}, \boldsymbol{p}_2, \boldsymbol{p}_1) \, d\boldsymbol{q} d\boldsymbol{p}_1 d\boldsymbol{p}_2$$

$$= - \int \frac{\partial \Phi(|\boldsymbol{q}|)}{\partial q_1^\alpha} \varphi(t, \xi, -\boldsymbol{q}, \boldsymbol{p}_2, \boldsymbol{p}_1) \, d\boldsymbol{q} d\boldsymbol{p}_1 d\boldsymbol{p}_2.$$

It is obvious that

$$\varphi(t, \xi, -(\boldsymbol{q}_1 - \boldsymbol{q}_2), \boldsymbol{p}_2, \boldsymbol{p}_1) = P_{1,2}\varphi(t, \xi, \boldsymbol{q}_1 - \boldsymbol{q}_2, \boldsymbol{p}_2, \boldsymbol{p}_1),$$

and therefore we can write

$$\int \frac{\partial \Phi_{1,2}}{\partial q_1^\alpha} f_2 \, d\boldsymbol{p}_1 d\boldsymbol{p}_2 d\boldsymbol{q}_2 = -\int \frac{\partial \Phi_{1,2}}{\partial q_1^\alpha} (P_{1,2}f_2) \, d\boldsymbol{p}_1 d\boldsymbol{p}_2 d\boldsymbol{q}_2,$$

so that

$$\int \frac{\partial \Phi_{1,2}}{\partial q_1^\alpha} f_2 \, d\boldsymbol{p}_1 d\boldsymbol{p}_2 d\boldsymbol{q}_2 = \frac{1}{2} \int \frac{\partial \Phi_{1,2}}{\partial q_1^\alpha} \{f_2 - P_{1,2}f_2\} \, d\boldsymbol{p}_1 d\boldsymbol{p}_2 d\boldsymbol{q}_2.$$

Further, we take into account the symmetry property (20) to obtain

$$\int \frac{\partial \Phi_{1,2}}{\partial q_1^\alpha} f_2 \, d\boldsymbol{p}_1 d\boldsymbol{p}_2 d\boldsymbol{q}_2 = \frac{\mu}{2} \sum_\beta \int \frac{\partial \Phi_{1,2}}{\partial q_1^\alpha} (\boldsymbol{q}_2 - \boldsymbol{q}_1)^\beta \frac{\partial f_2}{\partial \xi^\beta} \, d\boldsymbol{p}_1 d\boldsymbol{p}_2 d\boldsymbol{q}_2$$

$$+ \frac{\mu^2}{4} \sum_{\beta,\gamma} \int \frac{\partial \Phi_{1,2}}{\partial q_1^\alpha} (\boldsymbol{q}_2 - \boldsymbol{q}_1)^\beta (\boldsymbol{q}_2 - \boldsymbol{q}_1)^\gamma \frac{\partial^2 f_2}{\partial \xi^\beta \partial \xi^\gamma} \, d\boldsymbol{p}_1 d\boldsymbol{p}_2 d\boldsymbol{q}_2 + \mu^3 + \ldots \quad (25)$$

Note also that according to Equations (12) and (13) we have

$$\int p_1^\alpha p_1^\beta f_1 \, d\boldsymbol{p}_1 = \int (\boldsymbol{p}_1 - m\boldsymbol{u})^\alpha (\boldsymbol{p}_1 - m\boldsymbol{u})^\beta f_1 \, d\boldsymbol{p}_1 + m^2 \nu u^\alpha u^\beta \rho. \quad (26)$$

We substitute these relations into the right-hand side of Equation (24) to find

$$\frac{\partial (\rho u^\alpha)}{\partial t} = -\mu \sum_\beta \frac{\partial (\rho u^\alpha u^\beta)}{\partial \xi^\beta} - \mu \sum_\beta \frac{\partial T_{\alpha,\beta}}{\partial \xi^\beta}$$

$$+ \mu^2 \sum_{\beta,\gamma} \frac{\partial^2 T_{\alpha,\beta,\gamma}}{\partial \xi^\beta \partial \xi^\gamma} + \mu^3 + \ldots \quad (27)$$

where for brevity we have put

$$T_{\alpha,\beta} \equiv T_{\alpha,\beta}(f_1, f_2) = \frac{1}{m\nu^2} \int (\boldsymbol{p}_1 - m\boldsymbol{u})^\alpha (\boldsymbol{p}_1 - m\boldsymbol{u})^\beta f_1 \, d\boldsymbol{p}_1$$

$$+ \frac{1}{2m\nu^2} \int \frac{\partial \Phi(|\boldsymbol{q}_2 - \boldsymbol{q}_1|)}{\partial q_1^\alpha} (\boldsymbol{q}_2 - \boldsymbol{q}_1)^\beta f_2 \, d\boldsymbol{q}_2 d\boldsymbol{p}_1 d\boldsymbol{p}_2, \quad (28)$$

$$T_{\alpha,\beta,\gamma} \equiv T_{\alpha,\beta,\gamma}(f_1, f_2) =$$

$$+ \frac{1}{4m\nu^2} \int \frac{\partial \Phi(|\boldsymbol{q}_2 - \boldsymbol{q}_1|}{\partial q_1^\alpha} (\boldsymbol{q}_2 - \boldsymbol{q}_1)^\beta (\boldsymbol{q}_2 - \boldsymbol{q}_1)^\gamma f_2 \, d\boldsymbol{q}_2 d\boldsymbol{p}_1 d\boldsymbol{p}_2. \quad (29)$$

Finally, in view of the continuity equation (23), Equation (27) can be also represented in the form

$$\rho \frac{\partial(u^\alpha)}{\partial t} = -\mu \sum_\beta \rho u^\beta \frac{\partial(u^\alpha)}{\partial \xi^\beta} - \mu \sum_\beta \frac{\partial T_{\alpha,\beta}}{\partial \xi^\beta}$$

$$+ \mu^2 \sum_{\beta,\gamma} \frac{\partial^2 T_{\alpha,\beta,\gamma}}{\partial \xi^\beta \partial \xi^\gamma} + \mu^3 + \dots \quad (30)$$

At last, let us proceed to differentiate expression (14). If we notice that the following identity holds

$$\int f_1 \frac{\partial}{\partial t} |\boldsymbol{p}_1 - m\boldsymbol{u}|^2 \, d\boldsymbol{p}_1 = -2 \sum_\alpha \frac{\partial u^\alpha}{\partial t} \int f_1 (p_1^\alpha - mu^\alpha) \, d\boldsymbol{p}_1 = 0,$$

we obtain

$$\frac{\partial(\rho\varepsilon)}{\partial t} = \frac{1}{2m\nu} \int |\boldsymbol{p}_1 - m\boldsymbol{u}|^2 \frac{\partial f_1}{\partial t} \, d\boldsymbol{p}_1 + \frac{1}{2\nu^2} \int \Phi_{1,2} \frac{\partial f_2}{\partial t} \, d\boldsymbol{q}_2 d\boldsymbol{p}_1 d\boldsymbol{p}_2.$$

Consequently, making use of Equation (16) and (17) we find

$$\frac{\partial(\rho\varepsilon)}{\partial t} = -\frac{\mu}{2m^2\nu} \sum_\alpha \int |\boldsymbol{p}_1 - m\boldsymbol{u}|^2 \frac{\partial f_1}{\partial \xi^\alpha} \, d\boldsymbol{p}_1$$

$$+ \frac{1}{2m\nu^2} \sum_\alpha \int |\boldsymbol{p}_1 - m\boldsymbol{u}|^2 \frac{\partial \Phi_{1,2}}{\partial q_1^\alpha} \frac{\partial f_2}{\partial p_1^\alpha} \, d\boldsymbol{q}_2 d\boldsymbol{p}_1 d\boldsymbol{p}_2$$

$$+ \frac{1}{2\nu^2} \int \Phi_{1,2} \left\{ -\mu \sum_\alpha \frac{p^\alpha}{m} \frac{\partial f_2}{\partial \xi^\alpha} + \left[\frac{p_1^2 + p_2^2}{2m} + \Phi_{1,2}, \, f_2 \right] \right\} d\boldsymbol{q}_2 d\boldsymbol{p}_1 d\boldsymbol{p}_2$$

$$+ \frac{1}{2\nu^3} \int \Phi_{1,2} \left[\Phi_{1,3} + \Phi_{2,3}, \, f_3 \right] d\boldsymbol{q}_2 d\boldsymbol{q}_3 d\boldsymbol{p}_1 d\boldsymbol{p}_2 d\boldsymbol{p}_3.$$

Due to the identities

$$\int \Phi_{1,2} \left[\Phi_{1,3} + \Phi_{2,3}, \, f_3 \right] d\boldsymbol{p}_1 d\boldsymbol{p}_2 d\boldsymbol{p}_3 = 0,$$

$$\int \Phi_{1,2}\big[\Phi_{1,2},\, f_2\big]\, d\boldsymbol{p}_1 d\boldsymbol{p}_2 = 0$$

and in accordance with Equation (15)

$$\left[\frac{p_1^2 + p_2^2}{2m},\, f_2\right] = \frac{1}{m}\sum_\alpha \frac{\partial f_2}{\partial q_2^\alpha}(\boldsymbol{p}_1 - \boldsymbol{p}_2)^\alpha,$$

we can write

$$\begin{aligned}
\frac{\partial(\rho\varepsilon)}{\partial t} = &-\frac{\mu}{2m^2\nu}\sum_\alpha \int |\boldsymbol{p}_1 - m\boldsymbol{u}|^2 \frac{\partial f_1}{\partial \xi^\alpha}\, d\boldsymbol{p}_1 \\
&-\frac{1}{2m\nu^2}\sum_\alpha \int \Phi_{1,2}\, p_1^\alpha \frac{\partial f_2}{\partial p_1^\alpha}\, d\boldsymbol{q}_2 d\boldsymbol{p}_1 d\boldsymbol{p}_2 \\
&+\frac{1}{2m\nu^2}\sum_\alpha \int \bigg\{ \Phi_{1,2}\frac{\partial f_2}{\partial q_2^\alpha}(\boldsymbol{p}_1 - \boldsymbol{p}_2)^\alpha \\
&\qquad + |\boldsymbol{p}_1 - m\boldsymbol{u}|^2 \frac{\partial \Phi_{1,2}}{\partial q_1^\alpha}\frac{\partial f_2}{\partial p_1^\alpha} \bigg\}\, d\boldsymbol{q}_2 d\boldsymbol{p}_1 d\boldsymbol{p}_2
\end{aligned} \tag{31}$$

Note also that

$$\begin{aligned}
&\int \bigg\{ \Phi_{1,2}\frac{\partial f_2}{\partial q_2^\alpha}(\boldsymbol{p}_1 - \boldsymbol{p}_2)^\alpha + |\boldsymbol{p}_1 - m\boldsymbol{u}|^2 \frac{\partial \Phi_{1,2}}{\partial q_1^\alpha}\frac{\partial f_2}{\partial p_1^\alpha} \bigg\}\, d\boldsymbol{q}_2 d\boldsymbol{p}_1 d\boldsymbol{p}_2 \\
&= \int \bigg\{ -\frac{\partial \Phi_{1,2}}{\partial q_2^\alpha} f_2(\boldsymbol{p}_1 - \boldsymbol{p}_2)^\alpha - \frac{\partial|\boldsymbol{p}_1 - m\boldsymbol{u}|^2}{\partial p_1^\alpha}\frac{\partial \Phi_{1,2}}{\partial q_1^\alpha} f_2 \bigg\}\, d\boldsymbol{q}_2 d\boldsymbol{p}_1 d\boldsymbol{p}_2
\end{aligned} \tag{32}$$

With the aid of permutations and the symmetry conditions obtained above we can see that the following relations hold

$$\begin{aligned}
&\int \frac{\partial \Phi_{1,2}}{\partial q_1^\alpha}\big\{ 2mu^\alpha - p_1^\alpha - p_2^\alpha \big\} f_2\, d\boldsymbol{q}_2 d\boldsymbol{p}_1 d\boldsymbol{p}_2 \\
&= \int \frac{\partial \Phi_{1,2}}{\partial q_1^\alpha}\Big\{ mu^\alpha - \frac{p_1^\alpha + p_2^\alpha}{2} \Big\}(f_2 - P_{1,2}f_2)\, d\boldsymbol{q}_2 d\boldsymbol{p}_1 d\boldsymbol{p}_2 \\
&= \mu\sum_\beta \frac{\partial \Phi_{1,2}}{\partial q_1^\alpha}\Big\{ mu^\alpha - \frac{p_1^\alpha + p_2^\alpha}{2} \Big\}\frac{\partial f_2}{\partial \xi^\beta}(\boldsymbol{q}_2 - \boldsymbol{q}_1)^\beta\, d\boldsymbol{q}_2 d\boldsymbol{p}_1 d\boldsymbol{p}_2 \\
&\quad -\frac{\mu^2}{2}\sum_{\beta,\gamma} \frac{\partial \Phi_{1,2}}{\partial q_1^\alpha}\Big\{ mu^\alpha - \frac{p_1^\alpha + p_2^\alpha}{2} \Big\}\frac{\partial^2 f_2}{\partial \xi^\beta \partial \xi^\gamma}
\end{aligned}$$

$$\times (\boldsymbol{q}_2 - \boldsymbol{q}_1)^\beta (\boldsymbol{q}_2 - \boldsymbol{q}_1)^\gamma \, d\boldsymbol{q}_2 d\boldsymbol{p}_1 d\boldsymbol{p}_2 + \mu^3 + \dots$$

Thus Equations (31) and (32) yield

$$\frac{\partial(\rho\varepsilon)}{\partial t} = - \frac{\mu}{2m^2\nu} \sum_\alpha |\boldsymbol{p}_1 - m\boldsymbol{u}|^2 p_1^\alpha \frac{\partial f_1}{\partial \xi^\alpha} \, d\boldsymbol{p}_1$$

$$- \frac{\mu}{2m^2\nu} \sum_\alpha \int \Phi_{1,2} \, p_1^\alpha \frac{\partial f_2}{\partial \xi^\alpha} \, d\boldsymbol{q}_2 d\boldsymbol{p}_1 d\boldsymbol{p}_2$$

$$+ \frac{\mu}{2m^2\nu} \sum_{\alpha,\beta} \int \frac{\partial \Phi_{1,2}}{\partial q_1^\alpha} \left\{ mu^\alpha - \frac{p_1^\alpha + p_2^\alpha}{2} \right\} (\boldsymbol{q}_2 - \boldsymbol{q}_1)^\beta \frac{\partial f_2}{\partial \xi^\beta} \, d\boldsymbol{q}_2 d\boldsymbol{p}_1 d\boldsymbol{p}_2$$

$$- \frac{\mu^2}{4m\nu^2} \sum_{\alpha,\beta,\gamma} \frac{\partial \Phi_{1,2}}{\partial q_1^\alpha} \left\{ mu^\alpha - \frac{p_1^\alpha + p_2^\alpha}{2} \right\}$$

$$\times (\boldsymbol{q}_2 - \boldsymbol{q}_1)^\beta (\boldsymbol{q}_2 - \boldsymbol{q}_1)^\gamma \frac{\partial^2 f_2}{\partial \xi^\beta \partial \xi^\gamma} \, d\boldsymbol{q}_2 d\boldsymbol{p}_1 d\boldsymbol{p}_2 + \mu^3 + \dots$$

We note also that

$$\int |\boldsymbol{p}_1 - m\boldsymbol{u}|^2 p_1^\alpha \frac{\partial f_1}{\partial \xi^\alpha} \, d\boldsymbol{p}_1 = \frac{\partial}{\partial \xi^\alpha} \int |\boldsymbol{p}_1 - m\boldsymbol{u}|^2 p_1^\alpha f_1 \, d\boldsymbol{p}_1$$

$$+ 2 \sum_\beta m \frac{\partial u^\beta}{\partial \xi^\alpha} \int (\boldsymbol{p}_1 - m\boldsymbol{u})^\beta p_1^\alpha f_1 \, d\boldsymbol{p}_1,$$

$$\int \frac{\partial \Phi_{1,2}}{\partial q_1^\alpha} \left\{ mu^\alpha - \frac{p_1^\alpha + p_2^\alpha}{2} \right\} (\boldsymbol{q}_2 - \boldsymbol{q}_1)^\beta \frac{\partial f_2}{\partial \xi^\beta} \, d\boldsymbol{q}_2 d\boldsymbol{p}_1 d\boldsymbol{p}_2$$

$$= \frac{\partial}{\partial \xi^\beta} \int \frac{\partial \Phi_{1,2}}{\partial q_1^\alpha} \left\{ mu^\alpha - \frac{p_1^\alpha + p_2^\alpha}{2} \right\} (\boldsymbol{q}_2 - \boldsymbol{q}_1)^\beta f_2 \, d\boldsymbol{q}_2 d\boldsymbol{p}_1 d\boldsymbol{p}_2$$

$$- m \frac{\partial u^\alpha}{\partial \xi^\beta} \int \frac{\partial \Phi_{1,2}}{\partial q_1^\alpha} (\boldsymbol{q}_2 - \boldsymbol{q}_1)^\beta f_2 \, d\boldsymbol{q}_2 d\boldsymbol{p}_1 d\boldsymbol{p}_2,$$

and consequently

$$\frac{\partial(\rho\epsilon)}{\partial t} = - \frac{\mu}{2m^2\nu} \sum_\alpha \frac{\partial X_\alpha}{\partial \xi^\alpha} - \frac{\mu}{2m^2\nu} \sum_{\alpha,\beta} \frac{\partial u^\beta}{\partial \xi^\alpha} Y_{\alpha,\beta}$$

$$- \frac{\mu^2}{4m\nu^2} \sum_{\alpha,\beta,\gamma} \frac{\partial \Phi_{1,2}}{\partial q_1^\alpha} (\boldsymbol{q}_2 - \boldsymbol{q}_1)^\beta (\boldsymbol{q}_2 - \boldsymbol{q}_1)^\gamma$$

$$\times \left\{ mu^\alpha - \frac{p_1^\alpha + p_2^\alpha}{2} \right\} \frac{\partial^2 f_2}{\partial \xi^\beta \partial \xi^\gamma} \, d\boldsymbol{q}_2 d\boldsymbol{p}_1 d\boldsymbol{p}_2 + \mu^3 + \dots, \qquad (33)$$

where

$$X_\alpha = \frac{1}{m} \int |\boldsymbol{p}_1 - m\boldsymbol{u}|^2 p_1^\alpha f_1 \, d\boldsymbol{p}_1 + \frac{1}{\nu} \int \Phi_{1,2} p_1^\alpha f_1 \, d\boldsymbol{q}_2 d\boldsymbol{p}_1 d\boldsymbol{p}_2$$

$$- \frac{1}{\nu} \sum_\beta \int \frac{\partial \Phi_{1,2}}{\partial q_1^\beta} \left\{ mu^\beta - \frac{p_1^\beta + p_2^\beta}{2} \right\} (\boldsymbol{q}_2 - \boldsymbol{q}_1)^\alpha f_2 \, d\boldsymbol{q}_2 d\boldsymbol{p}_1 d\boldsymbol{p}_2,$$

$$Y_{\alpha,\beta} = 2 \int (\boldsymbol{p}_1 - m\boldsymbol{u})^\beta p_1^\alpha f_1 \, d\boldsymbol{p}_1 + \frac{m}{\nu} \int \frac{\partial \Phi_{1,2}}{\partial q_1^\beta} (\boldsymbol{q}_2 - \boldsymbol{q}_1)^\alpha f_2 \, d\boldsymbol{q}_2 d\boldsymbol{p}_1 d\boldsymbol{p}_2.$$

We make use of Equations (12)-(14) as well as the identity

$$\int (\boldsymbol{p}_1 - m\boldsymbol{u})^\beta u^\alpha f_1 \, d\boldsymbol{p}_1 = u^\alpha \int (\boldsymbol{p}_1 - m\boldsymbol{u})^\beta f_1 \, d\boldsymbol{p}_1 = 0,$$

to obtain

$$X_\alpha = 2m\nu u^\alpha \rho\varepsilon + \frac{1}{m} \int |\boldsymbol{p}_1 - m\boldsymbol{u}|^2 (\boldsymbol{p}_1 - m\boldsymbol{u})^\alpha f_1 \, d\boldsymbol{p}_1$$

$$+ \frac{1}{\nu} \int \Phi_{1,2} (\boldsymbol{p}_1 - m\boldsymbol{u})^\alpha f_2 \, d\boldsymbol{q}_2 d\boldsymbol{p}_1 d\boldsymbol{p}_2$$

$$- \frac{1}{\nu} \sum_\beta \int \frac{\partial \Phi_{1,2}}{\partial q_1^\beta} \left\{ mu^\beta - \frac{p_1^\beta + p_2^\beta}{2} \right\} (\boldsymbol{q}_2 - \boldsymbol{q}_1)^\alpha f_2 \, d\boldsymbol{q}_2 d\boldsymbol{p}_1 d\boldsymbol{p}_2.$$

If we put

$$S_{\alpha,\beta} = S_{\alpha,\beta}(f_1, f_2) = \frac{1}{m\nu} \int (p_1^\beta - mu^\beta)(p_1^\alpha - mu^\alpha) f_1 \, d\boldsymbol{p}_1$$

$$+ \frac{1}{2\nu^2} \int \frac{\partial \Phi_{1,2}}{\partial q_1^\beta} (q_2^\alpha - q_1^\alpha) f_2 \, d\boldsymbol{q}_2 d\boldsymbol{p}_1 d\boldsymbol{p}_2 = \frac{T_{\alpha,\beta}}{m}, \qquad (34)$$

$$S_\alpha = S_\alpha(f_1, f_2) = \frac{1}{2m^2\nu} \int |\boldsymbol{p}_1 - m\boldsymbol{u}|^2 (\boldsymbol{p}_1 - m\boldsymbol{u})^\alpha f_1 \, d\boldsymbol{p}_1$$

$$+ \frac{1}{m\nu^2} \int \Phi_{1,2} (p_1^\alpha - mu^\alpha)^2 f_2 \, d\boldsymbol{q}_2 d\boldsymbol{p}_1 d\boldsymbol{p}_2$$

$$+ \frac{1}{2m\nu^2} \sum_\beta \int \frac{\partial \Phi_{1,2}}{\partial q_1^\beta} \left\{ \frac{p_1^\beta + p_2^\beta}{2} - mu^\beta \right\} (\boldsymbol{q}_2 - \boldsymbol{q}_1)^\alpha f_2 \, d\boldsymbol{q}_2 d\boldsymbol{p}_1 d\boldsymbol{p}_2 \quad (35)$$

$$R = R(f_2) = \frac{1}{4m\nu^2} \sum_{\alpha,\beta,\gamma} \int \frac{\partial \Phi_{1,2}}{\partial q_1^\alpha} (\boldsymbol{q}_2 - \boldsymbol{q}_1)^\beta (\boldsymbol{q}_2 - \boldsymbol{q}_1)^\gamma$$

$$\times \left\{ mu^\alpha - \frac{p_1^\alpha + p_2^\alpha}{2} \right\} \frac{\partial^2 f_2}{\partial \xi^\beta \xi^\alpha} \, d\boldsymbol{q}_2 d\boldsymbol{p}_1 d\boldsymbol{p}_2, \quad (36)$$

we can write

$$\frac{\partial(\rho\varepsilon)}{\partial t} = -\mu \sum_\alpha \frac{\partial(u^\alpha \rho\varepsilon)}{\partial \xi^\alpha} - \mu \sum_\alpha \frac{\partial S_\alpha}{\partial \xi^\alpha}$$

$$- \mu \sum_{\alpha,\beta} \int \frac{\partial u^\beta}{\partial \xi^\alpha} S_{\alpha,\beta} - \mu^2 R + \mu^3 \ldots \quad (37)$$

instead of Equation (33).

5. Having determined the derivatives of the main hydrodynamics variables, we now return to Equation (16), which is to be solved taking into account the symmetry conditions (18) and (20), the normalization condition (21), and the correlation weakening condition (22). It should be emphasized that the usual expansion in powers of μ

$$f_s(t, \xi, \boldsymbol{q}_1, \ldots, \boldsymbol{q}_s, \boldsymbol{p}_1, \ldots, \boldsymbol{p}_s) = f_s^0(t, \xi, \boldsymbol{q}_1, \ldots, \boldsymbol{q}_s, \boldsymbol{p}_1, \ldots, \boldsymbol{p}_s)$$
$$+ \mu f_s^1(t, \xi, \boldsymbol{q}_1, \ldots, \boldsymbol{q}_s, \boldsymbol{p}_1, \ldots, \boldsymbol{p}_s) + \mu^2 + \ldots \quad (38)$$

cannot result in obtaining the equations of hydrodynamics, with the aid of which one could describe the evolution of hydrodynamics variables.

In fact, if we substitute Equation (38) into formulas (12)–(14), we find

$$\rho = \rho_0 + \mu\rho_1 + \ldots,$$
$$u^\alpha = u_0^\alpha + \mu u_1^\alpha + \ldots,$$
$$\varepsilon = \varepsilon_0 + \mu\varepsilon_1 + \ldots, \quad (39)$$

where ρ_0, u_0^α, and ε_0 are determined by Equations (12)-(14) where the quantities f^0 are substituted for f, and ρ_1, u_1^α, and ε_1 are determined by the same equations where f^1 are substituted for f, etc. On the other hand, if we substitute Equation (38) into Equation (16) and equate the coefficients of identical power of μ, we obtain, for instance, for the leading terms

$$\frac{\partial f_s^0}{\partial t} = [H_s; f_s^0] + \frac{1}{\nu} \int \left[\sum_{1 \le i \le s} \Phi_{i,s+1}; f_{s+1}^0 \right] d\boldsymbol{q}_{s+1} d\boldsymbol{p}_{s+1}$$

$$s = 1,\ 2,\ 3, \ldots \tag{40}$$

We note also that the symmetry conditions become

$$P_{ij} f_s^0 = f_s^0, \quad P_{1j} f^0 = f_s^0. \tag{41}$$

Since the function f_s^0 is spatially homogeneous with regard to $\boldsymbol{q}_1, \ldots, \boldsymbol{q}_s$ and there are no operators acting on the argument ξ, we find that the functions f_s^0 vary in time according to the same law as the function F_s does in the case of spatially homogeneous distribution. Therefore, the derivatives

$$\frac{\partial \rho_0}{\partial t}, \quad \frac{\partial u_0^\alpha}{\partial t}, \quad \frac{\partial \varepsilon_0}{\partial t}$$

must be identically equal to zero, since the formal expressions for them follow Equations (23), (27), and (37) in proceeding to the spatially homogeneous case, i.e. for $\mu = 0$. Thus, ρ_0, u_0^α, and ε_0 do not vary in time t. We find that in accordance with Equation (39) the variation of hydrodynamics functions is entirely due to the evolution of the correction terms proportional to the small parameter.

However, since one can make use of the asymptotic expansion (38) preserving one or two terms only until the correction terms are small compared with the leading ones. Therefore, these expansions are valid only for time intervals during which the quantities ρ, u^α, and ε do not vary significantly from their initial values. In other words, we are not able to derive the equations of hydrodynamics, i.e. equations which would describe the evolution of quantities ρ_0, u_0^α, and ε_0 for a sufficiently long period of time during which these quantities change considerably. Apparently, the difficulty of using the usual expansion is trivial enough and is connected with the appearance of secular term, as is usual in such cases.

In order to formulate a proper method of expansion similar to that used in nonlinear mechanics, we note that in the case under consideration one needs to know a particular solution $f_s(t, \xi, \boldsymbol{q}_1, \ldots, \boldsymbol{q}_s, \boldsymbol{p}_1, \ldots, \boldsymbol{p}_s)$ of the equations considered for which the derivatives $\partial f_s / \partial t$ are of the order of magnitude of μ from the beginning. Indeed, by introducing the small parameter μ and considering the expressions for F_s in the form of Equation (10), we actually perform the spatial averaging over a region the dimensions of which are sufficiently large by comparison with the molecular distances and investigate the solutions for which the derivative $\partial f_s / \partial t$ is proportional to μ and considered as averaged over time in a certain way. Then, in the first approximation, the functions

$$f_s = f_s^0 \tag{42}$$

satisfy the equations

$$\left[H_s; f_s^0\right] + \frac{1}{\nu} \int \left[\sum_{1 \le i \le s} \Phi_{i,s+1}; f_{s+1}^0\right] d\boldsymbol{q}_{s+1} d\boldsymbol{p}_{s+1} = 0, \tag{43}$$

the symmetry conditions (41), and the corresponding normalization and correlation weakening conditions. Note also that the functions f_s^0 are spatially homogeneous with regard to $\boldsymbol{q}_1, \ldots, \boldsymbol{q}_s$, since the above relations contain no operators acting on the variable ξ and this variable is contained in f_s^0 only as a parameter. Equations (43) are those very equations which determine the stationary spatially homogeneous distribution F_s. Suppose, as is valid for gases at usual densities, that there exists only one system of stationary solutions, namely that which corresponds to statistical equilibrium and which id fully determined by the values of the temperature, temperature, and average velocity vector.[q] Then we look for the expression of f_s^0 in the form

$$f_s^0 = \varphi_s(\xi, \boldsymbol{q}_1, \ldots, \boldsymbol{q}_s) \exp\left\{-\sum_{1 \le i \le s} \frac{|\boldsymbol{p}_i - \boldsymbol{k}|^2}{2m\theta}\right\}, \tag{44}$$

where $\boldsymbol{k}$ is a vector, and θ is the temperature. Since the function φ_s is spatially homogeneous, we have

$$f_1^0 = \varphi(\xi) \exp\left\{-\frac{|\boldsymbol{p}_1 - \boldsymbol{k}|^2}{2m\theta}\right\}. \tag{45}$$

---

[q]For amorphous solids such as glasses, there exist distributions which do not correspond to statistical equilibrium, since they have an infinitesimal relaxation velocity and, therefore, are stationary. On the other hand, for crystals which are not even in the state of statistical equilibrium, the probability density depends on the the location of the crystal lattice in space, and therefore it is necessary to fix a large number of parameters to determine the density.

Substituting expressions (44) and (45) into Equation (40) we obtain

$$\frac{\partial \varphi_s}{\partial q_1^\alpha} + \frac{1}{\theta} \frac{\partial U_s}{\partial q_1^\alpha} \varphi_s + \frac{1}{\theta} \frac{(2\pi m\theta)^{3/2}}{\nu} \int \frac{\partial \Phi_{1,s+1}}{\partial q_1^\alpha} \varphi_{s+1} \, dq_{s+1} = 0 \qquad (46)$$

where

$$U_s = \sum_{(1 \le i \le j \le s)} \Phi_{i,j}$$

and the functions φ_s are assumed to be symmetric with regard to the variables $q_1, \ldots, q_s$. Moreover, taking into account the correlation weakening condition for f_s^0 and formula (45) we see that the functions $\varphi_s(\xi, q_1, \ldots, q_s)$ tend to $\{\varphi(\xi)\}^s$ as the distance $q_1, \ldots, q_s$ tends to infinity.

In order to obtain the corresponding expressions for f_s^0 as well as for the coefficients of the expansion

$$f_s = f_s^0 + \mu f_s^1 + \cdots, \qquad (47)$$

we impose the following auxiliary condition: performing the expansion we should not expand the functions ρ, u^α, and ε in powers of μ. In other words, we determine the coefficients of expansion (47) so that the following relations hold

$$\rho = \frac{1}{\nu} \int f_1^0 \, dp_1, \quad m\rho u^\alpha = \frac{1}{\nu} \int p_1^\alpha f_1^0 \, dp_1,$$

$$\rho \varepsilon = \frac{1}{\nu} \int \frac{|p_1 - mu|^2}{2m} f_1^0 \, dp_1 + \frac{1}{2\nu^2} \int \Phi_{1,2} f_2^0 \, dq_2 dp_1 dp_2, \qquad (48)$$

and

$$\frac{1}{\nu} \int f_1^1 \, dp_1 = 0, \quad \frac{1}{\nu} \int p_1^\alpha f_1^1 \, dp_1 = 0,$$

$$\frac{1}{\nu} \int \frac{|p_1 - mu|^2}{2m} f_1^1 \, dp_1 + \frac{1}{2\nu^2} \int \Phi_{1,2} f_2^1 \, dq_2 dp_1 dp_2 = 0. \qquad (49)$$

Then Equations (44) and (45) imply that

$$k^\alpha = mu^\alpha, \quad \varphi(\xi) = \frac{\nu \rho}{(2\pi m\theta)^{3/2}}. \qquad (50)$$

Putting

$$\frac{\varphi_s(\xi, q_1, \ldots, q_s)}{(\varphi(\xi))^s} = \chi_s,$$

we find that the functions χ_s satisfy the equations

$$\frac{\partial \chi_s}{\partial q_1^\alpha} + \frac{1}{\theta}\frac{\partial U_s}{\partial q_1^\alpha}\chi_s + \frac{1}{\theta}\rho\int\frac{\partial \Phi_{1,s+1}}{\partial q_1^\alpha}\chi_{s+1}\,d\mathbf{q}_{s+1} = 0. \tag{51}$$

We note also that the functions χ_s are symmetric and spatially homogeneous with regard to $\mathbf{q}_1, \ldots, \mathbf{q}_s$. Moreover, $\chi_1 = 1$ while χ_s ($s = 2, 2, \ldots$) approaches unity when the distance between the molecules increase continuously. Thus, the functions χ_s are nothing but the distribution function for the coordinates for the molecule complex considered in the first chapter of the monograph [2] (in this monograph these functions are denoted by $F_s(\mathbf{q}_1, \ldots, \mathbf{q}_s)$). In the case under consideration, the particle number density $1/\nu$ is apparently replaced by ρ.

One can obtain explicit expressions for χ_s, for instance, with the aid of expansions in powers of ρ. In this case one does not need the explicit expressions for χ_s. It is only necessary that the quantities $\chi_2, \chi_3, \ldots$ be determined by the number ρ and that they be invariant with regard to translation and reflection transformations in the space of $\mathbf{q}$. Therefore, we put

$$\chi_s = \chi_s(\mathbf{q}_1, \ldots, \mathbf{q}_s, \rho),$$
$$\chi_1 = 1, \quad \chi_2 = g(|\mathbf{q}_1 - \mathbf{q}_2|, \rho). \tag{52}$$

We see that $g(r, \rho)$ is the usual molecular distribution function, which depends on the distance r between two molecules and on the average density ρ.

With the aid of the formulae obtained we write Equation (44) in the form

$$f_s^0 = \frac{(\nu\rho)^s}{(2\pi m\theta)^{3s/2}}\chi_s(\mathbf{q}_1, \ldots, \mathbf{q}_s, \rho)\exp\left\{-\sum_{1\le i\le s}\frac{|\mathbf{p}_i - m\mathbf{u}|^2}{2m\theta}\right\}. \tag{53}$$

We substitute this expression into formula (48) to obtain

$$\varepsilon = \frac{3}{2}\theta + \rho\int\Phi(|\mathbf{q}|)\,g(|\mathbf{q}|, \rho)\,d\mathbf{q} = \varepsilon(\rho, \theta). \tag{54}$$

It should be stressed that this expression is nothing but the thermodynamic average energy per molecule expressed in terms of the density and the temperature θ.

If one expresses the parameter θ in terms of ε and ρ with the aid of Equation (54), the expressions for f_0^s are completely determined by the values of ρ, $\boldsymbol{u}$, and ε as arbitrary functions of χ. Keeping in mind Equations (23), (30), and (37) we see that in the first approximation the derivatives $\partial\rho/\partial t$, $\partial\varepsilon/\partial t$, and $\partial\boldsymbol{u}/\partial t$ are equal to zero, and therefore ρ, $\boldsymbol{u}$, and ε can be considered as arbitrary constants with respect to the variable t.

6.   Let us now proceed to the final formulation of the corresponding method for asymptotic expansion of the function f_s in powers of the small parameter μ.

Making use of the methods of nonlinear mechanics, which are the corresponding generalization of Langrange's method of the variation of arbitrary constants, we look for an expression for f_s in the form

$$f_s = f_s^0(\xi, \boldsymbol{q}_1, \ldots, \boldsymbol{q}_s, \boldsymbol{p}_1, \ldots, \boldsymbol{p}_s | \rho, \boldsymbol{u}, \varepsilon)$$
$$+ \mu f_s^1(\xi, \boldsymbol{q}_1, \ldots, \boldsymbol{q}_s, \boldsymbol{p}_1, \ldots, \boldsymbol{p}_s | \rho, \boldsymbol{u}, \varepsilon) + \mu^2 + \ldots, \qquad (55)$$

where ρ, $\boldsymbol{u}$, and ε are solutions to the equations

$$\frac{\partial \rho}{\partial t} = \mu A_1(\xi | \rho, \boldsymbol{u}, \varepsilon) + \mu^2 A_2(\xi | \rho, \boldsymbol{u}, \varepsilon) + \mu^3 + \ldots,$$
$$\frac{\partial u^\alpha}{\partial t} = \mu B_1^\alpha(\xi | \rho, \boldsymbol{u}, \varepsilon) + \mu^2 B_2^\alpha(\xi | \rho, \boldsymbol{u}, \varepsilon) + \mu^3 + \ldots, \quad \alpha = 1,\ 2,\ 3,$$
$$\frac{\partial (\rho\varepsilon)}{\partial t} = \mu C_1(\xi | \rho, \boldsymbol{u}, \varepsilon) + \mu^2 C_2(\xi | \rho, \boldsymbol{u}, \varepsilon) + \mu^3 + \ldots. \qquad (56)$$

In this relations, f_s^0, f_s^1, $\ldots$; A_1, A_2, $\ldots$; B_1^α, B_2^α, $\ldots$; C_1, C_2, $\ldots$ are unknown functions of ρ, $\boldsymbol{u}$ and ε which depend on ξ and which are to be chosen in such a way that, after substituting the solutions of Equations (56) into expression (55), it satisfies Equations (16) and all the auxiliary conditions imposed on the functions f_s, including conditions (48) and (49). It is clear that expansions (55) and (56) are of a solely formal character and only the first one or two terms have practical significance. Therefore, the above formulae should be understood as those whose asymptotic approximation, obtained neglecting some power of the parameter μ, satisfy the corresponding equations up to this power.

If we take into account Equation (23) we find directly

$$A_1 = -\sum_\alpha \frac{\partial(\rho u^\alpha)}{\partial \xi^\alpha}, \quad A_2 = A_3 = \ldots = 0, \qquad (57)$$

i.e. the first equation of (56) is really the continuity equation. Then Equations (30) and (37) yield

$$B_1^\alpha = -\sum_\beta \rho u^\beta \frac{\partial u^\beta}{\partial \xi^\beta} - \sum_\beta \frac{\partial T_{\alpha,\beta}(f_1^0, f_2^0)}{\partial \xi^\beta},$$

$$B_2^\alpha = -\sum_\beta \frac{\partial T_{\alpha,\beta}(f_1^1, f_2^1)}{\partial \xi^\beta} + \sum_{\beta,\gamma} \frac{\partial^2 T_{\alpha,\beta,\gamma}(f_2^0)}{\partial \xi^\beta \partial \xi^\gamma}$$

$$C_1 = -\sum_\alpha \frac{\partial(\boldsymbol{u}^\alpha \rho \varepsilon)}{\partial \xi^\alpha} - \sum_\alpha \frac{\partial S_\alpha(f_1^0, f_2^0)}{\partial \xi^\alpha} - \sum_{\alpha,\beta} \frac{\partial u^\beta}{\partial \xi^\alpha} S_{\alpha,\beta}(f_1^0, f_2^0),$$

$$C_2 = -\sum_\alpha \frac{\partial S_\alpha(f_1^0, f_2^0)}{\partial \xi^\alpha} - \sum_{\alpha,\beta} \frac{\partial u^\beta}{\partial \xi^\alpha} S_{\alpha,\beta}(f_1^0, f_2^0) - R(f_2^0). \tag{58}$$

On the other hand, making use of the considerations of the preceding section it is not difficult to see that the first term in expansion (55) is determined by formula (53). Therefore the quantities B_1^α and C_1 can be easily found. Therefore, we are now able to show that the equations of the first approximation

$$\rho \frac{\partial u^\alpha}{\partial t} = \mu B_1^\alpha, \qquad \frac{\partial(\rho \varepsilon)}{\partial t} = \mu C_1, \tag{59}$$

which follow Equation (56) after neglecting the terms of the second order of smallness are the ordinary equations of the hydrodynamics of an ideal liquid.

Indeed, we substitute expressions (53) into Equation (58) and make use of the notations (28), (34), and (35) for $T_{\alpha,\beta}$, S_α, and $S_{\alpha,\beta}$ to obtain

$$B_1^\alpha = -\sum_\beta \rho u^\beta \frac{\partial u^\alpha}{\partial \xi^\alpha} - \frac{1}{m} \frac{\partial P}{\partial \xi^\alpha}$$

$$C_1 = -\sum_\alpha \frac{\partial(u^\alpha \rho \varepsilon)}{\partial \xi^\alpha} - -\sum_\alpha \frac{\partial u^\alpha}{\partial \xi^\alpha} P, \tag{60}$$

where

$$P = \rho \theta - \frac{\rho^2}{C} \int \frac{\partial \Phi(|\boldsymbol{q}|)}{\partial |\boldsymbol{q}|} g(|\boldsymbol{q}|, \rho) \, d\boldsymbol{q}. \tag{61}$$

Thus, the equations in the first approximation (59) being reconsidered in terms of the spatial variable $\boldsymbol{q} = \mu^{-1}\xi$ take the form

$$m\rho \left\{ \frac{\partial u^\alpha}{\partial t} + \sum_\beta \frac{\partial u^\alpha}{\partial q^\beta} u^\beta \right\} = -\frac{\partial P}{\partial q^\alpha}, \tag{62}$$

$$\frac{\partial(\rho\varepsilon)}{\partial t} + \sum_\alpha \frac{\partial(u^\alpha \rho\varepsilon)}{\partial q^\alpha} + P\sum_\alpha \frac{\partial u^\alpha}{\partial \xi^\alpha} = 0, \tag{63}$$

where $m\rho$ is the mass density. As we see (one may consult Equation (1.7) in the monograph [2]) the quantity P given by Equation (61) is the pressure, while relation (61) itself is the equation of state in the case when an explicit expression for the molecular distribution function g is known. Thus, Equation (62) is the ordinary equation of state of an ideal liquid.

To transform Equation (63) to the usual form we introduce the free energy of the system per molecule

$$\Psi = \Psi(\rho, \ \theta).$$

Since the quantities P and ε are determined with aid of the Gibbs canonical distribution, we have

$$P = \rho^2 \frac{\partial \Psi}{\partial \rho}, \quad \varepsilon = \Psi - \theta\frac{\partial \Psi}{\partial \theta}. \tag{64}$$

If we use the continuity equation, from Equation (63) we obtain

$$\rho\left(\frac{\partial\varepsilon}{\partial t} + \sum_\alpha u^\alpha \frac{\partial\varepsilon}{\partial q^\alpha}\right) + P\sum_\alpha \frac{\partial u^\alpha}{\partial q^\alpha} = 0,$$

or

$$\rho\frac{\partial\varepsilon}{\partial\rho}\left(\frac{\partial\rho}{\partial t} + \sum_\alpha u^\alpha \frac{\partial\rho}{\partial q^\alpha}\right) + \rho\frac{\partial\varepsilon}{\partial\theta}\left(\frac{\partial\theta}{\partial t} + \sum_\alpha u^\alpha \frac{\partial\theta}{\partial q^\alpha}\right) + P\sum_\alpha \frac{\partial u^\alpha}{\partial q^\alpha} = 0.$$

We then use the continuity equation again to find

$$\rho\frac{\partial\varepsilon}{\partial\theta}\left(\frac{\partial\theta}{\partial t} + \sum_\alpha u^\alpha \frac{\partial\theta}{\partial q^\alpha}\right) + \left(P - \rho^2\frac{\partial\varepsilon}{\partial\rho}\right)\sum_\alpha \frac{\partial u^\alpha}{\partial q^\alpha} = 0. \tag{65}$$

After determining the entropy

$$S = -\frac{\partial\Psi}{\partial\theta}, \tag{66}$$

Equation (64) yields

$$P - \rho^2\frac{\partial\varepsilon}{\partial\rho} = -\rho^2\theta\frac{\partial S}{\partial\rho}, \quad \rho\frac{\partial\varepsilon}{\partial\theta} = \rho\theta\frac{\partial S}{\partial\theta}.$$

Hence, Equation (65) becomes

$$\frac{\partial S}{\partial \theta}\left(\frac{\partial \theta}{\partial t} + \sum_\alpha u^\alpha \frac{\partial \theta}{\partial q^\alpha}\right) - \rho\frac{\partial S}{\partial \rho}\sum_\alpha \frac{\partial u^\alpha}{\partial q^\alpha} = 0.$$

or

$$\frac{\partial S}{\partial \theta}\left(\frac{\partial \theta}{\partial t} + \sum_\alpha u^\alpha \frac{\partial \theta}{\partial q^\alpha}\right) + \frac{\partial S}{\partial \rho}\left(\frac{\partial \rho}{\partial t} + \sum_\alpha \frac{\partial u^\alpha}{\partial q^\alpha}\right) = 0.$$

Equation (63) can be written finally as follows:

$$\frac{\partial S}{\partial t} + \sum_\alpha u^\alpha \frac{\partial S}{\partial q^\alpha} = 0. \tag{67}$$

We see that this is the usual adiabatic equation (see, for instance, paper [3]).

Thus, we have shown that the equations of the first approximation are the ordinary equations of hydrodynamics for an ideal liquid. It is also possible to show that the equations of the second order, which are obtained from Equation (56) by neglecting terms of the order of magnitude μ^3, are the equations of a viscous liquid taking into account heat transfer processes.

References

1. Bogolubov, N.N. (1946) *Journal of Experimental and Theoretical Physics*, **16**, 681.
   Bogolubov, N.N. (1946) *Journal of Experimental and Theoretical Physics*, **16**, 691,
2. Bogolubov, N.N. (1946) Problems of a Dynamical Theory in Statistical Physics , Moscow, Gostekhizdat.
3. Landau, L.D., Lifshitz, E.M. (1944) *Mechanics of Continuous Media* Moscow, Gostekhizdat.

CHAPTER 4

ON THE HYDRODYNAMICS OF A SUPERFLUID LIQUID

Introduction

The object of these lectures is to derive the hydrodynamic equations for a superfluid liquid from the equations of motion of a system of identical Bose particles, and to obtain in this way a "hydrodynamic approximation" for the Green Functions. To simplify the presentation we shall consider our system to be an ideal liquid.

We shall consider a system of identical Bose particles with pair interactions, with a Hamiltonian which in the representation of second quantization has the the form:[r]

$$H = + \frac{1}{2m} \int \nabla \psi^\dagger(t,r) \nabla \psi(t,r) \, d\vec{r} - \lambda \int \psi^\dagger(t,r) \psi(t,r) \, d\vec{r}$$
$$+ \frac{1}{2} \int \Phi(r - r') \psi^\dagger(t,r) \psi^\dagger(t,r') \psi(t,r') \psi(t,r) \, d\vec{r} \, d\vec{r}'$$
$$+ \int \{ \eta(t,r) \psi^\dagger(t,r) + \eta^*(t,r) \psi(t,r) + U(t,r) \psi^\dagger(t,r) \psi(t,r) \} \, d\vec{r}.$$

Here λ is a constant, and $\psi^\dagger(t,r)$, $\psi(t,r)$ are Bose operators in the Heisenberg representation, with the usual commutation relations. The reader will have noticed that in addition to the usual terms, our Hamiltonian includes additional terms corresponding to "particle sources" $\eta(t,r)$, $\eta^*(t,r)$ and an external field $U(t,r)$; η, η^* and U are given c-number functions of $\vec{r}$ and t. The introduction of these "external sources" is necessary because we

---

[r]We use a system of units in which $\hbar = 1$. When $\vec{r}$ is the argument of a function we write it simply as r.

intend to obtain an expression for the Green's function by varying the usual
hydrodynamic averages with respect to them.

1.   Preliminary Identities

To derive the hydrodynamic equations we shall need a set of identities for
the time derivatives of the following "local" quantities:

$$\rho(t,r) = \langle \psi^\dagger(t,r)\psi(t,r) \rangle, \quad \phi(t,r) = \langle \psi(t,r) \rangle,$$

$$j_\alpha(t,r) = \frac{i}{2}\left\langle \frac{\partial \psi^\dagger(t,r)}{\partial r_\alpha}\psi(t,r) - \psi^\dagger(t,r)\frac{\partial \psi(t,r)}{\partial r_\alpha} \right\rangle, \quad (\alpha = 1,2,3)$$

$$\rho(t,r)\epsilon(t,r) = -\frac{1}{4m}\langle (\Delta\psi^\dagger(t,r))\psi(t,r) + \psi^\dagger(t,r)(\Delta\psi(t,r)) \rangle$$

$$+ \frac{1}{2}\int \Phi(r - r')\langle \psi^\dagger(t,r)\psi^\dagger(t,r')\psi(t,r')\psi(t,r) \rangle \, d\vec{r}', \quad (1.1)$$

which represents respectively the mean particle number density (ρ), the mean
current $(\vec{j})$ and the mean energy per particle (ϵ). In equation (1.1) the
pointed brackets denote an average taken with respect to some statistical
operator; in general the latter need not correspond to statistical equilibrium.

To calculate the time derivatives of the above quantities we use the
equations of motion, which for our hamiltonian have the form:

$$i\frac{\partial \psi(t,r)}{\partial t} = -\lambda\psi(t,r) - \frac{\Delta}{2m}\psi(t,r)$$

$$+ \int \Phi(r - r')\psi^\dagger(t,r')\psi(t,r') \, d\vec{r}' \, \psi(t,r)$$

$$+ U(t,r)\psi(t,r) + \eta(t,r),$$

$$i\frac{\partial \psi^\dagger(t,r)}{\partial t} = \lambda\psi^\dagger(t,r) + \frac{\Delta}{2m}\psi^\dagger(t,r)$$

$$- \psi^\dagger(t,r)\int \Phi(r - r')\psi^\dagger(t,r')\psi(t,r') \, d\vec{r}'$$

$$- U(t,r)\psi^\dagger(t,r) - \eta^*(t,r). \quad (1.2)$$

Differentiating the first equations (1.1), we get

$$\frac{\partial \rho(t,r)}{\partial t} = \left\langle \frac{\partial \psi^\dagger(t,r)}{\partial t}\psi(t,r) + \psi^\dagger(t,r)\frac{\partial \psi(t,r)}{t} \right\rangle$$

$$= \frac{i}{2m}\langle -(\Delta\psi^\dagger(t,r))\psi(t,r) + \psi^\dagger(t,r)(\Delta\psi(t,r))\rangle$$

$$+ i\int \Phi(r-r')\langle \psi^\dagger(t,r)\psi^\dagger(t,r')\psi(t,r')\psi(t,r)$$

$$- \psi^\dagger(t,r)\psi^\dagger(t,r')\psi(t,r')\psi(t,r)\rangle\, d\vec{r}'$$

$$+ i\eta^*(t,r)\phi(t,r) - i\eta(t,r)\phi^*(t,r).$$

But since

$$\sum_\alpha \frac{\partial}{\partial r_\alpha}\left\langle \frac{\partial\psi^\dagger(t,r)}{\partial r_\alpha}\psi(t,r) - \psi^\dagger(t,r)\frac{\partial\psi(t,r)}{\partial r_\alpha}\right\rangle$$

$$= \langle (\Delta\psi^\dagger(t,r))\psi(t,r) - \psi^\dagger(t,r)(\Delta\psi(t,r))\rangle$$

and the term containing Φ is identical zero, we get as our final result:

$$m\frac{\partial\rho(t,r)}{\partial t} + \sum_\alpha \frac{\partial j_\alpha(t,r)}{\partial r_\alpha} = im[\eta^*(t,r)\phi(t,r) - \eta(t,r)\phi^*(t,r)]. \qquad (1.3)$$

In exactly the same way we get for the current density:[s]

$$\frac{\partial j_\alpha}{\partial t} = \frac{1}{2}\left\langle \left(\frac{\partial}{\partial r_\alpha}i\frac{\partial\psi^\dagger}{\partial t}\right)\psi + \frac{\partial\psi^\dagger}{\partial r_\alpha}i\frac{\partial\psi}{\partial t} - i\frac{\partial\psi^\dagger}{\partial t}\frac{\partial\psi}{\partial r_\alpha} - \psi^\dagger\left(\frac{\partial}{\partial r_\alpha}i\frac{\partial\psi}{\partial t}\right)\right\rangle$$

$$= \frac{1}{4m}\left\langle \left(\frac{\partial}{\partial r_\alpha}\Delta\psi^\dagger\right)\psi - \frac{\partial\psi^\dagger}{\partial r_\alpha}(\Delta\psi) - (\Delta\psi^\dagger)\frac{\partial\psi}{\partial r_\alpha} + \psi^\dagger\left(\frac{\partial}{\partial r_\alpha}\Delta\psi\right)\right\rangle$$

$$- \frac{1}{2}\left\langle \left(\frac{\partial}{\partial r_\alpha}\int \Phi(r-r')\psi^\dagger(t,r)\psi^\dagger(t,r')\psi(t,r')\, d\vec{r}'\right)\psi\right.$$

$$- \frac{\partial\psi^\dagger}{\partial r_\alpha}\int \Phi(r-r')\psi^\dagger(t,r')\psi(t,r')\, d\vec{r}'\psi(t,r)$$

$$+ \psi^\dagger\left(\frac{\partial}{\partial r_\alpha}\int \Phi(r-r')\psi^\dagger(t,r')\psi(t,r')\, d\vec{r}'\psi(t,r)\right)$$

$$\left. - \int \Phi(r-r')\psi^\dagger(t,r)\psi^\dagger(t,r')\psi(t,r')\, d\vec{r}'\frac{\partial\psi(t,r)}{\partial r_\alpha}\right\rangle$$

$$+ \frac{1}{2}\left\langle \frac{\partial\psi^\dagger}{\partial r_\alpha}\eta - \frac{\partial\eta^*}{\partial r_\alpha}\psi + \eta^*\frac{\partial\psi}{\partial r_\alpha} - \psi^\dagger\frac{\partial\eta}{\partial r_\alpha}\right\rangle - \langle\psi^\dagger\psi\rangle\frac{\partial U}{\partial r_\alpha}$$

$$+ \frac{1}{2}\left\langle \left(\frac{\partial}{\partial r_\alpha}\lambda\psi^\dagger\right)\psi - \frac{\partial\psi^\dagger}{\partial r_\alpha}\lambda\psi - \lambda\psi^\dagger\frac{\partial\psi}{\partial r_\alpha} + \psi^\dagger\left(\frac{\partial}{\partial r_\alpha}\lambda\psi\right)\right\rangle.$$

---

[s]From now on we omit the arguments t, r of ψ, $\psi^\dagger$, erc., where there is no risk of ambiguity.

Using the definitions (1.1) and simplifying the expressions somewhat, we find

$$\frac{\partial j_\alpha}{\partial t} = \frac{1}{4m}\frac{\partial}{\partial r_\alpha}\Delta\rho - \frac{1}{2m}\sum_\beta \left\langle \frac{\partial \psi^\dagger}{\partial r_\alpha}\frac{\partial \psi}{\partial r_\beta} + \frac{\partial \psi^\dagger}{\partial r_\beta}\frac{\partial \psi}{\partial r_\alpha}\right\rangle$$

$$- \int \frac{\partial \Phi(r-r')}{\partial r_\alpha}\langle \psi^\dagger(t,r)\psi^\dagger(t,r')\psi(t,r')\psi(t,r)\rangle\, d\vec{r}'$$

$$+ \rho\frac{\partial}{\partial r_\alpha}(\lambda - U) + \frac{1}{2}\left(\frac{\partial \phi^*}{\partial r_\alpha}\eta + \frac{\partial \phi}{\partial r_\alpha}\eta^* - \phi^*\frac{\partial \eta}{\partial r_\alpha} - \phi\frac{\partial \eta^*}{\partial r_\alpha}\right).$$

It is convenient to introduce a new quantity $D_t(r, r - r')$

$$\langle \psi^\dagger(t,r)\psi^\dagger(t,r')\psi(t,r')\psi(t,r)\rangle \equiv D_t(r, r' - r) = D_t(r', r - r')$$

$$= \frac{1}{2}\{D_t(r', r - r') + D_t(r, r' - r)\}.$$

Using this quantity and changing the variable of integration to $\vec{R} = \vec{r} - \vec{r}'$, we can write the equation for the time derivative of j in the form:

$$\frac{\partial j_\alpha}{\partial t} = \frac{1}{4m}\frac{\partial}{\partial r_\alpha}\Delta\rho - \frac{1}{2m}\sum_\beta \left\langle \frac{\partial \psi^\dagger}{\partial r_\alpha}\frac{\partial \psi}{\partial r_\beta} + \frac{\partial \psi^\dagger}{\partial r_\beta}\frac{\partial \psi}{\partial r_\alpha}\right\rangle$$

$$- \int \frac{\partial \Phi(\vec{R})}{\partial R_\alpha}\{D_t(r, -R) + D_t(r - R, R\}\, d\vec{R}$$

$$+ \rho\frac{\partial}{\partial r_\alpha}(\lambda - U) + \frac{1}{2}\left(\frac{\partial \phi^*}{\partial r_\alpha}\eta + \frac{\partial \phi}{\partial r_\alpha}\eta^* - \phi^*\frac{\partial \eta}{\partial r_\alpha} - \phi\frac{\partial \eta^*}{\partial r_\alpha}\right). \qquad (1.4)$$

Lastly, we have the identity

$$\frac{\partial(\rho\epsilon)}{\partial t} = -\frac{1}{m}\left\langle \left(\Delta\frac{\partial \psi^\dagger}{\partial t}\right)\psi + (\Delta\psi^\dagger)\frac{\partial \psi}{\partial t} + \frac{\partial \psi^\dagger}{\partial t}(\Delta\psi) + \psi^\dagger\left(\Delta\frac{\partial \psi}{\partial t}\right)\right\rangle$$

$$+ \frac{1}{2}\int \Phi(r-r')\frac{\partial}{\partial t}\langle \psi^\dagger(t,r)\psi^\dagger(t,r')\psi(t,r')\psi(t,r)\rangle\, d\vec{r}'.$$

We proceed to calculate the last term:

$$i\frac{\partial}{\partial t}\langle \psi^\dagger(t,r)\psi^\dagger(t,r')\psi(t,r')\psi(t,r)\rangle$$

$$= \frac{1}{2m}\langle ((\Delta\psi^\dagger(t,r))\psi^\dagger(t,r')\psi(t,r')\psi(t,r)\rangle + \frac{1}{2m}\langle \psi^\dagger(t,r)(\Delta\psi^\dagger(t,r'))$$

$$\times \psi(t,r')\psi(t,r)\rangle - \frac{1}{2m}\langle \psi^\dagger(t,r)\psi^\dagger(t,r')(\Delta\psi(t,r'))\psi(t,r)\rangle$$

$$+ \frac{1}{2m}\langle \psi^\dagger(t,r)\psi^\dagger(t,r')\psi(t,r')(\Delta\psi(t,r))\rangle - \int \Phi(r-r')\langle \psi^\dagger(t,r)$$

$$\times \psi^\dagger(t,r_1)\psi(t,r_1)\psi^\dagger(t,r')\psi(t,r')\psi(t,r)\rangle \, d\vec{r}'' + \int \Phi(r-r_1)$$

$$\times \langle \psi^\dagger(t,r)\psi^\dagger(t,r')\psi(t,r')\psi^\dagger(t,r_1)\psi(t,r_1)\psi(t,r)\rangle \, d\vec{r}' + \int \Phi(r'-r_1)$$

$$\times \langle \psi^\dagger(t,r)\psi^\dagger(t,r')\psi^\dagger(t,r_1)\psi(t,r_1)\psi(t,r')\psi(t,r)\rangle \, d\vec{r}' + \int \Phi(r'-r_1)$$

$$\times \langle \psi^\dagger(t,r)\psi^\dagger(t,r')\psi^\dagger(t,r_1)\psi(t,r_1)\psi(t,r')\psi(t,r)\rangle \, d\vec{r}'$$

$$+ \eta(r,t)\langle \psi^\dagger(t,r)\psi^\dagger(t,r')\psi(t,r')\rangle - \eta^*(t,r)\langle \psi^\dagger(t,r')\psi(t,r)\psi(t,r)\rangle$$

$$+ \eta(t,r')\langle \psi^\dagger(t,r)\psi^\dagger(t,r')\psi(t,r)\rangle - \eta^*(t,r')\langle \psi^\dagger(t,r)\psi(t,r')\psi(t,r)\rangle.$$

Using the fact that $\psi^\dagger(t,r)\psi(t,r)$ commutes with $\psi^\dagger(t,r')\psi(t,r')$ we can write this equation as:

$$\frac{\partial}{\partial t}\langle \psi^\dagger(t,r)\psi^\dagger(t,r')\psi(t,r')\psi(t,r)\rangle$$

$$= \frac{i}{2m}\langle \psi^\dagger(t,r)\psi^\dagger(t,r')(\Delta\psi(t,r'))\psi(t,r) + \psi^\dagger(t,r)\psi^\dagger(t,r')\psi(t,r')$$

$$\times (\Delta\psi(t,r)) - (\Delta\psi^\dagger(t,r)\psi^\dagger(t,r')\psi(t,r')\psi(t,r) - \psi^\dagger(t,r)$$

$$\times (\Delta\psi^\dagger(t,r'))\psi(t,r')\psi(t,r)\rangle + i\eta^*(t,r)\langle \psi^\dagger(t,r')\psi(t,r')\psi(t,r)\rangle$$

$$+ i\eta^*(t,r')\langle \psi^\dagger(t,r)\psi(t,r')\psi(t,r)\rangle - i\eta(t,r)\langle \psi^\dagger(t,r)$$

$$\times \psi^\dagger(t,r')\psi(t,r')\rangle - i\eta(t,r')\langle \psi^\dagger(t,r)\psi^\dagger(t,r')\psi(t,r)\rangle.$$

Substituting this into equation for $\partial(\rho\epsilon)/\partial t$ and using the equations of motion, we find:

$$\frac{\partial(\rho\epsilon)}{\partial t} = \frac{i}{8m^2}\langle ((\Delta\Delta\psi^\dagger)\psi - (\Delta\psi^\dagger)(\Delta\psi) + (\Delta\psi^\dagger)(\Delta\psi) - \psi^\dagger(\Delta\Delta\psi)$$

$$+ \frac{i}{4m}\langle -\left(\Delta\int \Phi(r-r')\psi^\dagger(t,r)\psi^\dagger(t,r')\psi(t,r') \, d\vec{r}'\right)\psi(t,r)$$

$$+ \int \Phi(r-r')[(\Delta\psi^\dagger(t,r))\psi^\dagger(t,r')\psi(t,r')\psi(t,r) - \psi^\dagger(t,r)$$

$$\times \psi^\dagger(t,r')\psi(t,r')(\Delta\psi(t,r))]d\vec{r}' + \psi^\dagger(t,r)\left(\Delta\int \Phi(r-r')\psi^\dagger(t,r')\right.$$

$$\times \psi(t,r')\psi(t,r) \, d\vec{r}'\right)\rangle + \frac{i}{2m}\int \Phi(r-r')\langle \psi^\dagger(t,r)\psi^\dagger(t,r')$$

$$\times \, (\Delta\psi(t,r'))\psi(t,r) + \psi^\dagger(t,r)\psi^\dagger(t,r')\psi(t,r')(\Delta\psi(t,r))$$

$$- \, (\Delta\psi^\dagger(t,r))\psi^\dagger(t,r')\psi(t,r')\psi(t,r) - \psi^\dagger(t,r)(\Delta\psi^\dagger(t,r'))$$

$$\times \, \psi(t,r')\psi(t,r')\rangle d\vec{r}' + \frac{i}{2}\int \Phi(r-r')\{\eta^*(t,r')\langle\psi^\dagger(t,r)\psi(t,r')$$

$$\times \, \psi(t,r)\rangle + \eta^*(t,r)\langle\psi^\dagger(t,r')\psi(t,r')\psi(t,r)\rangle - \eta(t,r)$$

$$\times \, \langle\psi^\dagger(t,r)\psi^\dagger(t,r')\psi(t,r')\rangle - \eta(t,r')\langle\psi^\dagger(t,r)\psi^\dagger(t,r')$$

$$\times \, \psi(t,r)\rangle\}d\vec{r}' - \frac{i}{4m}\langle(\Delta\eta^*)\psi - \eta^*\Delta\psi - (\Delta\psi^\dagger)\eta - \psi^\dagger(\Delta\eta)\rangle$$

$$- \frac{i}{2m}\sum_\alpha\Big[\frac{\partial}{\partial r_\alpha}(U-\lambda)\Big\langle\frac{\partial\psi^\dagger}{\partial r_\alpha}\psi - \psi^\dagger\frac{\partial\psi}{\partial r_\alpha}\Big\rangle\Big]. \tag{1.5}$$

Now we notice that

$$\sum_\beta\frac{\partial}{\partial r_\beta}\Big\{\Big(\frac{\partial}{\partial r_\beta}\delta\psi^\dagger\Big)\psi - \Delta\psi^\dagger\frac{\partial\psi}{\partial r_\beta} - \psi^\dagger\frac{\partial}{\partial r_\beta}\Delta\psi + \frac{\Delta\psi^\dagger}{\partial r_\beta}\Delta\psi\Big\}$$

$$= (\Delta\Delta\psi^\dagger)\psi + \psi^\dagger(\Delta\Delta\psi)$$

$$= \sum_\beta\frac{\partial^2}{\partial r_\beta^2}\{(\Delta\psi^\dagger)\psi - \psi^\dagger(\Delta\psi)\} - 2\sum_\beta\frac{\partial}{\partial r_\beta}\Big\{(\Delta\psi^\dagger)\frac{\partial\psi}{\partial r_\beta} - \frac{\partial\psi^\dagger}{\partial r_\beta}(\Delta\psi)\Big\}.$$

We can also transform the term containing Φ; we have

$$- \Big(\Delta\int\Phi(r-r')\psi^\dagger(t,r)\psi^\dagger(t,r')\psi(t,r')\,d\vec{r}'\Big)\psi(t,r)$$

$$+ \, (\psi^\dagger(t,r))\int\Phi(t-r')\psi^\dagger(t,r')\psi(t,r')\,d\vec{r}'\,\psi(t,r)$$

$$+ \, \psi^\dagger(t,r)\Big(\Delta\int\Phi(r-r')\psi^\dagger(t,r')\psi(t,r')\,d\vec{r}'\Big)\psi(t,r)$$

$$- \int\Phi(r-r')\psi^\dagger(t,r)\psi^\dagger(t,r')\psi(t,r')\,d\vec{r}'\,(\Delta\psi(t,r))$$

$$= -2\sum_\beta\int\frac{\partial\Phi(r-r')}{\partial r_\beta}\frac{\partial\psi^\dagger(t,r)}{\partial r_\beta}\psi^\dagger(t,r')\psi(t,r')\,d\vec{r}'\,\psi(t,r)$$

$$+ \, 2\sum_\beta\int\frac{\partial\Phi(r-r')}{\partial r_\beta}\psi^\dagger(t,r)\psi^\dagger(t,r')\psi(t,r')\frac{\partial\psi(t,r)}{\partial r_\beta}$$

and also

$$\int\Phi(r-r')\psi^\dagger(t,r)\psi^\dagger(t,r')\psi(t,r')\,d\vec{r}'\,\Delta\psi(t,r)$$

$$- \int \Phi(r - r')(\Delta \psi^\dagger(t, r)) \psi^\dagger(t, r') \psi(t, r') \, d\vec{r}' \, \psi(t, r)$$

$$= \sum_\beta \int \Phi(r - r') \frac{\partial}{\partial r_\beta} \left\{ \psi^\dagger(t, r) \psi^\dagger(t, r') \psi(t, r') \frac{\partial \psi(t, r)}{\partial r_\beta} \right.$$

$$\left. - \frac{\partial \psi^\dagger(t, r)}{\partial r_\beta} \psi^\dagger(t, r') \psi(t, r') \psi(t, r) \right\} d\vec{r}'$$

$$= \sum_\beta \frac{\partial}{\partial r_\beta} \int \Phi(r - r') \left\{ \psi^\dagger(t, r) \psi^\dagger(t, r') \psi(t, r') \frac{\partial \psi(t, r)}{\partial r_\beta} \right.$$

$$\left. - \frac{\partial \psi^\dagger(t, r)}{\partial r_\beta} \psi^\dagger(t, r') \psi(t, r') \psi(t, r) \right\} d\vec{r}' - \sum_\beta \int \frac{\partial \Phi(r - r')}{\partial r_\beta}$$

$$\left\{ \psi^\dagger(t, r) \psi^\dagger(t, r') \psi(t, r') \frac{\partial \psi(t, r)}{\partial r_\beta} - \frac{\partial \psi^\dagger(t, r)}{\partial r_\beta} \psi^\dagger(t, r') \right.$$

$$\left. \times \psi(t, r') \psi(t, r) \right\} d\vec{r}'.$$

We also get in a similar way

$$\int \Phi(r - r') \psi^\dagger(t, r) \{ \psi^\dagger(t, r')(\Delta \psi(t, r')) - (\Delta \psi^\dagger(t, r')) \psi(t, r') \} \psi(t, r) d\vec{r}'$$

$$= \sum_\beta \int \frac{\partial \Phi(r - r')}{\partial r_\beta} \psi^\dagger(t, r) \left\{ \psi^\dagger(t, r') \frac{\partial \psi^\dagger(t, r')}{\partial r'_\beta} - \frac{\partial \psi^\dagger(t, r')}{\partial r'_\beta} \psi(t, r) \right\} d\vec{r}'.$$

Substituting of these equations into (1.5) finally gives us

$$\frac{\partial (\rho \epsilon)}{\partial t} = \frac{i}{8m^2} \Delta \langle (\Delta \psi^\dagger) \psi - \psi^\dagger (\Delta \psi) \rangle + \frac{i}{4m} \sum_\beta \frac{\partial}{\partial r_\beta} \left\{ -\frac{1}{m} \right.$$

$$\times \left\langle (\Delta \psi^\dagger) \frac{\partial \psi}{\partial r_\beta} - \frac{\partial \psi^\dagger}{\partial r_\beta} (\Delta \psi) \right\rangle + \int \Phi(r - r') \left\langle \psi^\dagger(t, r) \psi^\dagger(t, r') \right.$$

$$\times \left. \psi(t, r') \frac{\partial \psi(t, r)}{\partial r_\beta} - \frac{\partial \psi^\dagger(t, r)}{\partial r_\beta} \psi^\dagger(t, r') \psi(t, r') \psi(t, r) \right\rangle d\vec{r}' \right\}$$

$$+ \frac{i}{4m} \sum_\beta \int \frac{\partial \Phi(r - r')}{\partial r_\beta} \left\langle \psi^\dagger(t, r) \psi^\dagger(t, r') \psi(t, r') \frac{\partial \psi(t, r)}{\partial r_\beta} \right.$$

$$- \frac{\partial \psi^\dagger(t, r)}{\partial r_\beta} \psi^\dagger(t, r') \psi(t, r') \psi(t, r) + \psi^\dagger(t, r) \left[\psi^\dagger(t, r') \frac{\partial \psi^\dagger(t, r')}{\partial r'_\beta} \right.$$

$$-\frac{\partial \psi^\dagger(t,r')}{\partial r'_\beta}\psi(t,r)\Big]\psi(t,r)\Big\rangle d\vec{r}' - \frac{1}{m}\sum_\beta j_\beta \frac{\partial}{\partial r_\beta}(U-\lambda)$$

$$+\frac{i}{4m}(\eta\Delta\phi^* - \eta^*\Delta\phi + \phi^*\Delta\eta - \phi\Delta\eta^*) + \frac{i}{2}\int \Phi(r-r')$$

$$\times \{\eta^*(t,r)\langle\psi^\dagger(t,r')\psi(t,r')\psi(t,r)\rangle + \eta^*(t,r')\langle\psi^\dagger(t,r)\psi(t,r')$$

$$\times \psi(t,r)\rangle - \eta(t,r')\langle\psi^\dagger(t,r)\psi^\dagger(t,r')\psi(t,r)\rangle - \eta(t,r)$$

$$\times \langle\psi^\dagger(t,r)\psi^\dagger(t,r')\psi(t,r')\rangle\} d\vec{r}'.$$

If we introduce the quantity

$$G_t^{(\alpha)}(r,r-r') = \frac{i}{4m}\Big\langle \psi^\dagger(t,r')\Big(\psi^\dagger(t,r)\frac{\partial\psi(t,r)}{\partial r_\alpha}$$

$$-\frac{\partial\psi^\dagger(t,r)}{\partial r_\alpha}\psi(t,r)\Big)\psi(t,r')\Big\rangle \tag{1.6}$$

and go over to a new variable $\vec{R} = \vec{r}' - \vec{r}$ in some of the integrals, the identity (1.6) takes the form

$$\frac{\partial(\rho\epsilon)}{\partial t} = \frac{i}{8m^2}\Delta\langle(\Delta\psi^\dagger)\psi - \psi^\dagger(\Delta\psi)\rangle + \sum_\beta \frac{\partial}{\partial r_\beta}\Big\{\frac{i}{4m^2}$$

$$\times \Big\langle\frac{\partial\psi^\dagger}{\partial r_\beta}(\Delta\psi) - (\Delta\psi^\dagger)\frac{\partial\psi}{\partial r_\beta}\Big\rangle + \int \Phi(R)G_t^{(\beta)}(r,R)\,d\vec{R}\Big\} + \sum_\beta \int \frac{\partial\Phi(R)}{\partial R_\beta}$$

$$\times \{G_t^{(\beta)}(r,-R) + G_t^{(\beta)}(r-R,R)\}\,d\vec{R} - \frac{1}{m}\sum_\beta j_\beta\frac{\partial}{\partial r_\beta}(U-\lambda)$$

$$+\frac{i}{4m}(\eta\Delta\phi^* - \eta^*\Delta\phi + \phi^*\Delta\eta - \phi\Delta\eta^*) + \frac{i}{2}\int \Phi(r-r')$$

$$\times \{\eta^*(t,r)\times\langle\psi^\dagger(t,r')\psi(t,r')\psi(t,r)\rangle + \eta^*(t,r')\langle\psi^\dagger(t,r)\psi(t,r')\psi(t,r)\rangle$$

$$-\eta(t,r')\langle\psi^\dagger(t,r)\psi^\dagger(t,r')\psi(t,r)\rangle - \eta(t,r)\langle\psi^\dagger(t,r)\psi^\dagger(t,r')\psi(t,r')\rangle\}\,d\vec{r}'.$$

$$\tag{1.7}$$

2. Hydrodynamic Equations for a Normal Liquid

We shall now proceed to derive the hydrodynamic equations for a normal (non-superfluid) liquid. Actually, this problem has already been treated in the work of K. P. Gurov [1], and the only reason for dwelling on it here is that is will form the basis of a subsequent generalization to the superfluid case. For the purpose of this section the sources are unimportant and so we shall take

$$\eta = \eta^\dagger = 0.$$

Consider, first of all, the statistical equilibrium state of the normal liquid, which is characterized by the usual parameters: the particle number density ρ, temperature θ and velocity $\vec{v}$ characterizing the motion of the liquid as a whole. The dependence on the velocity $\vec{v}$ is trivial; by using the transformation of the field operators

$$\psi \to \psi \, e^{im\vec{v}\vec{r}}$$

we can express the mean values in the state with velocity $\vec{v}$

$$\langle \ldots \rangle_{\rho,\theta,v}$$

in terms of the mean values

$$\langle \ldots \rangle_{\rho,\theta,0}$$

in the statistical equilibrium state of the liquid at rest. For example

$$\vec{j} = m\rho\vec{v},$$

$$\epsilon = \epsilon^0 + \frac{mv^2}{2} = \epsilon(\rho,\theta) + \frac{mv^2}{2} \qquad (2.1)$$

where $\varepsilon(\rho,\theta)$ is the mean energy per particle in the statistical equilibrium state of the liquid at rest.

In what follows we shall have to deal with averages of the general type:

$$\mathscr{U} = \langle (\mathscr{D}_1 \psi^\dagger(t,r))(\mathscr{D}_2 \psi(t,r)) \rangle_{\rho,\theta,v}$$
$$\mathscr{B} = \langle (\mathscr{D}_1 \psi^\dagger(t,r))(\mathscr{D}_2 \psi^\dagger(t,r'))(\mathscr{D}_3 \psi(t,r'))(\mathscr{D}_4 \psi(t,r)) \rangle_{\rho,\theta,v}$$

where the $\mathscr{D}_\nu$ are linear combinations of constants and differential operators with respect to the spatial variables (cf., for example, the expressions for D_t and $G_t^{(\alpha)}$; in particular, $\mathscr{D}_\nu$ may be simply equal to unity.

In view of the total spatial homogeneity of the state of statistical equilibrium, it is easy to se that

$$\mathscr{U} = \mathscr{U}(\rho,\theta,v),$$

$$\mathscr{B} = \mathscr{B}(\rho, \theta, v/r - r')$$
$$U(t, r) = U = \text{const},$$

where neither $\mathscr{U}$ nor $\mathscr{B}$ can depend on the value of U; this follows from the fact that all dependence on U can be eliminated by the gauge transformation

$$\psi \to e^{iUt}\psi$$

while $\mathscr{U}$ and $\mathscr{B}$ are invariant under this transformation.

We now proceed to consider the type of non-equilibrium processes discussed in hydrodynamics, namely those for which non-equilibrium quantities of the type

$$\mathscr{U}(t, r) = \langle(\mathscr{D}_1\psi^\dagger(t, r))(\mathscr{D}_2\psi(t, r))\rangle_{\rho,\theta,v}$$
$$\mathscr{B}(t, r - r') = \langle(\mathscr{D}_1\psi^\dagger(t, r))(\mathscr{D}_2\psi^\dagger(t, r'))(\mathscr{D}_3\psi(t, r'))(\mathscr{D}_4\psi(t, r))\rangle_{\rho,\theta,v} \quad (2.2)$$

and, of course, also the external field $U(t, r)$ us sufficiently slowly varying under translations in space and time. We shall assume that the deviations form statistical equilibrium are asymptotically damped out, so that we can define, at least as an order of magnitude, some relaxation time T and mean free path l. The non-equilibrium processes to be considered, that, are those for which the quantities (2.2), and $U(t, r)$, change sufficiently slowly under the translations

$$t \to t + t_0; \quad r \to r + t_0, \quad r' \to r' + r_0; \quad |t_0| \approx T; \quad |r_0| \approx l,$$

and there are only small deviations from "local statistical equilibrium" in the neighborhood of any given point. To put it more precisely, we assume that quantities of the type (2.2) differ by a sufficiently small amount from the corresponding equilibrium values

$$\mathscr{U}(\rho(t, r), \theta(t, r), v(t, r)), \quad \mathscr{B}(\rho(t, r), \theta(t, r), v(t, r)/r - r') \quad (2.3)$$

and that the difference tends to zero with vanishing gradients of ρ, θ, $\vec{v}$ and u. In equation (2.3) $\theta(t, r)$ and $\vec{v}(t, r)$ are defined (implicitly) by equations (2.1) in terms of the current j and mean energy ε at the point in question.

The assumption that for processes with sufficiently small gradients of ρ, θ, $\vec{v}$ and u the liquid in each small region is in a state of approximate local statistical equilibrium is an essential condition for the transition to the hydrodynamics equations to be possible. For instance, if we were to consider a dynamical system with no interactions at all (i.e., a system if non-interacting bosons) then of course we could consider processes characterized by arbitrary small gradients of ρ, j, ϵ, etc., and could formally introduce local values of $\vec{v}$ and θ. However, in this case quantities of the type (2.2) could not, in general, be expressed even approximately in terms of their "quasi-equilibrium" values (2.3). This is what makes the hydrodynamics equations inapplicable to a system of completely non-interacting particles. We might also remark that it is essential to the usual (Hilbert-Chapman-Enskog) derivation of the hydrodynamic equations for gases from the Boltzmann equation that this equations has solutions representing states close to local equilibrium, so that an expansion in powers of the gradients is legitimate.

After these preliminary remarks we return to our assumption about the nature of the nonequilibrium process considered. To formulate it in a form suitable for calculation it is convenient to introduce a small parameter μ which measures the "slowness" of the process and the deviation from isotropy. We shall represent averages of the type (2.2), and also the external field, in the form

$$U(t,r) = \tilde{U}(\mu t, \mu r) = U(\tau, \xi),$$
$$\mathscr{U}(t,r) = \tilde{\mathscr{U}}(\mu t, \mu r; \mu) = \tilde{U}(\tau, \xi; \mu),$$
$$\mathscr{B}(t,r) = \tilde{\mathscr{B}}(\mu t, \mu r, r - r'; \mu) = \tilde{\mathscr{B}}(\tau, \xi, R, \mu),$$
$$t = \mu t, \quad \vec{\xi} = \mu \vec{r}, \quad \vec{R} = \vec{r} - \vec{r}'. \tag{2.4}$$

In particular, we write

$$\rho(t,r) = \tilde{\rho}(\tau, \xi), \quad \vec{j}(t,r) = \vec{\tilde{j}}(\tau, \xi), \quad \epsilon(t,r) = \tilde{\varepsilon}(\tau, \xi).$$

The representation (2.4) is simply a formalization of our assumption about the slowness of the time rate of change and the weakness of the deviation from local isotropy.

To formulate the assumption of weak deviation from local statistical equilibrium, we write

$$\tilde{\mathscr{U}}(\tau, \xi; \mu) = \tilde{\mathscr{U}}^{(0)}(\tau, \xi) + \mu \tilde{\mathscr{U}}^{(1)}(\tau, \xi) + \mu^2 \tilde{U}^{(2)}(\tau, \xi) + \ldots,$$
$$\tilde{\mathscr{B}}(\tau, \xi, R; \mu) = \tilde{\mathscr{B}}^{(0)}(\tau, \xi, R) + \mu \tilde{\mathscr{B}}^{(1)}(\tau, \xi, R) + \mu^2 \tilde{\mathscr{B}}^{(2)}(\tau, \xi, R) + \ldots,$$
$$\tilde{\mathscr{U}}^{(0)}(\tau, \xi) = \tilde{\mathscr{U}}(\tilde{\rho}(\tau, \xi), \tilde{\theta}(\tau, \xi), \tilde{v}(\tau, \xi)),$$

$$\tilde{\mathscr{B}}^{(0)}(\tau, \xi, R) = \tilde{\mathscr{B}}(\tilde{\rho}(\tau, \xi), \tilde{\theta}(\tau, \xi), \tilde{v}(\tau, \xi), R), \tag{2.5}$$

where $\tilde{\tilde{\theta}}$ and $\tilde{\tilde{v}}$ are defined by the relations

$$\tilde{\tilde{v}} = \frac{1}{m\tilde{\rho}}\vec{\tilde{j}},$$

$$\tilde{\tilde{\epsilon}} = \epsilon(\tilde{\rho}, \tilde{\theta}) + \frac{m\tilde{v}^2}{2} \tag{2.6}$$

and postulate that $\tilde{\mathscr{U}}^{(1)}$, $\tilde{\mathscr{B}}^{(1)}$ are linear in the gradients of $\tilde{\rho}$, $\tilde{\theta}$, $\tilde{v}$, $\tilde{u}$ etc. Therefore, (2.5) is equivalent to the assumption that quantities of the types $\tilde{\mathscr{U}}$, $\tilde{\mathscr{B}}$ may be expanded in powers of the gradients of ρ, $\vec{v}$, θ and u. (Actually the possibility of such an expansion in intimately connected with the validity of a Boltzmann-like kinetic equation).

Now we may proceed directly to the derivation of the hydrodynamic equations. We shall have to deal with functions[t] $\mathscr{D}_t(r, R)$, $G_t^{(\alpha)}(r, R)$ belonging to the general type $\mathscr{B}$. Let us write:

$$\mathscr{D}_t(r, R) = \tilde{\mathscr{D}}_\tau(\xi, R; \mu),$$
$$G_t^{(\alpha)}(r, R) = \tilde{G}_\tau^{(\alpha)}(\xi, R; \mu),$$

or in abbreviation notation:

$$\mathscr{D}_t(r, R) = \tilde{\mathscr{D}}_\tau(\xi, R),$$
$$G_t^{(\alpha)}(r, R) = \tilde{G}_\tau^{(\alpha)}(\xi, R).$$

Then, going over from the variables (t, r) to our new variables (τ, ξ) in the identities of the preceding section (equation (1.3), (1.4), and (1.7)) we get:

$$m\frac{\partial\tilde{\rho}}{\partial\tau} + \sum_\beta \frac{\partial\tilde{j}(\tau_\beta, \xi)}{\partial\xi_\beta} = 0, \tag{2.7}$$

$$\frac{\partial\tilde{j}_\alpha(\tau, \xi)}{\partial\tau} = \frac{1}{2m}\sum_\beta \frac{\partial}{\partial\xi_\beta}\left\{\delta_{\alpha\beta}\mu^2\frac{\Delta\xi}{2}\tilde{\rho}(\tau, \xi) - \left\langle\frac{\partial\psi^\dagger}{\partial r_\alpha}\frac{\partial\psi}{\partial r_\beta} + \frac{\partial\psi^\dagger}{\partial r_\beta}\frac{\partial\psi}{\partial r_\alpha}\right\rangle\right\}$$

---

[t]Translator's Note: D_t is the same D_t introduced in section 1.

$$-\frac{1}{2\mu}\int\frac{\partial\Phi(R)}{\partial R_\alpha}\{\tilde{\tilde{\mathscr{D}}}_\tau(\xi,-R)+\tilde{\tilde{\mathscr{D}}}_\tau(\xi-\mu R,R)\}\,d\vec{R}$$

$$-\tilde{\tilde{\rho}}(\tau,\xi)\frac{\partial}{\partial\xi_\alpha}\tilde{\tilde{U}}(\tau,\xi),\tag{2.8}$$

$$\frac{\partial(\tilde{\tilde{\rho}}\tilde{\tilde{\epsilon}})}{\partial\tau}=\frac{i\mu}{8m^2}\Delta_\xi\langle(\Delta\psi^\dagger)\psi-\psi^\dagger(\Delta\psi)\rangle$$

$$+\sum_\beta\frac{\partial}{\partial\xi_\beta}\left\{\frac{i}{4m^2}\left\langle\frac{\partial\psi^\dagger}{\partial r_\beta}\Delta\psi-(\Delta\psi^\dagger)\frac{\partial\psi}{\partial r_\beta}\right\rangle+\int\Phi(R)G_\tau^{(\beta)}(\xi,R)\,d\vec{R}\right\}$$

$$+\frac{1}{\mu}\sum_\beta\int\frac{\partial\Phi(R)}{\partial R_\beta}\{\tilde{\tilde{G}}_\tau^{(\beta)}(\xi,-R)+\tilde{\tilde{G}}_\tau^{(\beta)}(\xi-\mu R,R)\}\,d\vec{R}$$

$$+\frac{1}{m}\sum_\beta\tilde{\tilde{j}}_\beta\frac{\partial\tilde{\tilde{U}}}{\partial\xi_\beta}.\tag{2.9}$$

In equation (2.8) the quantity $\tilde{\tilde{\mathscr{D}}}_\tau(\xi-\mu R,R)$ is multiplied by the factor $\partial\Phi(R)/\partial R_\alpha$, which decreases rapidly for large R. Expanding $\tilde{\tilde{\mathscr{D}}}$ in powers of μR, we get for the value of the relevant term in (2.8)

$$\int\frac{\partial\Phi(R)}{\partial R_\alpha}\left\{\tilde{\tilde{\mathscr{D}}}_\tau(\xi,-R)+\tilde{\tilde{\mathscr{D}}}_\tau(\xi,R)-\mu\sum_\beta R_\beta\frac{\partial\tilde{\tilde{\mathscr{D}}}_\tau(\xi,R)}{\partial\xi_\beta}\right.$$

$$\left.+\frac{\mu^2}{2}\sum_{\beta,\gamma}R_\alpha R_\gamma\frac{\partial^2\tilde{\tilde{\mathscr{D}}}_\tau(\xi,R)}{\partial\xi_\beta\partial\xi_\gamma}+0((\mu r)^3)\right\}d\vec{R}$$

$$=-\mu\sum_\beta\int\frac{\partial\Phi(R)}{\partial R_\alpha}R_\beta\frac{\partial\tilde{\tilde{\mathscr{D}}}_\tau(\xi,R)}{\partial\xi_\beta}\,d\vec{R}+0(\mu^3).$$

Here we have used the fact that $\partial\Phi(R)/\partial R_\alpha$ is an odd function of R, and also that $\tilde{\tilde{\mathscr{D}}}_\tau(\xi,R)$ is an even function of R to zeroth order in μ. Thus, keeping only terms of order μ, we get the equation

$$\frac{\partial\tilde{\tilde{j}}_\alpha}{\partial\tau}=\frac{1}{2m}\sum_\beta\frac{\partial}{\partial\xi_\beta}\left\{-\left\langle\frac{\partial\psi^\dagger}{\partial r_\alpha}\frac{\partial\psi}{\partial r_\beta}+\frac{\partial\psi^\dagger}{\partial r_\beta}\frac{\partial\psi}{\partial r_\alpha}\right\rangle\right\}$$

$$+ \frac{1}{2} \sum_{\beta} \frac{\partial}{\partial \xi_\beta} \int \frac{\partial \Phi(R)}{\partial R_\alpha} R_\beta \tilde{\tilde{\mathscr{D}}}_\tau(\xi, R) \, d\vec{R} - \tilde{\rho} \frac{\partial \tilde{\tilde{U}}}{\partial \xi_\alpha}$$

or

$$\frac{\partial \tilde{\tilde{j}}_\alpha}{\partial \tau} = \sum_{\beta} \frac{\partial T_{\alpha\beta}}{\partial \xi_\beta} - \tilde{\tilde{\rho}} \frac{\partial \tilde{\tilde{U}}}{\partial \xi_\alpha} \tag{2.10}$$

where

$$T_{\alpha\beta}(\tau, \xi) = - \frac{1}{2m} \left\langle \frac{\partial \psi^\dagger}{\partial r_\alpha} \frac{\partial \psi}{\partial r_\beta} + \frac{\partial \psi^\dagger}{\partial r_\beta} \frac{\partial \psi}{\partial r_\alpha} \right\rangle$$
$$+ \frac{1}{2} \int \frac{\partial \Phi(R)}{\partial R_\alpha} R_\beta \tilde{\tilde{\mathscr{D}}}_\tau(\xi, R) \, d\vec{R}. \tag{2.11}$$

Performing an analogous expansion in powers of μ for the quantities $\tilde{\tilde{G}}_\tau^{(\beta)}(\xi - \mu R, R)$ in equation (2.9), we get to the first order in μ:

$$\frac{\partial(\tilde{\tilde{\rho}}\tilde{\tilde{e}})}{\partial \tau} = \frac{i}{8m^2} \mu \Delta_\xi \langle (\Delta \psi^\dagger)\psi - \psi^\dagger(\Delta\psi) \rangle + \sum_{\beta} \frac{\partial I_\beta}{\partial \xi_\beta} - \frac{1}{m} \sum_{\beta} \tilde{\tilde{j}}_\beta \frac{\partial \tilde{\tilde{U}}}{\partial \xi_\beta}. \tag{2.12}$$

Here I_α is given by the expression

$$I_\alpha = - \frac{i}{4m^2} \left\langle (\Delta\psi^\dagger) \frac{\partial \psi}{\partial r_\alpha} - \frac{\partial \psi^\dagger}{\partial r_\alpha} \Delta\psi \right\rangle - \int \Phi(R) G_\tau^{(\alpha)}(\xi, R) \, d\vec{R}$$
$$- \sum_{\beta} \int \frac{\partial \Phi(R)}{\partial R_\beta} R_\alpha \tilde{\tilde{G}}_\tau^{(\beta)}(\xi, R) \, d\vec{R}$$
$$+ \frac{\mu}{2} \sum_{\beta,\gamma} \frac{\partial}{\partial \xi_\gamma} \int \frac{\partial \Phi(R)}{\partial R_\beta} R_\alpha R \tilde{\tilde{G}}_\tau^{(\beta)}(\xi, R) \, d\vec{R}. \tag{2.13}$$

Now, the quantities $T_{\alpha\beta}$, I_α are composed of terms of the types $\mathscr{U}$ and $\mathscr{B}$ [cf. (2.2)]. and therefore we may expand them according to (2.5); in fact we get

$$T_{\alpha\beta} = T_{\alpha\beta}^{(0)} + \mu T_{\alpha\beta}^{(1)} + \dots$$
$$I_\alpha = I_\alpha^{(0)} + \mu I_\alpha^{(1)} + \dots \tag{2.14}$$

$$T_{\alpha\beta}^{(0)} = T_{\alpha\beta}(\tilde{\rho}, \tilde{\tilde{\theta}}, \tilde{v}); \quad I_\alpha^{(0)} = I_\alpha(\tilde{\rho}, \tilde{\tilde{\theta}}, \tilde{v}) \tag{2.15}$$

$T_{\alpha\beta}(\tilde{\rho}, \tilde{\theta}, \tilde{v})$ and $I_\alpha(\tilde{\rho}, \tilde{\theta}, \tilde{v})$ are to be obtained from expressions (2.11) and (2.13) respectively by replacing the averages

$$\langle \ldots \rangle$$

by the corresponding equilibrium averages

$$\langle \ldots \rangle_{\rho,\theta,v}.$$

Taking terms of order μ into account in (2.14) would give us the viscous-liquid approximation. In as much as this work is restricted to the ideal-liquid approximation, it will be sufficient to consider only the terms $T_{\alpha\beta}^{(0)}$ and $I_\alpha^{(0)}$.

Let us put these terms in a more convenient form. First of all, it is advantageous to express the average

$$\langle \ldots \rangle_{\rho,\theta,v}$$

in terms of

$$\langle \ldots \rangle_{\rho,\theta,0}.$$

This we can do by performing the transformation

$$\psi \to \psi e^{im\vec{v}\vec{r}}.$$

We find

$$\left\langle \frac{\partial \psi^\dagger}{\partial r_\alpha} \frac{\partial \psi}{\partial r_\beta} + \frac{\partial \psi^\dagger}{\partial r_\beta} \frac{\partial \psi}{\partial r_\alpha} \right\rangle_{\rho,\theta,v}$$

$$= \left\langle \left(\frac{\partial \psi^\dagger}{\partial r_\alpha} - imv_\alpha \psi^\dagger \right) \left(\frac{\partial \psi}{\partial r_\beta} + imv_\beta \psi \right) + \left(\frac{\partial \psi^\dagger}{\partial r_\beta} - imv_\beta \psi^\dagger \right) \right.$$

$$\left. \times \left(\frac{\partial \psi}{\partial r_\alpha} + imv_\alpha \psi \right) \right\rangle_{\rho,\theta,v} + \left\langle \frac{\partial \psi^\dagger}{\partial r_\alpha} \frac{\partial \psi}{\partial r_\beta} + \frac{\partial \psi^\dagger}{\partial r_\beta} \frac{\partial \psi}{\partial r_\alpha} \right\rangle_{\rho,\theta,0} + 2m^2 v_\alpha v_\beta \rho,$$

since terms containing only one derivative vanish in the equilibrium state with $\vec{v} = 0$ because of the reflection invariance of this state. Therefore, noticing that for $\vec{v} = 0$ $T_{\alpha\beta}$ can only be an isotropic tensor, we get

$$T_{\alpha\beta}(\rho, \theta, v) = -m\rho v_\alpha v_\beta - \delta_{\alpha\beta} \mathscr{P}(\rho, \theta), \tag{2.16}$$

where for arbitrary $\alpha \, (= 1, 2, 3)$

$$-\mathscr{P}(\rho, \theta) = -\frac{1}{m} \left\langle \frac{\partial \psi^\dagger}{\partial r_\alpha} \frac{\partial \psi}{\partial r_\alpha} \right\rangle_{\rho,\theta,0} + \frac{1}{2} \int \frac{\partial \Phi(R)}{\partial R_\alpha} R_\alpha \, \mathscr{D}(R|\rho, \theta) \, d\vec{R}, \tag{2.17}$$

where

$$\mathscr{D}(r - r'|\rho, \theta) = \langle \psi^\dagger(r)\psi^\dagger(r')\psi(r')\psi(r)\rangle_{\rho,\theta,0}.$$

We shall now prove that if $F(\rho, \theta)$ is the free energy per particle and N is the total number of particles,

$$\mathscr{P}(\rho, \theta) = -\frac{\partial NF\left(\frac{N}{V}, \theta\right)}{\partial V} = \rho^2 \frac{\partial F(\rho, \theta)}{\partial \rho}, \tag{2.18}$$

i.e., that $\mathscr{P}$ is just the usual thermodynamic pressure. To establish this, let us suppose we subject the volume to a linear transformation along, say, the α-th coordinate axis: $r_\alpha \to Lr_\alpha$, $r_\beta \to r_\beta$ for $\alpha \neq \beta$. Let H_L be the Hamiltonian of the "stretched" system:

$$H_L = \frac{1}{2m}\int \psi^\dagger(r)p_L^2\psi(r)\,d\vec{r} + \frac{1}{2}\int \Phi_L(r - r')\psi^\dagger(r)\psi^\dagger(r')\psi(r')\psi(r)\,d\vec{r}\,d\vec{r}',$$

where

$$p_L^2 = -\frac{1}{L^2}\frac{\partial^2}{\partial r_\alpha^2} - \sum_{\alpha \neq \beta}\frac{\partial^2}{\partial r_\beta^2},$$

and $\Phi_L(R)$ is obtained from $\Phi(R)$ by the substitution $R_\alpha \to LR_\alpha$, $R_\beta \to R_\beta$ for $\alpha \neq \beta$. Then

$$\frac{\partial NF}{\partial V} = \frac{1}{V}\left\langle\left(\frac{\partial H_L}{\partial L}\right)_{L=1}\right\rangle_{\rho,\theta,0},$$

or explicitly

$$\begin{aligned}
\frac{\partial NF}{\partial V} &= \frac{1}{m}\left\langle\psi^\dagger(r)\frac{\partial^2}{\partial r_\alpha^2}\psi(r)\right\rangle_{\rho,\theta,0} + \frac{1}{2}\int\frac{\partial\Phi(R)}{\partial R_\alpha}R_\alpha\,\mathscr{D}(R|\rho,\theta)\,d\vec{R} \\
&= -\frac{1}{m}\left\langle\frac{\partial\psi^\dagger}{\partial r_\alpha}\frac{\partial\psi}{\partial r_\alpha}\right\rangle_{\rho,\theta,0} + \frac{1}{2}\int\frac{\partial\Phi(R)}{\partial R_\alpha}R_\alpha\,\mathscr{D}(R|\rho,\theta)\,d\vec{R},
\end{aligned}$$

which proves the relation (2.18).

Substituting of (2.14), (2.15), and (2.16) in (2.10) finally gives:

$$\frac{\partial\tilde{\tilde{j}}_\alpha}{\partial\tau} = \sum_\beta\frac{\partial}{\partial\xi_\beta}\left(m\tilde{\tilde{\rho}}\tilde{v}_\alpha\tilde{v}_\beta + \delta_{\alpha\beta}\mathscr{P}(\tilde{\rho},\tilde{\theta})\right) - \tilde{\rho}\frac{\partial\tilde{U}}{\partial\xi_\alpha}. \tag{2.19}$$

A similar transformation may be carried out for the energy density equation (2.12). Notice fist of all that

$$G^{(\alpha)}(r - r'|\rho, \theta, v) =$$

$$= \frac{i}{4m} \langle \psi^\dagger(r')[\psi^\dagger(r)imv_\alpha\psi(r) + \psi^\dagger(r)imv_\alpha\psi(r)]\psi(r')\rangle_{\rho,\theta,0}$$

$$= -\frac{v_\alpha}{2}\mathscr{D}(r - r'|\rho,\theta).$$

Now we transform the first term in expression (2.13) for $I_\alpha^{(0)}$ in the same way as we did the corresponding terms for $T_{\alpha\beta}^{(0)}$, remembering that the average is taken over a state of statistical equilibrium:

$$-\frac{i}{4m}\Big\langle\sum_\beta\Big(\frac{\partial^2\psi^\dagger}{\partial r_\beta^2}\frac{\partial\psi}{\partial r_\alpha} + \frac{\partial\psi^\dagger}{\partial r_\alpha}\frac{\partial^2\psi}{\partial r_\beta\,\partial r_\beta}\Big)\Big\rangle_{\rho,\theta,v}$$

$$= -\frac{m}{2}v_\alpha\sum_\beta v_\beta^2\langle\psi^\dagger\psi\rangle_{\rho,\theta,0} - \frac{1}{2m}\sum_\beta v_\beta\Big\langle\frac{\partial\psi^\dagger}{\partial r_\alpha}\frac{\partial\psi}{\partial r_\beta} + \frac{\partial\psi^\dagger}{\partial r_\beta}\frac{\partial\psi}{\partial r_\alpha}\Big\rangle_{\rho,\theta,0}$$

$$+ \frac{1}{4m}v_\alpha\sum_\beta\frac{\partial}{\partial r_\beta}\Big\langle\frac{\partial\psi^\dagger}{\partial r_\beta}\psi + \psi^\dagger\frac{\partial\psi}{\partial r_\beta}\Big\rangle_{\rho,\theta,0} - \frac{v_\alpha}{2m}\sum_\beta\Big\langle\frac{\partial\psi^\dagger}{\partial r_\beta}\frac{\partial\psi}{\partial r_\beta}\Big\rangle_{\rho,\theta,0}$$

$$= -\rho v_\alpha\frac{mv^2}{2} - \frac{v_\alpha}{m}\Big\langle\frac{\partial\psi^\dagger}{\partial r_\alpha}\frac{\partial\psi}{\partial r_\alpha}\Big\rangle_{\rho,\theta,0} - \frac{v_\alpha}{2m}\sum_\beta\Big\langle\frac{\partial\psi^\dagger}{\partial r_\beta}\frac{\partial\psi}{\partial r_\beta}\Big\rangle_{\rho,\theta,0}$$

We thereby obtain:

$$I_\alpha(\rho,\theta,v) = -\rho v_\alpha\frac{mv^2}{2} - \frac{v_\alpha}{m}\Big\langle\frac{\partial\psi^\dagger}{\partial r_\alpha}\frac{\partial\psi}{\partial r_\alpha}\Big\rangle_{\rho,\theta,0} - \frac{v_\alpha}{2m}\sum_\beta\Big\langle\frac{\partial\psi^\dagger}{\partial r_\beta}\frac{\partial\psi}{\partial r_\beta}\Big\rangle_{\rho,\theta,0}$$

$$-\frac{v_\alpha}{2}\int\Phi(R)\mathscr{D}(R|\rho,\theta)\,d\vec{R} + \frac{v_\alpha}{2}\int\frac{\partial\Phi(R)}{\partial R_\alpha}R_\alpha\mathscr{D}(R|\rho,\theta)\,d\vec{R}.$$

On the other hand,

$$\frac{1}{2m}\sum_\beta\Big\langle\frac{\partial\psi^\dagger}{\partial r_\beta}\frac{\partial\psi}{\partial r_\beta}\Big\rangle_{\rho,\theta,0} + \frac{1}{2}\int\Phi(R)\mathscr{D}(R|\rho,\theta)\,d\vec{R}$$

$$= \rho\epsilon(\rho,\theta) = \rho\Big\{F(\rho,\theta) - \theta\frac{\partial F(\rho,\theta)}{\partial\theta}\Big\},$$

and therefore, using (2.17), we get

$$I_\alpha(\rho,\theta,v) = -v_\alpha\Big\{\rho\Big(\epsilon(\rho,\theta) + \frac{mv^2}{2}\Big) + \mathscr{P}(\rho,\theta)\Big\}.$$

Equation (2.12) therefore leads to the following expression in our approximation

$$\frac{\partial(\tilde{\rho}\tilde{\epsilon})}{\partial\tau} = -\sum_\alpha \frac{\partial}{\partial\xi_\alpha}\tilde{v}_\alpha\left\{\tilde{\rho}\left(\epsilon(\tilde{\rho},\tilde{\theta}) + \frac{m\tilde{v}^2}{2}\right) + \mathscr{P}(\tilde{\rho},\tilde{\theta})\right\} - \frac{1}{m}\sum_\alpha \tilde{j}_\alpha \frac{\partial\tilde{U}}{\partial\xi_\alpha}, \quad (2.20)$$

where the definition of the local velocity and temperature is given by (2.6).

Finally, let us introduce the entropy

$$S(\rho,\theta) = -\frac{\partial F(\rho,\theta)}{\partial\theta}.$$

Then by combining the equations obtained above we can easily get in place of (2.20) an equation for the entropy density:

$$\frac{\partial\tilde{s}}{\partial\tau} + \sum_\alpha \tilde{v}_\alpha \frac{\partial\tilde{s}}{\partial\xi_\alpha} = 0. \quad (2.21)$$

In equations (2.7), (2.19) and (2.21) we can now go back to the original variables t, r; obviously this is achieved simply by substituting t, r for τ, ξ everywhere and removing the $\approx$ signs. Thus we finally arrive at the usual hydrodynamic equations for a normal liquid.

3. Hydrodynamic Equations for a Superfluid

A superfluid liquid is characterized by the presence of a non-vanishing expectation value $\langle\psi(r,t)\rangle = \phi(t,r) \neq 0$, even for vanishingly small source terms. Thus the equations of motion for the local quantities already considered must be supplemented by an equation of motion for the quantity ϕ. Such an equation may be obtained by taking expectation values in the equation of motion for ψ; it reads

$$i\frac{\partial\phi(t,r)}{\partial t} = -\lambda\phi(t,r) - \frac{\Delta}{2m}\phi(t,r)$$

$$+ \int \Phi(r-r')\langle\psi^\dagger(t,r')\psi(t,r')\psi(t,r)\rangle\, d\vec{r}' + U(t,r)\phi(t,r) + \eta(t,r).$$

In this equation ϕ is complex. For our purpose it is more convenient do deal with two real quantities, the modulus and phase of ϕ: $\phi = a \exp(i\chi)$. We obtain equations for a and χ by direct substitution into equation for ϕ

$$\frac{\partial \chi}{\partial t} = \lambda + \frac{\Delta a}{2ma} - \frac{1}{2m}\left(\frac{\partial \chi}{\partial \vec{r}}\right)^2 - U(t,r) + \frac{\zeta^* + \zeta}{2a}$$
$$- \frac{1}{2a^2}\int \Phi(R)\{X_t(r,R) + X^*(r,R)\}\,d\vec{R}, \qquad (3.1)$$

where

$$X_t(r, r' - r) = \langle \psi^\dagger(t,r')\psi(t,r')\psi(t,r)\rangle \psi^*(t,r),$$
$$\chi(t,r) = \eta(t,r)\,e^{-i\chi(t,r)},$$

and for $a(t,r)$:

$$i\frac{\partial a^2}{\partial t} = -\frac{i}{m}\left\{a^2\Delta\chi + 2\sum_\beta \frac{\partial \chi}{\partial r_\beta}\frac{\partial a}{\partial r_\beta}a\right\}$$
$$+ \int \Phi(R)\{X_t(r,R) - X^*(r,R)\}\,d\vec{R} + a(\zeta - \zeta^*). \qquad (3.2)$$

In what follows we shall consider frequently use the superfluid velocity $v_s^{(\alpha)} = \frac{1}{m}\frac{\partial \chi}{\partial r_\alpha}$, which, according to (3.1), satisfies the equation

$$m\frac{\partial v_s^{(\alpha)}}{\partial t} = \frac{\partial}{\partial r}\left\{\frac{\Delta a}{2ma_\alpha} - \frac{mv_s^2}{2} - U - \frac{\zeta + \zeta^*}{2a}\right.$$
$$\left. - \frac{1}{2a^2}\int \Phi(R)\{X_t(r,R) - X^*(r,R)\}\,d\vec{R}\right\}. \qquad (3.3)$$

Consider first of all a statistical equilibrium state of the superfluid liquid with $u = 0$, $\eta = 0$. Whereas the statistical equilibrium state of a normal liquid is characterized by ρ, θ and a single velocity $\vec{v}$, we now have states in general must be described by two *two* velocities, e.g. the velocity of the condensate $\vec{v}_s$ and that of the "normal component" $\vec{v}_n$. We shall denote averages taken over such a statistical equilibrium state by the symbol

$$\langle \ldots \rangle_{v_s, v_n}$$

suppressing the indices ρ, θ for the sake of conciseness. We shall find it convenient to express this type of average in terms of the averages

$$\langle \ldots \rangle_{v'_s,0}; \quad \text{where} \quad \vec{v}'_s = \vec{v}_s - \vec{v}_n;$$

corresponding to a state with zero "normal" velocity. To do so we can obviously perform a Galilean transformation on the field operators:

$$\psi \to \psi \exp(im\vec{v}_n \vec{r}).$$

Let us start, then, by considering the state characterized by ρ, θ and $\vec{v}_s$ with $\vec{v}_n = 0$. In this case:

$$a = \text{conts} = a(\rho, \theta, u); \quad F = F(\rho, \theta, u); \quad u^2 = \frac{v_s^2}{2},$$

and the chemical potential has the form

$$\frac{\partial(FN)}{\partial N} = F + \rho \frac{\partial F}{\partial \rho} = \Lambda(\rho, \theta, u).$$

Let us introduce the current j by the equation

$$j_\alpha = \rho \frac{\partial F}{\partial v_s^{(\alpha)}} = \rho \frac{\partial F}{\partial u} v_s^{(\alpha)},$$

and put

$$\rho \frac{\partial F}{\partial u} = \rho_s m,$$

so that

$$j_\alpha = m \rho_s v_s^{(\alpha)}.$$

Clearly $\rho_s \le \rho$.

It will now be shown that the definition of j just introduces is equivalent to the usual definition. This may conveniently be done by transforming to a system of coordinates moving with velocity $\vec{v}_s$ and then taking all averages only over states with $\vec{v}_s = 0$, $\vec{v}_n = -\vec{v}_s$; this maneuver will not of course change the numerical volume of the averaged quantity[u]. In the new

---

[u]Translator's Note: We are simply calculating the old quantities by going to a new representation.

coordinate system the Hamiltonian, as a function of the new operators ψ, has the form

$$H = \frac{1}{2m} \int \sum_\alpha \left(\frac{\partial \psi^\dagger}{\partial r_\alpha} - imv_s^{(\alpha)} \psi^\dagger \right) \left(\frac{\partial \psi}{\partial r_\alpha} + imv_s^{(\alpha)} \psi \right) d\vec{r}$$

$$+ \int \Phi(r - r') \psi^\dagger(r) \psi^\dagger(r') \psi(r') \psi(r) \, d\vec{r} \, d\vec{r}'.$$

Calculating $\rho \dfrac{\partial F}{\partial v_s^{(\alpha)}}$ in this representation, we find

$$\rho \frac{\partial F}{\partial v_s^{(\alpha)}} = \frac{1}{V} \left\langle \frac{\partial H}{\partial v_s^{(\alpha)}} \right\rangle_{0,-v_s}$$

$$= \frac{i}{2V} \int \left\langle \left(\frac{\partial \psi^\dagger}{\partial r_\alpha} - imv_s^{(\alpha)} \psi^\dagger \right) \psi - \psi^\dagger \left(\frac{\partial \psi}{\partial r_\alpha} + imv_s^{(\alpha)} \psi \right) \right\rangle_{0,-v_s} d\vec{r}$$

$$= \frac{i}{2V} \int \left\langle \frac{\partial \psi^\dagger}{\partial r_\alpha} \psi - \psi^\dagger \frac{\partial \psi}{\partial r_\alpha} \right\rangle_{v_s,0} d\vec{r} = \frac{i}{2} \left\langle \frac{\partial \psi^\dagger}{\partial r_\alpha} \psi - \psi^\dagger \frac{\partial \psi}{\partial r_\alpha} \right\rangle_{v_s,0} = j_\alpha.$$

The last expression is just the usual definition of j.

Going back now to (3.1), we observe that in a state of thermodynamic equilibrium $X = X(r - r'|\rho, \theta, v_s)$. Equation (3.1) gives us:[v]

$$\frac{1}{2a^2} \int \Psi(R) \{ X(R|\rho, \theta, v_s) + X^*(R|\rho, \theta, v_s) \} d\vec{R} = \Lambda(\rho, \theta, u) - \frac{mv_s^2}{2}.$$

Hence, using (3.2) (and remembering that $a = \text{const.}$, $\zeta = 0$), we get

$$\frac{1}{a^2} \int \Psi(R) X(R|\rho, \theta, v_s) \, d\vec{R} = \frac{1}{a^2} \int \Psi(R) X^*(R|\rho, \theta, v_s) \, d\vec{R}$$

$$= \Lambda(\rho, \theta, u) - \frac{mv_s^2}{2}. \tag{3.4}$$

Now we proceed to calculate the tensor [cf. (1.4)]

$$T_{\alpha\beta}(\rho, \theta, v_s) = -\frac{1}{2m} \left\langle \frac{\partial \psi^\dagger}{\partial r_\alpha} \frac{\partial \psi}{\partial r_\beta} + \frac{\partial \psi^\dagger}{\partial r_\beta} \frac{\partial \psi}{\partial r_\alpha} \right\rangle_{v_s,0}$$

$$+ \frac{1}{2} \int \frac{\partial \Phi(R)}{\partial R_\alpha} R_\beta \, \mathscr{D}(R|\rho, \theta, v_s) \, d\vec{R}$$

[v]Translator's Note: The constant λ in equations (1.3), (3.1) is here identified with the chemical potential $\Lambda(\rho, \theta, u)$.

in the equilibrium state $(v_s, 0)$. From general symmetry consideration it follows that, since the only direction characterized the system is that of $\vec{v}_s$, $T_{\alpha\beta}$ is of the form

$$T_{\alpha\beta} = A(\rho, \theta, u)v_s^{(\alpha)}v_s^{(\beta)} + \delta_{\alpha\beta}B(\rho, \theta, u),$$

where A and B are scalars. As proved above (section 2) the diagonal element $T_{\alpha\alpha}$ is equal to the derivative

$$\frac{1}{V}\frac{\partial(NF)}{\partial L}\bigg|_{L=1},$$

where $F(L)$ is the free energy of the system when "stretched" by a factor L along the α axis. Therefore,

$$\rho\frac{\partial F}{\partial L}\bigg|_{L=1} = T_{\alpha\alpha} = A(v_s^{(\alpha)})^2 + B.$$

On the other hand

$$\rho\frac{\partial F}{\partial L}\bigg|_{L=1} = \rho^2\frac{\partial F}{\partial\rho} - \rho\frac{\partial F}{\partial L}(v_s^{(\alpha)})^2 = -\mathscr{P}(\rho, \theta, u) - m\rho_s(v_s^{(\alpha)})^2.$$

Therefore,

$$A = -m\rho_s, \quad B = -\mathscr{P}(\rho, \theta, u),$$

and so, finally:

$$T_{\alpha\beta} = -m\rho_s v_s^{(\alpha)}v_s^{(\beta)} - \delta_{\alpha\beta}\mathscr{P}(\rho, \theta, u).$$

Now consider states characterized by two velocities $\vec{v}_s$, $\vec{v}_n$. Notice first of all that the expressions

$$X(r - r'|\rho, \theta, v_s, v_n) = \langle\psi^\dagger(r')\psi(r')\psi(r)\rangle_{v_s,v_n}\langle\psi^\dagger(r)\rangle_{v_s,v_n},$$

$$\mathscr{D}(r - r'|\rho, \theta, v_s, v_n) = \langle\psi^\dagger(r)\psi^\dagger(r')\psi(r')\psi(r)\rangle_{v_s,v_n}$$

are invariant with respect to Galilean transformations

$$\psi \to \psi\,\exp(im\vec{v}\,\vec{r}),$$

and therefore can be functions only of the relative velocity of superfluid and normal components:

$$X(r - r'|\rho, \theta, v_s, s_n) = X(r - r'|\rho, \theta, v_s - v_n),$$

$$\mathscr{D}(r - r'|\rho, \theta, v_s, s_n) = \mathscr{D}(r - r'|\rho, \theta, v_s - v_n).$$

Consider the expression for j_α:

$$
\begin{aligned}
j_\alpha =& \frac{i}{2}\left\langle \frac{\partial \psi^\dagger}{\partial r_\alpha}\psi - \psi^\dagger \frac{\partial \psi}{\partial r_\alpha}\right\rangle_{v_s, v_n}\\
=& \frac{i}{2}\left\langle \left(\frac{\partial \psi^\dagger}{\partial r_\alpha} - imv_n^{(\alpha)}\psi^\dagger\right)\psi - \psi^\dagger\left(\frac{\partial \psi}{\partial r_\alpha} + imv_n^{(\alpha)}\psi\right)\right\rangle_{v_s - v_n, 0}\\
=& mv_n^{(\alpha)}\rho + m\rho_s(v_s^{(\alpha)} - v_n^{(\alpha)}) = m\rho_s v_s^{(\alpha)} + m\rho_n v_n^{(\alpha)}.
\end{aligned}
$$

In the last expression we have introduced the "density of the normal component" $\rho_n = \rho - \rho_s$. Now let us calculate the stress tensor:

$$
\begin{aligned}
T_{\alpha\beta}(\rho, \theta, v_s, v_n) =& \frac{1}{2}\int \frac{\partial \Phi(R)}{\partial R_\alpha} R_\beta\, \mathscr{D}(R|\rho, \theta, v_s, v_n)\, d\vec{R}\\
& - \frac{1}{2m}\left\langle\left(\frac{\partial \psi^\dagger}{\partial r_\alpha} - imv_n^{(\alpha)}\psi^\dagger\right)\left(\frac{\partial \psi}{\partial r_\beta} + imv_n^{(\beta)}\psi\right)\right.\\
& \left. + \left(\frac{\partial \psi^\dagger}{\partial r_\beta} - imv_n^{(\beta)}\psi^\dagger\right)\left(\frac{\partial \psi}{\partial r_\alpha} + imv_n^{(\alpha)}\psi\right)\right\rangle_{v_s - v_n, 0}\\
=& -\frac{1}{2m}\left\langle\frac{\partial \psi^\dagger}{\partial r_\alpha}\frac{\partial \psi}{\partial r_\beta} + \frac{\partial \psi^\dagger}{\partial r_\beta}\frac{\partial \psi}{\partial r_\alpha}\right\rangle_{v_s - v_n, 0} + \frac{1}{2}\int \frac{\partial \Phi(R)}{\partial R_\alpha}\\
& \times R_\beta\, \mathscr{D}(R|\rho, \theta, v_s - v_n)\, d\vec{R} + \frac{iv_n^{(\alpha)}}{2}\left\langle\psi^\dagger\frac{\partial \psi}{\partial r_\beta} - \frac{\partial \psi^\dagger}{\partial r_\beta}\psi\right\rangle_{v_s - v_n, 0}\\
& - \frac{iv_n^{(\beta)}}{2}\left\langle\frac{\partial \psi^\dagger}{\partial r_\beta}\psi - \psi^\dagger\frac{\partial \psi}{\partial r_\beta}\right\rangle_{v_s - v_n, 0} - m\rho v_n^{(\alpha)}v_n^{(\beta)}\\
=& -\delta_{\alpha\beta}\mathscr{P}(\rho, \theta, u) - m\rho_s v_s^{(\alpha)}v_s^{(\beta)} - m\rho_n v_n^{(\alpha)}v_n^{(\beta)}. \qquad (3.5)
\end{aligned}
$$

Here and subsequently $u = \dfrac{(v_s - v_n)^2}{2}$.

Next we transform the expression for the average energy:

$$
\begin{aligned}
\rho\epsilon(\rho, \theta, v_s, v_n) =& \frac{1}{2}\int \Phi(R)\, \mathscr{D}(R|\rho, \theta, v_s - v_n, 0)\, d\vec{R}\\
& - \frac{1}{4m}\left\langle\left[\left(\frac{\partial}{\partial \vec{r}} - im\vec{v}_n\right)^2\psi\right]\psi + \psi^\dagger\left[\left(\frac{\partial}{\partial \vec{r}} + im\vec{v}_n\right)^2\psi\right]\right\rangle_{v_s - v_n, 0}\\
=& \rho E(\rho, \theta, u) + \frac{m\rho v_n^2}{2} + m\rho_s(\vec{v}_s - \vec{v}_n)\vec{v}_n. \qquad (3.6)
\end{aligned}
$$

Here, as usual, $E(\rho, \theta, u) = F - \theta \dfrac{\partial F}{\partial \theta}$.

We also need to find an expression for the vector I_α in the statistical equilibrium state specified by $\vec{v}_n$ and $\vec{v}_s$. Since

$$
\begin{aligned}
G^{(\alpha)}&(r - r'|\rho, \theta, v_s, v_n) \\
&= \frac{i}{4m} \left\langle \psi^\dagger(t, r') \left\{ \psi^\dagger(t, r) \frac{\partial \psi(t, r)}{\partial r_\alpha} - \frac{\partial \psi^\dagger(t, r)}{\partial r_\alpha} \psi(t, r) \right\} \psi(t, r') \right\rangle_{v_s, v_n} \\
&= \frac{i}{4m} \left\langle \psi^\dagger(t, r') \left\{ \psi^\dagger(t, r) \left(\frac{\partial \psi(t, r)}{\partial r_\alpha} + imv_n^{(\alpha)} \psi(t, r) \right) \right. \right. \\
&\qquad \left. \left. - \left(\frac{\partial \psi^\dagger(t, r)}{\partial r_\alpha} - imv_n^{(\alpha)} \psi^\dagger(t, r) \right) \psi(t, r) \right\} \psi(t, r') \right\rangle_{v_s - v_n, 0} \\
&= G^{(\alpha)}(r - r'|\rho, \theta, v_s - v_n) - \frac{v^{(\alpha)}}{2} \mathscr{D}(r - r'|\rho, \theta, v_s - v_n)
\end{aligned}
$$

we get for I_α itself

$$
\begin{aligned}
I_\alpha(\rho, \theta, v_s, v_n) &= -\frac{i}{4m^2} \left\langle (\Delta \psi^\dagger) \frac{\partial \psi}{\partial r_\alpha} - \frac{\partial \psi^\dagger}{\partial r_\alpha} \Delta \psi \right\rangle_{v_s, v_n} \\
&\quad + \int \Phi(R) G^{(\alpha)}(R|\rho, \theta, v_s, v_n) \, d\vec{R} - \sum_\beta \int \frac{\partial \Phi(R)}{\partial R_\beta} R_\alpha \, G^{(\beta)}(R|\rho, \theta, v_s, v_n) \, d\vec{R} \\
&= -\frac{i}{4m^2} \left\langle \left[\left(\frac{\partial}{\partial \vec{r}} - im\vec{v}_n \right)^2 \psi^\dagger \right] \left(\frac{\partial \psi}{\partial r_\alpha} + imv_n^{(\alpha)} \psi \right) - \left(\frac{\partial \psi^\dagger}{\partial r_\alpha} + imv_n^{(\alpha)} \psi^\dagger \right) \right. \\
&\quad \left. \times \left[\left(\frac{\partial}{\partial \vec{r}} - im\vec{v}_n \right)^2 \psi \right] \right\rangle_{v_s, v_n, 0} + \int \Phi(R) G^{(\alpha)}(R|\rho, \theta, v_s - v_n) \, d\vec{R} \\
&\quad - \frac{v^{(\alpha)}}{2} \int \Phi(R) \, \mathscr{D}(R|\rho, \theta, v_s - v_n) \, d\vec{R} \\
&\quad - \sum_\beta \int \frac{\partial \Phi(R)}{\partial R_\beta} R_\alpha \, G^{(\beta)}(R|\rho, \theta, v_s - v_n) \, d\vec{R} \\
&\quad + \frac{1}{2} \sum_\beta \int \frac{\partial \Phi(R)}{\partial R_\beta} R_\alpha v_n^{(\beta)} \mathscr{D}(R|\rho, \theta, v_s - v_n) \, d\vec{R} \\
&= \frac{i}{4m^2} \left\langle (\Delta \psi^\dagger) \frac{\partial \psi}{\partial r_\alpha} - \frac{\partial \psi^\dagger}{\partial r_\alpha} \Delta \psi \right\rangle_{v_s - v_n, 0} + \int \Phi(R) G^{(\alpha)}(R|\rho, \theta, v_s - v_n) \, d\vec{R} \\
&\quad - \sum_\beta \int \frac{\partial \Phi(R)}{\partial R_\beta} R_\alpha \, G^{(\beta)}(R|\rho, \theta, v_s - v_n) \, d\vec{R}
\end{aligned}
$$

$$+ \frac{1}{4m} v_n^{(\alpha)} \langle (\Delta \psi^\dagger) \psi + \psi^\dagger (\Delta \psi) \rangle_{v_s - v_n, 0} - \frac{1}{2m} \sum_\beta v_n^{(\beta)}$$

$$\times \left\langle \frac{\partial \psi^\dagger}{\partial r_\beta} \frac{\partial \psi}{\partial r_\alpha} + \frac{\partial \psi^\dagger}{\partial r_\alpha} \frac{\partial \psi}{\partial r_\beta} \right\rangle_{v_s - v_n, 0} + \frac{1}{2} \sum_\beta v_n^{(\beta)} \int \frac{\partial \Phi(R)}{\partial R_\beta}$$

$$\times R_\alpha \mathscr{D}(R|\rho, \theta, v_s - v_n) \, d\vec{R} - \frac{v_n^{(\alpha)}}{2} \int \Phi(R) \mathscr{D}(R|\rho, \theta, v_s - v_n) \, d\vec{R}$$

$$- \frac{m v_n^2}{2} \rho v_n^{(\alpha)} - \frac{i}{2} v_n^{(\alpha)} \sum_\beta v_n^{(\beta)} \left\langle \frac{\partial \psi^\dagger}{\partial r_\beta} \psi - \psi^\dagger \frac{\partial \psi}{\partial r_\beta} \right\rangle_{v_s - v_n, 0}$$

$$+ \frac{i}{4} v_n^2 \left\langle \psi^\dagger \frac{\partial \psi}{\partial r_\beta} - \frac{\partial \psi^\dagger}{\partial r_\beta} \psi \right\rangle_{v_s - v_n, 0}.$$

Using the definition of $I_\alpha(\rho, \theta, v_s - v_n, 0)$ formulae (3.6) and (3.5) and the definition of j_α we get:

$$\begin{aligned}
I_\alpha(\rho, \theta, v_s, v_n) =\,& I_\alpha(\rho, \theta, v_s - v_n, 0) - v_n^{(\alpha)} \rho E(\rho, \theta, u) \\
& + \sum_\beta v_n^{(\beta)} T_{\alpha\beta}(\rho, \theta, v_s - v_n, 0) - \sum_\beta v_n^{(\alpha)} v_n^{(\beta)} m \rho_s (v_s^{(\beta)} - v_n^{(\beta)}) \\
& - \frac{m v_n^2}{2} \rho_s (v_s^{(\alpha)} - v_n^{(\alpha)}) - \frac{m v_n^2}{2} v_n^{(\alpha)} \rho \\
=\,& I_\alpha(\rho, \theta, v_s - v_n, 0) - v_n^{(\alpha)} \Big[\rho E + \frac{m v_n^2}{2} \rho + \mathscr{P} \\
& + m \rho_s \vec{v}_n (\vec{v}_s - \vec{v}_n) - m \rho_s (v_s^{(\alpha)} - v_n^{(\alpha)}) \Big(\vec{v}_s \, \vec{v}_n - \frac{v_n^2}{2} \Big) \Big]. \quad (3.7)
\end{aligned}$$

Let us write

$$I_\alpha(\rho, \theta, v_s - v_n, 0) = [-\Lambda(\rho, \theta, u) \rho_s + A(\rho, \theta, u)](v_s^{(\alpha)} - v_n^{(\alpha)}). \quad (3.8)$$

Below it will be shown that $A \equiv 0$.

So far we have assumed that $U = 0$. We should, of course, get the same expressions for j_α, X, $T_{\alpha\beta}$ and I_α in the case $U = \text{const.}$, since all these expressions are invariant under the gauge transformation

$$\psi \to \psi \, e^{-iUt}.$$

After these preliminary transformations of the equilibrium quantities we can go on to discuss the derivation of the hydrodynamic equations for a superfluid. We shall use a natural generalization of the procedure expounded in section 2.

We introduce, as on page 133, a small parameter and put

$$\xi = \mu \vec{r}, \quad \tau = \mu t.$$

We then write

$$\rho(t, r) = \tilde{\rho}(\tau, \xi), \quad \vec{j}(t, r) = \tilde{\vec{j}}(\tau, \xi), \quad \epsilon(t, r) = \tilde{\epsilon}(\tau, \xi),$$

and also

$$U(t, r) = \tilde{U}(\tau, \xi),$$

$$\vec{v}_s(t, r) = i \frac{\frac{\partial}{\partial \vec{r}} \langle \psi(t, r) \rangle}{m \langle \psi(t, r) \rangle} = \tilde{\vec{v}}_s(\tau, \xi),$$

$$a(t, r) = \sqrt{\langle \psi^\dagger(t, r) \rangle \langle \psi(t, r) \rangle} = |\langle \psi(t, r) \rangle| = \tilde{a}(\tau, \xi) \neq 0.$$

It should be emphasized that the assumption that $\vec{v}_s$ is a slowly varying function of t and r (which obviously implies that the denominator in the expression for $\vec{v}_s$ does not vanish, i.e., $a \neq 0$) restricts us to considering irrotational flow:

$$\text{rot } \vec{v}_s = 0.$$

The case when $\langle \psi(t, r) \rangle$ can vanish has been investigated by S. V. Iordanskii [2]. To obtain an expression for the Green's function we actually only need the hydrodynamic equations in the "acoustic" case, when all the quantities introduced above differ from their equilibrium values by a vanishingly small amount. In this case we have

$$\vec{v}_s = \delta \vec{v}_s = \frac{-i}{m \langle \psi \rangle_{v_s, v_n}} \left\langle \frac{\partial \phi}{\partial \vec{r}} \right\rangle, \quad a = a_0 + \delta a,$$

and hence everywhere

$$\text{rot } \vec{v}_s = 0, \quad a \neq 0.$$

Thus, in calculating the Green's functions we have the condition of irrotational flow fulfilled automatically.

Notice also that since the gradients with respect to t and $\vec{r}$ of ρ, v_s etc., are quantities of first order in μ, equation (1.3) implies that we must take η (or equivalently, ζ) proportional to μ. Let us therefore write

$$\zeta(t, r) = \mu \tilde{\zeta}(\tau, \xi).$$

To simplify the notation we shall from here on omit the index $\approx$ on functions of τ and ξ, provided there is no danger of confusion.

Now, starting from the expressions for $\rho(\tau, \xi)$, $\vec{j}(\tau, \xi)$, $\epsilon(\tau, \xi)$ and $\vec{v}_s(\tau, \xi)$ (all of which are defined as microscopic averages) we can introduce quantities $\theta(\tau, \xi)$ and $\vec{v}_n(\tau, \xi)$, which together with $\rho(\tau, \xi)$ and $\vec{v}_s(\tau, \xi)$ describe the "local quasi-equilibrium state". Thus, we take the functions

$$\epsilon(\rho, \theta, v_s, v_n); \quad \rho_s(\rho, \theta, u) = \frac{1}{m} \rho \frac{\partial F(\rho, \theta, u)}{\partial u}; \quad u = \frac{(v_n - v_s)^2}{2};$$
$$\rho_n(\rho, \theta, u) = \rho - \rho_s(\rho, \theta, u),$$

which express ϵ, ρ_s and ρ in statistical equilibrium, and define functions

$$\theta(\tau, \xi), \quad \vec{v}_n(\tau, \xi) \tag{3.9}$$

by the relation

$$\epsilon(\tau, \xi) = \epsilon(\rho, \theta, v_s, v_n),$$
$$\vec{j}(\tau, \xi) = \rho_s(\rho, \theta, u)\vec{v}_s + \rho_n(\rho, \theta, u)\vec{v}_n. \tag{3.10}$$

Once we have introduced the functions (3.9) in this way, we can also define quantities $\rho_s(\tau, \xi)$, $\rho_n(\tau, \xi)$ by the relations

$$\rho_s(\tau, \xi) = \rho_s(\rho, \theta, u), \quad \rho_n(\tau, \xi) = \rho_n(\rho, \theta, u).$$

Now we formulate our assumption that for the non-equilibrium process considered the local state of the liquid is only slightly different from the local quasi-equilibrium state. Following the procedure of section 2, we write

$$a(\tau, \xi) = a(\rho, \theta, u) + \mu a^{(1)}(\tau, \xi) + \ldots,$$
$$X_\tau(\xi, R) = X(R|\rho, \theta, v_s, v_n) + \mu X^{(1)}(\tau, \xi) + \ldots$$
$$= X(R|\rho, \theta, v_s - v_n) + \mu X^{(1)}(\tau, \xi) + \ldots,$$
$$T_{\alpha\beta}(\tau, \xi) = T_{\alpha\beta}(\rho, \theta, v_s, v_n) + \mu T_{\alpha\beta}^{(1)}(\tau, \xi) + \ldots,$$
$$I_\alpha(\tau, \xi) = I_\alpha(\rho, \theta, v_s, v_n) + \mu I_\alpha^{(1)}(\tau, \xi) + \ldots \tag{3.11}$$

and substitute these expansions, along with (3.10), in equations (1.3), (1.4), (1.7)^w, and (3.3), after expressing the latter in terms of variables τ, ξ. In

---

^{w}We transform equations (1.4) and (1.7) into forms analogous to (2.10) and (2.12)

accordance with our ideal-liquid approximation we shall neglect correction terms of order μ.

Consider, for example, equation (3.3). We have

$$m\frac{\partial v_s^{(\alpha)}}{\partial \tau} = \frac{\partial}{\partial \xi_\alpha}\left\{\mu^2\frac{\Delta\xi a}{2ma} - \frac{mv_s^2}{2} - U - \mu\frac{\zeta^* + \zeta}{2a}\right.$$
$$\left. - \frac{1}{2a^2}\int\Phi(R)[X_\tau(\xi, R) + X_\tau^*(\xi, R)]\,d\vec{R}\right\},$$

and therefore, in our approximation,

$$m\frac{\partial v_s^{(\alpha)}}{\partial \tau} = -\frac{\partial}{\partial \xi_\alpha}\left\{\frac{mv_s^2}{2} + U + \frac{1}{2a^2(\rho, \theta, u)}\int\Phi(R)[X(R|\rho, \theta, v_s - v_n)\right.$$
$$\left. + X^*(R|\rho, \theta, v_s - v_n)]\,d\vec{R}\right\}.$$

Hence, using (3.4), we finally get:

$$m\frac{\partial v_s^{(\alpha)}}{\partial \tau} = -\frac{\partial}{\partial \xi_\alpha}\left\{\frac{mv_s^2}{2} - \frac{m}{2}(v_s - v_n)^2 + \Lambda(\rho, \theta, u) + U\right\}$$
$$u = \frac{(v_s - v_n)^2}{2}. \tag{3.12}$$

In the same way we get from (1.3), (1.4), (1.7), and (3.8)

$$\frac{\partial\rho}{\partial\tau} + \sum_\alpha\frac{\partial(\rho_s v_s^{(\alpha)} + \rho_n v_n^{(\alpha)})}{\partial\xi_\alpha} = ia(\zeta^* - \zeta), \tag{3.13}$$

$$\frac{\partial(\rho_s v_s^{(\alpha)} + \rho_n v_n^{(\alpha)})}{\partial\tau} = \sum_\beta\frac{\partial T_{\alpha\beta}(\rho, \theta, v_s, v_n)}{\partial\xi_\beta} + imv_s^{(\alpha)}(\zeta^* - \zeta)a - \rho\frac{\partial U}{\partial\xi_\alpha}, \tag{3.14}$$

$$\frac{\partial\rho\epsilon(\rho, \theta, v_s, v_n)}{\partial\tau} \doteq \sum_\beta\frac{\partial I_\beta(\rho, \theta, v_s, v_n)}{\partial\xi_\beta} - \sum_\beta(\rho_s v_s^{(\alpha)} + \rho_n v_n^{(\alpha)})\frac{\partial U}{\partial\xi_\beta}$$
$$+ ia(\zeta^* - \zeta)\left\{\frac{mv_s^2}{2} + \Lambda(\rho, \theta, \mu) - \frac{(v_s - v_n)^2}{2}m\right\}. \tag{3.15}$$

Substituting in equations (3.14) and (3.15) expressions (3.5)-(3.8) for $T_{\alpha\beta}(\rho, \theta, v_s, v_n)$, $\epsilon(\rho, \theta, v_s, v_n)$ and $I_\beta(\rho, \theta, v_s, v_n)$ we get:

$$m\left\{\frac{\partial(\rho_s v_s^{(\alpha)} + \rho_n v_n^{(\alpha)})}{\partial\tau} + \sum_\beta \frac{\partial(\rho_s v_s^{(\alpha)} v_s^{(\beta)} + \rho_n v_n^{(\alpha)} v_n^{(\beta)})}{\partial\xi_\alpha}\right\}$$

$$= -\frac{\partial\mathscr{P}}{\partial\xi_\alpha} - \rho\frac{\partial U}{\partial\xi_\alpha} + imav_s^{(\alpha)}(\zeta^* - \zeta), \qquad (3.16)$$

$$\frac{\partial}{\partial\tau}\left[\rho E(\rho, \theta, u) + \frac{m\rho v_n^2}{2} + m\rho_s(\vec{v}_s - \vec{v}_n)\vec{v}_n\right] + \sum_\beta \frac{\partial}{\partial\xi_\beta}$$

$$\times \left\{v_n^{(\beta)}\left[\frac{\rho m v_n^2}{2} + \rho_s m\vec{v}_n(\vec{v}_s - \vec{v}_n) + \rho E + \mathscr{P}\right] + (v_s^{(\beta)} - v_n^{(\beta)})\rho_s\right.$$

$$\times \left[\Lambda + m\left(\vec{v}_s\vec{v}_n - \frac{v_n^2}{2}\right)\right]\right\} + \sum_\beta \frac{\partial U}{\partial\xi_\beta}(\rho_s v_s + \rho_n v_n) - ia(\zeta^* - \zeta)$$

$$\times \left[\Lambda + m\left(\vec{v}_s\vec{v}_n - \frac{v_n^2}{2}\right)\right] = -\sum_\beta \frac{\partial}{\partial\xi_\beta}(v_s^{(\beta)} - v_n^{(\beta)})A. \qquad (3.17)$$

Now we go back in equations (3.12), (3.13), (3.16), and (3.17) from the auxiliary variables τ, $\vec{\xi}$ to our original variables $t, \vec{r}$, This gives us the following system of hydrodynamic equations:

$$m\frac{\partial v_s^{(\alpha)}}{\partial t} + \frac{\partial}{\partial r_\alpha}\left\{m\left(v_s v_n - \frac{v_n^2}{2}\right) + \Lambda + U\right\} = 0, \qquad (3.18a)$$

$$\frac{\partial\rho}{\partial t} + \sum_\beta \frac{\partial(\rho_s v_s^{(\alpha)} + \rho_n v_n^{(\alpha)})}{\partial r_\beta} - ia(\zeta^* - \zeta) = 0, \qquad (3.18b)$$

$$m\left\{\frac{\partial(\rho_s v_s^{(\alpha)} + \rho_n v_n^{(\alpha)})}{\partial t} + \sum_\beta \frac{\partial(\rho_s v_s^{(\alpha)} v_s^{(\beta)} + \rho_n v_n^{(\alpha)} v_n^{(\beta)})}{\partial r_\beta}\right\}$$

$$+ \frac{\partial\mathscr{P}}{\partial r_\alpha} + \rho\frac{\partial U}{\partial r_\alpha} - imav_s^{(\alpha)}(\zeta^* - \zeta) = 0, \qquad (3.18c)$$

$$\frac{\partial}{\partial t}\left[\rho E + \frac{m\rho v_n^2}{2} + m\rho_s(\vec{v}_s - \vec{v}_n)\vec{v}_n\right]$$

$$+ \sum_{\beta} \frac{\partial}{\partial r_\beta} \left\{ v_n^{(\beta)} \left[\frac{\rho m v_n^2}{2} + \rho_s m \vec{v}_n (\vec{v}_s - \vec{v}_n) + \rho E + \mathscr{P} \right] \right.$$

$$\left. + (v_s^{(\beta)} - v_n^{(\beta)}) \rho_s \left[\Lambda + m \left(\vec{v}_s \vec{v}_n - \frac{v_n^2}{2} \right) \right] \right\} + \sum_{\beta} \frac{\partial U}{\partial r_\beta} (\rho_s v_s + \rho_n v_n)$$

$$- i a (\zeta^* - \zeta) \left[\Lambda + m \left(\vec{v}_s \vec{v}_n - \frac{v_n^2}{2} \right) \right] = - \sum_{\beta} \frac{\partial}{\partial r_\beta} (v_s^{(\beta)} - v_n^{(\beta)}) A. \qquad (3.18d)$$

We remind the reader that in these equations

$$a = a(\rho, \theta, u), \quad A = A(\rho, \theta, u), \quad u = \frac{(v_s - v_n)^2}{2}, \quad \rho_s = \frac{1}{m} \rho \frac{\partial F(\rho, \theta, u)}{\partial u}$$

$$\rho_n = \rho - \rho_s; \quad E = F(\rho, \theta, u) - \theta \frac{\partial F(\rho, \theta, u)}{\partial \theta};$$

$$\mathscr{P} = \rho^2 \frac{\partial F(\rho, \theta, u)}{\partial \rho}, \quad \Lambda = F(\rho, \theta, u) + \rho \frac{\partial F(\rho, \theta, u)}{\partial \rho},$$

$$\zeta(t, r) = \eta(t, r) e^{-i\chi(t,r)} = \eta \, e^{-im\Theta},$$

where $\Theta(t, r)$ is the velocity potential for superfluid flow:

$$v_s^{(\alpha)} = \frac{\partial \Theta}{\partial r_\alpha}.$$

Equation (3.18d) can be simplified by introducing the entropy

$$S = - \frac{\partial F}{\partial \theta}.$$

In fact, using the other equations of the group (3.18) we can transform the first term of (3.18d) as follows

$$\frac{\partial}{\partial t} \left[\rho E + \frac{m \rho v_n^2}{2} + m \rho_s \vec{v}_n (\vec{v}_s - \vec{v}_n) \right]$$

$$= \frac{\partial}{\partial t} \left[\rho \left(F - \theta \frac{\partial F}{\partial \theta} \right) - \frac{m \rho v_n^2}{2} + m \vec{v}_n (\rho_s \vec{v}_s + \rho_n \vec{v}_n) \right]$$

$$= \frac{\partial \rho}{\partial t} \left(F + \rho \frac{\partial F}{\partial \rho} \right) - \theta \frac{\partial}{\partial t} \left(\rho \frac{\partial F}{\partial \theta} \right) + m \rho_s (v_s^{(\alpha)} - v_n^{(\alpha)}) \frac{\partial (v_s^{(\alpha)} - v_n^{(\alpha)})}{\partial t}$$

$$- \frac{\partial \rho}{\partial t} \frac{m v_n^2}{2} - (\rho_s + \rho_n) m v_n^{(\alpha)} \frac{\partial v_n^{(\alpha)}}{\partial t} + \frac{\partial}{\partial t} m v_n^{(\alpha)} (\rho_s v_s^{(\alpha)} + \rho_n v_n^{(\alpha)})$$

$$= \frac{\partial \rho}{\partial t}\left(\Lambda - \frac{mv_n^2}{2}\right) - \theta\frac{\partial}{\partial t}\left(\rho\frac{\partial F}{\partial \theta}\right) + m\rho_s(v_s^{(\alpha)} - v_n^{(\alpha)})\frac{\partial(v_s^{(\alpha)} - v_n^{(\alpha)})}{\partial t}$$

$$- \frac{\partial \rho}{\partial t}\frac{mv_n^2}{2} - (\rho_s + \rho_n)mv_n^{(\alpha)}\frac{\partial v_n^{(\alpha)}}{\partial t} + \frac{\partial}{\partial t}\,mv_n^{(\alpha)}(\rho_s v_s^{(\alpha)} + \rho_n v_n^{(\alpha)})$$

$$= \frac{\partial \rho}{\partial t}\left(\Lambda - \frac{mv_n^2}{2}\right) - \theta\frac{\partial}{\partial t}\left(\rho\frac{\partial F}{\partial \theta}\right) + m\rho_s(v_s^{(\alpha)} - v_n^{(\alpha)})\frac{\partial(v_s^{(\alpha)} - v_n^{(\alpha)})}{\partial t}$$

$$- (\rho_s + \rho_n)mv_n^{(\alpha)}\frac{\partial v_n^{(\alpha)}}{\partial t} + (\rho_s v_s^{(\alpha)} + \rho_n v_n^{(\alpha)})m\frac{\partial v_n^{(\alpha)}}{\partial t}$$

$$- mv_n^{(\alpha)}\frac{\partial}{\partial r_\beta}[v_n^{(\alpha)}(\rho_s v_s^{(\beta)} + \rho_n v_n^{(\beta)}) + \rho_s v_s^{(\beta)}(v_s^{(\alpha)} - v_n^{(\alpha)})] - v_n^{(\alpha)}\frac{\partial \mathscr{P}}{\partial r_\alpha}$$

$$- (\rho_s + \rho_n)\,v_n^{(\alpha)}\frac{\partial U}{\partial r_\alpha} + imv_n^{(\alpha)}v_s^{(\alpha)}a(\zeta^* - \zeta)$$

$$= \frac{\partial \rho}{\partial t}\left(\Lambda - \frac{mv_n^2}{2}\right) - \theta\frac{\partial}{\partial t}\left(\rho\frac{\partial F}{\partial \theta}\right) + m\rho_s(v_s^{(\alpha)} - v_n^{(\alpha)})\frac{\partial v_s^{(\alpha)}}{\partial t}$$

$$- mv_n^{(\alpha)}v_n^{(\alpha)}\frac{\partial}{\partial r_\beta}(\rho_s v_s^{(\beta)} + \rho_n v_n^{(\beta)}) - mv_n^{(\alpha)}(\rho_s v_s^{(\beta)} + \rho_n v_n^{(\beta)})\frac{\partial v_n^{(\alpha)}}{\partial r_\beta}$$

$$- mv_n^{(\alpha)}(v_s^{(\beta)} - v_n^{(\beta)})\frac{\partial \rho_s v_n^{(\alpha)}}{\partial r_\beta} - m\rho_s v_n^{(\alpha)}v_s^{(\beta)}\frac{\partial(v_s^{(\alpha)} - v_n^{(\alpha)})}{\partial r_\beta}$$

$$- v_n^{(\alpha)}\frac{\partial \mathscr{P}}{\partial r_\alpha} - (\rho_s + \rho_n)\,v_n^{(\alpha)}\frac{\partial U}{\partial r_\alpha} + imv_n^{(\alpha)}v_s^{(\alpha)}a(\zeta^* - \zeta)$$

$$= -\theta\frac{\partial}{\partial t}\left(\rho\frac{\partial F}{\partial \theta}\right) - \left(\Lambda + \frac{mv_n^2}{2}\right)\left[\frac{\partial}{\partial r_\beta}(\rho_s v_s^{(\beta)} + \rho_n v_n^{(\beta)}) - ia(\zeta^* - \zeta)\right]$$

$$+ m\rho_s(v_s^{(\alpha)} - v_n^{(\alpha)})\frac{\partial v_s^{(\alpha)}}{\partial t} - m\rho_n v_n^{(\alpha)}v_n^{(\beta)}\frac{\partial v_n^{(\alpha)}}{\partial r_\beta} - m\rho_s v_s^{(\beta)}v_n^{(\alpha)}\frac{\partial v_s^{(\alpha)}}{\partial r_\beta}$$

$$- mv_n^{(\alpha)}(v_s^{(\alpha)} - v_n^{(\alpha)})\frac{\partial \rho_s v_n^{(\beta)}}{\partial r_\beta} + ima(\zeta^* - \zeta)(v_n^{(\alpha)}v_s^{(\alpha)} - v_n^2)$$

$$- v_n^{(\alpha)}\frac{\partial \mathscr{P}}{\partial r_\alpha} - (\rho_s + \rho_n)\,v_n^{(\alpha)}\frac{\partial U}{\partial r_\alpha}$$

Also, the term $\dfrac{\partial}{\partial r_\beta}(v_n^{(\beta)}\rho E)$ in (3.18d) can be rewritten

$$\frac{\partial}{\partial r_\beta}(v_n^{(\beta)}\rho E) = \Lambda\frac{\partial \rho v_n^{(\beta)}}{\partial r_\beta} - \mathscr{P}\frac{\partial v_n^{(\beta)}}{\partial r_\beta} - \theta\frac{\partial}{\partial r_\beta}\left(\rho v_n^{(\beta)}\frac{\partial F}{\partial \theta}\right)$$

$$+ m\rho_s v_n^{(\beta)}(v_s^{(\alpha)} - v_n^{(\alpha)})\frac{\partial(v_s^{(\alpha)} - v_n^{(\alpha)})}{\partial r_\beta}.$$

Now we rearrange the left-hand side of the (transformed) equation (3.18d), grouping together similar terms:

$$-\theta\frac{\partial}{\partial t}\left(\rho\frac{\partial F}{\partial \theta}\right) - \theta\frac{\partial}{\partial r_\beta}\left(\rho v_n^{(\beta)}\frac{\partial F}{\partial \theta}\right)$$

$$+ \rho_s(v_s^{(\alpha)} - v_n^{(\alpha)})\left[m\frac{\partial v_s^{(\alpha)}}{\partial t} + \frac{\partial \Lambda}{\partial r_\alpha} + m\left(\vec{v}_s\vec{v}_n - \frac{v_n^2}{2}\right) + U\right]$$

$$- \left[\rho_s(v_s^{(\alpha)} - v_n^{(\alpha)})\frac{\partial U}{\partial r_\alpha} + (\rho_s + \rho_n)v_n^{(\alpha)}\frac{\partial U}{\partial r_\alpha} - (\rho_s v_s^{(\alpha)} + \rho_n v_n^{(\alpha)})\frac{\partial U}{\partial r_\alpha}\right]$$

$$+ \Lambda\left[\frac{\partial \rho}{\partial t} + \frac{\partial \rho v_n^{(\beta)}}{\partial r_\beta} + \frac{\partial \rho_s(v_s^{(\beta)} - v_n^{(\beta)})}{\partial r_\beta} - ia(\zeta^* - \zeta)\right]$$

$$+ ia(\zeta^* - \zeta)\left[\Lambda + m(v_n^{(\beta)}v_s^{(\beta)} - v_n^2) - \Lambda - m\left(v_n^{(\beta)}v_s^{(\beta)} - \frac{v_n^2}{2}\right) + \frac{mv_n^2}{2}\right]$$

$$- \frac{mv_n^2}{2}\frac{\partial}{\partial r_\beta}\left[\rho_s v_s^{(\beta)} + \rho_n v_n^{(\beta)} - 2\rho_s v_s^{(\beta)} - \rho_n v_n^{(\beta)} - \rho_s v_n^{(\beta)} + 2\rho_s v_n^{(\beta)} + \rho_s v_s^{(\beta)}\right.$$

$$\left. - \rho_s v_n^{(\beta)}\right] - mv_n^{(\alpha)}v_s^{(\alpha)}\frac{\partial}{\partial r_\beta}\left[\rho_s v_s^{(\beta)} - \rho_s v_n^{(\beta)} - \rho_s(v_s^{(\beta)} - v_n^{(\beta)})\right]$$

$$- m\frac{\partial v_n^{(\beta)}}{\partial r_\beta}\left[\rho_n v_n^{(\alpha)}v_n^{(\beta)} + \rho_n v_n^{(\beta)}(v_s^{(\alpha)} - v_n^{(\alpha)}) - \rho_s v_n^{(\beta)}v_n^{(\alpha)} - \rho_n v_n^{(\beta)}v_n^{(\alpha)}\right.$$

$$\left. - \rho_s v_n^{(\beta)}v_s^{(\alpha)} + 2\rho_s v_n^{(\beta)}v_n^{(\alpha)}\right] - m\frac{\partial v_s^{(\beta)}}{\partial r_\beta}\left[\rho_s v_n^{(\alpha)}v_s^{(\beta)} + \rho_s v_n^{(\beta)}(v_s^{(\alpha)} - v_n^{(\alpha)})\right.$$

$$\left. - \rho_s v_n^{(\alpha)}v_n^{(\beta)}\right] + \frac{\partial(\mathscr{P}v_n^{(\beta)})}{\partial r_\beta} - v_n^{(\beta)}\frac{\partial \mathscr{P}}{\partial r_\beta} - \mathscr{P}\frac{\partial v_n^{(\beta)}}{\partial r_\beta}$$

$$= m\rho_s v_n^{(\alpha)}v_n^{(\beta)}\left(\frac{\partial v_n^{(\alpha)}}{\partial r_\beta} - \frac{\partial v_n^{(\beta)}}{\partial r_\alpha}\right) + \theta\left[\frac{\partial(\rho S)}{\partial t} + \frac{\partial}{\partial r_\beta}(\rho v_n^{(\beta)} S)\right]$$

$$= \theta\left[\frac{\partial(\rho S)}{\partial t} + \frac{\partial}{\partial r_\beta}(\rho v_n^{(\beta)} S)\right].$$

Thus we form the following equation for the entropy density

$$\theta\left[\frac{\partial(\rho S)}{\partial t} + \sum_\beta \frac{\partial(\rho v_n^{(\beta)} S)}{\partial r_\beta}\right] = \sum_\beta \frac{\partial}{\partial r_\beta}\left[(v_n^{(\beta)} - v_s^{(\beta)})A(\rho, \theta, u)\right]. \qquad (3.19)$$

We can use the equations obtained above to prove that

$$A(\rho, \theta, u) = 0. \tag{3.20}$$

Consider the state of statistical equilibrium in the absence of an external field U, but with η a given function of $\vec{r}$:

$$U = 0, \quad \zeta = \zeta(r), \quad \theta = \text{const}, \quad \rho = \rho(r), \quad \vec{v}_n = 0, \quad \vec{v}_s = \vec{v}_s(r).$$

Then we get from (3.19)

$$\sum_{\beta} \frac{\partial}{\partial r_\beta} v_s^{(\beta)} A\left(\rho, \theta, \frac{v_s^2}{2}\right) = 0, \tag{3.21}$$

and from the continuity equation (3.18b)

$$\sum_{\beta} \frac{\partial}{\partial r_\beta}(\rho_s v_s^{(\beta)}) = ia(\zeta^* - \zeta). \tag{3.22}$$

Moreover, we get from (3.18b)

$$\frac{\partial}{\partial r_\alpha} \Lambda\left(\rho, \theta, \frac{v_s^2}{2}\right) = 0. \tag{3.23}$$

Remembering that

$$\vec{v}_s = \frac{\partial \Theta}{\partial \vec{r}},$$

we see at once the we have three equations[(3.21)-(3.23)], for two unknown functions; this overdetermines the problem. This is the basis of our assertion (3.20), as we shall now see.

Consider first the case when η, $\vec{v}_s$ and the deviation of ρ from a constant value are vanishingly small. Then, neglecting terms of second order in these quantities, we get

$$\sum_{\beta} \frac{\partial v_s^{(\beta)}}{\partial r_\beta} = \frac{ia}{\rho_s}(\zeta^* - \zeta); \quad A(\rho, \theta, 0) \sum_{\beta} \frac{\partial v_s^{(\beta)}}{\partial r_\beta} = 0,$$

whence obviously

$$A(\rho, \theta, 0) = 0.$$

Now consider the more general case, when

$$\vec{v}_s(r) = \vec{w} + \delta\vec{v}_s, \quad \rho = \rho + \delta\rho(r); \quad \vec{w} \neq 0,$$

and $\delta\vec{v}_s$, $\delta\rho$ and ζ are vanishingly small. Then we get from (3.21) and (3.22)

$$A\sum_\beta \frac{\partial \delta v_s^{(\beta)}}{\partial r_\beta} + \sum_\beta w^{(\beta)}\left(\frac{\partial A}{\partial \rho}\frac{\partial \delta\rho}{\partial r_\beta} + \frac{\partial A}{\partial u}\sum_\gamma w^{(\gamma)}\frac{\partial \delta v_s^{(\gamma)}}{\partial r_\beta}\right) = 0,$$

$$\rho_s\sum_\beta \frac{\partial \delta v_s^{(\beta)}}{\partial r_\beta} + \sum_\beta w^{(\beta)}\left(\frac{\partial \rho_s}{\partial \rho}\frac{\partial \delta\rho}{\partial r} + \frac{\partial \rho_s}{\partial u}\sum_\gamma w^{(\gamma)}\frac{\partial \delta v_s^{(\gamma)}}{\partial r_\beta}\right) = ia(\zeta^* - \zeta).$$

$$\text{(3.24)}$$

Let us suppose

$$\zeta(r) = e^{i\vec{k}\vec{r}}\delta G; \quad \delta G = \text{const.}$$

Then obviously we can put

$$\delta\vec{v}_s = e^{i\vec{k}\vec{r}}\delta\vec{v} + e^{-i\vec{k}\vec{r}}\delta\vec{v}^*,$$

$$\delta\rho = e^{i\vec{k}\vec{r}}\delta\Gamma + e^{-i\vec{k}\vec{r}}\delta\Gamma^*,$$

where $\delta\vec{v}$ and $\delta\Gamma$ are infinitesimal constants.

Let us choose the arbitrary vector $\vec{k}$ in such a way that

$$\vec{k}\vec{w} = 0.$$

With this choice of $\vec{k}$ we get

$$\sum_\beta w^{(\beta)}\frac{\partial \delta\rho}{\partial r_\beta} = 0, \quad \sum_\beta w^{(\beta)}\frac{\partial \delta v_s^{(\gamma)}}{\partial r_\beta} = 0,$$

and therefore (3.24) takes the form

$$A\sum_\beta \frac{\partial \delta v_s^{(\beta)}}{\partial r_\beta} = 0, \quad \rho_s\sum_\beta \frac{\partial \delta v_s^{(\beta)}}{\partial r_\beta} = ia(\zeta^* - \zeta).$$

from which follows equation (3.20).

We may now write in place of equation (3.19)

$$\frac{\partial(\rho S)}{\partial t} + \sum_\beta \frac{\partial(\rho v_n^{(\beta)} S)}{\partial r_\beta} = 0. \tag{3.25}$$

Thus equations (3.18a)-(3.18c) and (3.25) constitute the hydrodynamics equations for a superfluid liquid in our ideal-liquid approximation.

It should be remarked that for the case $u = 0$, $\eta = 0$ this system of equations was first obtained by L. D. Landau [3] from phenomenological considerations.

4.   Variational Equations and Green's Functions

Let us now use equations (3.18), (3.25) to consider the case of infinitesimal derivation from the statistical equilibrium state of the liquid at rest. Put

$$\rho = \rho^0 + \delta\rho, \quad \vec{v}_s = \delta\vec{v}_s, \quad \vec{v}_n = \delta\vec{n}_s, \quad S = S^0 + \delta S, \quad U = \delta U, \quad \eta = \delta\eta$$

thus making a transition from the hydrodynamic equations to the linearized "acoustic" equations. Since in this case $\eta = \zeta$ up to terms of second order in the infinitesimals, the linearized equations take the form (we omit the superscript 0 on ρ and S)

$$\frac{\partial\delta\rho}{\partial t} + \rho_s\sum_\beta\frac{\partial v_s^{(\beta)}}{\partial r_\beta} + \rho_n\sum_\beta\frac{\partial v_n^{(\beta)}}{\partial r_\beta} = i\sqrt{\rho_0}(\eta^* - \eta),$$

$$m\rho_s\frac{\partial v_s^{(\beta)}}{\partial t} + m\rho_n\frac{\partial v_n^{(\beta)}}{\partial t} = -\frac{\partial\delta\mathscr{P}}{\partial r_\alpha} - \rho\frac{\partial U}{\partial r_\alpha},$$

$$m\frac{\partial v_s^{(\beta)}}{\partial t} = -\frac{\partial(\delta\Lambda + U)}{\partial r_\alpha},$$

$$\rho\frac{\partial\delta S}{\partial t} + S\frac{\partial\delta\rho}{\partial t} + \rho S\sum_\beta\frac{\partial v_n^{(\beta)}}{\partial r_\beta} = 0, \tag{4.1}$$

where

$$\sqrt{\rho_0} = \langle\psi\rangle_{0,0} = \langle\psi^\dagger\rangle_{0,0},$$

$$\delta\Lambda = \frac{\partial\Lambda}{\partial\rho}\delta\rho + \frac{\partial\Lambda}{\partial\theta}\delta\theta = -S\delta\theta + \frac{1}{\rho}\delta\mathscr{P}.$$

Before going on to examine the system of equations (4.1), we shall elucidate the connection between the various quantities involved and the

Green's functions. The source term in the original Hamiltonian may be regarded as a variation (perturbation) of the Hamiltonian, i.e. we may write $H = H_0 + \delta H$, where

$$\delta H = \int \{\psi(r,t)\eta^*(r,t) + \psi^\dagger(r,t)\eta(r,t)\}\, d\vec{r}.$$

(Here and subsequently we put $u = 0$). Let us put

$$\eta(r,t) = e^{-i\omega t + \varepsilon t + i\vec{k}\vec{r}}\eta_k + e^{i\omega t + \varepsilon t - i\vec{k}\vec{r}}\eta_{-k},$$

$$\eta^*(r,t) = e^{-i\omega t + \varepsilon t + i\vec{k}\vec{r}}\eta^*_{-k} + e^{i\omega t + \varepsilon t - i\vec{k}\vec{r}}\eta^*_k, \qquad (4.2)$$

where $\varepsilon > 0$, $\varepsilon \to 0$. (This means we switch on δH adiabatically).

Consider the variations

$$\delta\phi = \delta\langle\psi(r,t)\rangle = e^{i\omega t + \varepsilon t}\,\delta\mathcal{U}(r) + e^{i\omega t + \varepsilon t}\,\delta\mathcal{B}(r)$$

$$\delta\phi^* = \delta\langle\psi^\dagger(r,t)\rangle = e^{-i\omega t + \varepsilon t}\,\delta\mathcal{B}^*(r) + e^{i\omega t + \varepsilon t}\,\delta\mathcal{U}^*(r).$$

A general theorem on the variation of averages under the action of a perturbation switched on adiabatically (see e.g. reference [4]) tells us that if we consider only the component of δH proportional to $e^{i\vec{k}\vec{r}}$, then

$$\delta\mathcal{U}(r) = 2\pi \int e^{i\vec{k}\vec{r}\,'}\{\ll \psi(r);\psi(r') \gg_{\omega+i\varepsilon} \eta^*_{-k}+ \ll \psi(r);\psi^\dagger(r') \gg_{\omega+i\varepsilon} \eta_k\}\, d\vec{r}\,',$$

$$\delta\mathcal{B}^*(r) = 2\pi \int e^{i\vec{k}\vec{r}\,'}\{\ll \psi^\dagger(r);\psi(r') \gg_{\omega+i\varepsilon} \eta^*_{-k}+ \ll \psi^\dagger(r);\psi^\dagger(r') \gg_{\omega+i\varepsilon} \eta_k\}\, d\vec{r}\,',$$

$$\ll \psi(r);\psi(r') \gg_E = \frac{1}{2\pi}\int \ll a_q;a_{-q} \gg_E\, e^{i\vec{q}(\vec{r}-\vec{r}\,')}\, d\vec{q},$$

$$\ll \psi^\dagger(r);\psi(r') \gg_E = \frac{1}{2\pi}\int \ll a^\dagger_{-q};a_{-q} \gg_E\, e^{i\vec{q}(\vec{r}-\vec{r}\,')}\, d\vec{q},$$

$$\ll \psi^\dagger(r);\psi^\dagger(r') \gg_E = \frac{1}{2\pi}\int \ll a^\dagger_{-q};a^\dagger_q \gg_E\, e^{i\vec{q}(\vec{r}-\vec{r}\,')}\, d\vec{q},. \qquad (4.2')$$

Here $E = \omega + i\varepsilon$, and $\ll a_q;a_{-q} \gg$, etc., denote the Fourier components of the corresponding retarded Green's functions. Using (4.2) and (4.2') we find

$$\delta\mathcal{U}(r) = 2\pi\{\ll a_k;a_{-k} \gg_E \eta^*_{-k}+ \ll a_k;a^\dagger_k \gg \eta_k\}\, e^{i\vec{k}\vec{r}} = \delta\phi_k\, e^{i\vec{k}\vec{r}},$$

$$\delta \mathscr{B}^*(r) = 2\pi \{\ll a^\dagger_{-k}; a_k \gg_E \eta^*_{-k} + \ll a_k; a^\dagger_k \gg \eta_k\} e^{i\vec{k}\vec{r}} = \delta\phi^*_{-k} e^{i\vec{k}\vec{r}}.$$

In terms of the Fourier components $\delta\phi_k$, $\delta\phi^*_k$ the variations of ϕ, ϕ^* can be written in the form

$$\delta\phi(t, r) = e^{-i\omega t + \varepsilon t + i\vec{k}\vec{r}} \delta\phi_k + +e^{i\omega t + \varepsilon t - i\vec{k}\vec{r}} \delta\phi_{-k},$$

$$\delta\phi^*(t, r) = e^{-i\omega t + \varepsilon t + i\vec{k}\vec{r}} \delta\phi^*_{-k} + +e^{i\omega t + \varepsilon t - i\vec{k}\vec{r}} \delta\phi^*_k.$$

The quantity $\delta\phi(t, r)$ does not appear explicitly in equations (4.1), but it is connected with $\vec{v}_s(t, r)$. Using the definition of $\vec{v}_s$, it is easy to obtain the expression

$$v^{(\alpha)}_s(t, r) = \frac{i}{2m\sqrt{\rho_0}} \left(\frac{\partial \delta\phi^*(t, r)}{\partial r_\alpha} - \frac{\partial \delta\phi(t, r)}{\partial r_\alpha}\right).$$

We introduce the Fourier components of $\vec{v}_s$:

$$v^{(\alpha)}_s(t, r) = e^{-i\omega t + \varepsilon t + i\vec{k}\vec{r}} v^{(\alpha)}_s(k) + e^{i\omega t + \varepsilon t - i\vec{k}\vec{r}} v^{(\alpha)}_s(-k),$$

for which, from the expression for $v^{(\alpha)}_s$ and the definition of $\delta\phi_k$, we have

$$v^{(\alpha)}_s(k) = \frac{\pi}{m\sqrt{\rho_0}} k^{(\alpha)} \{\ll a_k - a^\dagger_{-k}, a_{-k} \gg_E \eta^*_{-k} + \ll a_k - a^\dagger_{-k}, a^\dagger_k \gg_E \eta_k\}.$$

$$(4.3)$$

Moreover, by definition:

$$a^2(t, r) = \phi^*(t, r)\phi(t, r)$$

whence we get:

$$\delta a = \frac{1}{2}[\delta\phi^*(t, r) + \delta\phi(t, r)] = \frac{\partial a}{\partial \rho} \delta\rho + \frac{\partial a}{\partial \theta} \delta\theta.$$

Then going over once again to Fourier components and using the definition of $\delta\phi_k$, we get

$$\delta a(t, r) = e^{-i\omega t + \varepsilon t + i\vec{k}\vec{r}} \delta a_k + e^{i\omega t + \varepsilon t - i\vec{k}\vec{r}} \delta a_{-k},$$

$$\delta a_k = \pi \{\ll a_k + a^\dagger_{-k}, a_{-k} \gg_E \eta^*_{-k} + \ll a_k + a^\dagger_{-k}, a^\dagger_k \gg_E \eta_k\}. \quad (4.4)$$

Formulae (4.3) and (4.4) connect the hydrodynamic quantities obtained from equation (4.1) with the Green's functions. It should be noted that since the hydrodynamic equations are valid only for "slow" changes of the hydrodynamic variables, the connection is asymptotic in nature. being true only for $k \ll 1/l$ and $E \ll 1/T$, where l is the mean free path and T the relaxation time.

In order to find the solutions of the equations appropriate to this limit and so obtain asymptotic expressions for the Green's functions, we need only substitute equations (4.2) for η and η^* in (4.1) (as in normal acoustic theory) and seek a solution proportional to η_k and η^*_{-k}. Rewriting equation (4.1) (with $\delta U = U = 0$) as an equation for the Fourier components, we get:

$$-E\delta\rho(k) + \rho_s \sum_\beta k^{(\beta)} v_s^{(\beta)}(k) + \rho_n \sum_\beta k^{(\beta)} v_n^{(\beta)}(k) = \sqrt{\rho_0}(\eta^*_{-k} - \eta_k), \quad (4.5a)$$

$$mE[\rho_s v_s^{(\beta)}(k) + \rho_n v_n^{(\beta)}(k)] = k^{(\alpha)}\left[\frac{\partial \mathscr{P}}{\partial \rho}\delta\rho(k) + \frac{\partial \mathscr{P}}{\partial \theta}\delta\theta(k)\right], \quad (4.5b)$$

$$mE v_s^{(\alpha)}(k) = k^{(\alpha)}\left[-S\delta\theta + \frac{1}{\rho}\frac{\partial \mathscr{P}}{\partial \rho}\delta\rho(k) + \frac{1}{\rho}\frac{\partial \mathscr{P}}{\partial \theta}\delta\theta(k)\right], \quad (4.5c)$$

$$E[\rho\delta S(k) + S\delta\rho(k)] = \rho S \sum_\beta k^{(\beta)} v_n^{(\beta)}(k), \quad (4.5d)$$

$$\delta S(k) = \frac{\partial S}{\partial \theta}\delta\theta + \frac{\partial S}{\partial \rho}\delta\rho(k). \quad (4.5e)$$

Let us start examining some limiting cases. Consider first the case $E = 0$. Then it follows from (4.5b) and (4.5c) that $\delta\rho(k) = \delta\theta(k) = 0$, and hence from (4.5d) that

$$\sum_\beta k^{(\beta)} v_n^{(\beta)}(k) = 0.$$

Then we get from (4.5a)

$$\rho_s \sum_\beta k^{(\beta)} v_s^{(\beta)}(k) = \sqrt{\rho_0}(\eta^*_{-k} - \eta_k),$$

or using (4.3)

$$\frac{\pi \rho_s k^2}{m\rho_0}\{\ll a_k - a^\dagger_{-k}, a_{-k} \gg_{E=0} \eta^*_{-k} + \ll a_k - a^\dagger_{-k}, a^\dagger_k \gg_{E=0} \eta_k\} = \eta^*_{-k} - \eta_k.$$

Equating the coefficients of η^*_{-k} and η_k on two sides of this equation, we get

$$\ll a_k - a^\dagger_{-k}, a_{-k} \gg_{E=0} = \frac{m\rho_0}{\pi\rho_s} \frac{1}{k^2},$$

$$\ll a_k - a^\dagger_{-k}, a^\dagger_k \gg_{E=0} = -\frac{m\rho_0}{\pi\rho_s} \frac{1}{k^2}. \tag{4.6}$$

Also we have

$$\delta a = \frac{\partial a}{\partial \rho} \delta\rho + \frac{\partial a}{\partial \theta} \delta\theta,$$

or, according to (4.4),

$$\ll a_k + a^\dagger_{-k}, a_{-k} \gg_{E=0} \eta^*_{-k} + \ll a_k + a^\dagger_{-k}, a^\dagger_k \gg_{E=0} \eta_k = 0.$$

Hence,

$$\ll a_k; a_{-k} \gg_{E=0} = - \ll a^\dagger_{-k}; a_{-k} \gg_{E=0},$$

$$\ll a_k; a^\dagger_k \gg_{E=0} = - \ll a^\dagger_{-k}; a^\dagger_k \gg_{E=0} . \tag{4.7}$$

substituting these relations in (4.6), we find

$$\ll a_k; a_{-k} \gg_{E=0} - \ll a^\dagger_{-k}; a_k \gg_{E=0} = 2 \ll a_k; a_{-k} \gg_{E=0} = \frac{m\rho_0}{\pi\rho_s} \frac{1}{k^2}$$

$$\ll a_k; a^\dagger_k \gg_{E=0} - \ll a^\dagger_{-k}; a^\dagger_k \gg_{E=0} = 2 \ll a_k; a^\dagger_k \gg_{E=0} = -\frac{m\rho_0}{\pi\rho_s} \frac{1}{k^2}.$$

Thus we have obtained the "$1/k^2$ theorem" [5] for the Green's function, with the coefficient explicit. It is interesting to note that the coefficient contains not only ρ_s but also the actual density of the condensate particles ρ_0.

Next, consider the special case $\theta = 0$. Strictly speaking, the hydrodynamic equations can have only a formal meaning in this case, since the relaxation time as $\theta \to 0$ becomes very long. However, we shall consider this case formally in order to see the results given by our formulae in this limit. All expressions are now considerably simplified, since

$$\rho_s = \rho, \quad \rho_n = 0, \quad S = 0, \quad \frac{\partial S}{\partial \rho} = 0, \quad \frac{\partial S}{\partial \theta} = 0, \quad \frac{\partial \mathscr{P}}{\partial \theta} = 0,$$

$$\delta\Lambda = \frac{1}{\rho}\delta\mathscr{P} = \frac{1}{\rho}\left(\frac{\partial\mathscr{P}}{\partial\rho}\right)_{\theta=0}\delta\rho.$$

Equations (4.5) give:

$$- E\delta\rho(k) + \rho_s \sum_\beta k^{(\beta)} v_s^{(\beta)}(k) = \sqrt{\rho_0}(\eta_{-k}^* - \eta_k),$$

$$Em\rho v_s^{(\alpha)}(k) = k^{(\alpha)} \left(\frac{\partial \mathscr{P}}{\partial \rho}\right)_{\theta=0} \delta\rho(k).$$

The second equation can be written:

$$\delta\rho(k) = \frac{E\rho}{k^2 c^2} \sum_\beta k^{(\beta)} v_s^{(\beta)}(k), \quad c^2 = \frac{1}{m}\left(\frac{\partial \mathscr{P}}{\partial \rho}\right).$$

Substituting this expression into the first equation, we get

$$\sum_\beta k^{(\beta)} v_s^{(\beta)}(k) = \frac{k^2 \sqrt{\rho_0} c^2}{\rho(E^2 - c^2 k^2)}(\eta_k - \eta_{-k}^*)$$

$$= \frac{\pi k^2}{m\sqrt{\rho_0}}\{\ll a_k - a_{-k}^\dagger, a_{-k} \gg_{E=0} \eta_{-k}^* + \ll a_k - a_{-k}^\dagger, a_k^\dagger \gg_{E=0} \eta_k\}.$$

Since $\delta\rho(k)$ contains not only first-order infinitesimals but also an extra small factor E, and we are interested in the asymptotic form of the Green's functions for $E, k \to 0$, equation (4.7) is still valid to lowest order, and so, equating the coefficients of η_k and η_{-k}^*, we get:

$$\ll a_k; a_{-k} \gg \approx -\frac{\rho_0 c^2 m}{2\pi\rho(E^2 - c^2 k^2)},$$

$$\ll a_k; a_k^\dagger \gg \approx \frac{\rho_0 c^2 m}{2\pi\rho(E^2 - c^2 k^2)}.$$

Finally, consider the general case $(\theta \neq 0)$. From (4.5b) and (4.5c) we have

$$mE[\rho_s v_s^{(\alpha)}(k) + \rho_n v_n^{(\alpha)}(k)] = k^{(\alpha)} \delta\mathscr{P},$$

$$mE(\rho_s + \rho_n) v_s^{(\alpha)}(k) = -k^{(\alpha)} \rho S \delta\theta(k) + k^{(\alpha)} \delta\mathscr{P}.$$

Subtraction of the second equation from the first gives:

$$mE\rho_n[v_n^{(\alpha)}(k) - v_s^{(\alpha)}(k)] = \rho S k^{(\alpha)} \delta\theta(k). \tag{4.8}$$

Equation (4.5a) can be rewritten as

$$-E\delta\rho(k) + \rho \sum_{\beta} k^{(\beta)} v_s^{(\beta)}(k) + \rho_n \sum_{\beta} k^{(\beta)} [v_n^{(\beta)}(k) - v_s^{(\beta)}(k)] = \sqrt{\rho_0}(\eta_{-k}^* - \eta_k).$$

(4.9)

Using this equation to eliminate $\delta\rho(k)$ from (4.5d), we get:

$$E\rho\delta S(k) + S\rho_s \sum_{\beta} k^{(\beta)} v_s^{(\beta)}(k) + S\rho_n \sum_{\beta} k^{(\beta)} v_n^{(\beta)}(k)$$
$$- S\sqrt{\rho_0}(\eta_{-k}^* - \eta_k) = S(\rho_s + \rho_n) \sum_{\beta} k^{(\beta)} v_s^{(\beta)}(k).$$

Then, expressing $\delta S(k)$ in terms of $\delta\theta(k)$ and $\delta\rho(k)$ [by equation (4.5e)] and again eliminating $\delta\rho(k)$, we find

$$E\rho\left(\frac{\partial S}{\partial \theta}\right)_{\rho} \delta\theta(k) = \sqrt{\rho_0}(\eta_{-k}^* - \eta_k)\left(S + \rho\left(\frac{\partial S}{\partial \rho}\right)_{\theta}\right)$$
$$- \rho^2\left(\frac{\partial S}{\partial \rho}\right)_{\theta} \sum_{\beta} k^{(\beta)} v_s^{(\beta)}(k) + \left(S\rho_s - \rho_n\rho\left(\frac{\partial S}{\partial \rho}\right)_{\theta}\right) \sum_{\beta} k^{(\beta)} [v_n^{(\beta)}(k) - v_s^{(\beta)}(k)],$$

or, using (4.8)

$$E\delta\theta(k) = - \frac{\rho\left(\frac{\partial S}{\partial \rho}\right)_{\theta}}{\left(\frac{\partial S}{\partial \theta}\right)_{\rho}} \sum_{\beta} k^{(\beta)} v_s^{(\beta)} + \frac{S\rho_s - \rho_n\rho\left(\frac{\partial S}{\partial \rho}\right)_{\theta}}{\rho\left(\frac{\partial S}{\partial \theta}\right)_{\rho}} \frac{\rho Sk^2}{mE\rho_n} \delta\theta(k)$$
$$+ \sqrt{\rho_0} \frac{S + \rho\left(\frac{\partial S}{\partial \rho}\right)_{\theta}}{\rho\left(\frac{\partial S}{\partial \theta}\right)_{\rho}} (\eta_{-k}^* - \eta_k).$$

Therefore, we finally get

$$\delta\theta(k) = - \frac{\rho\left(\frac{\partial S}{\partial \rho}\right)_{\theta} mE}{\left(\frac{\partial S}{\partial \theta}\right)_{\rho} \left[mE^2 - \frac{S\frac{\rho_s}{\rho_n} - \rho\left(\frac{\partial S}{\partial \rho}\right)_{\theta}}{\left(\frac{\partial S}{\partial \theta}\right)_{\rho}}\right]} \sum_{\beta} k^{(\beta)} v_s^{(\beta)}$$

$$+ \frac{\sqrt{\rho_0}\left(S + \rho\left(\frac{\partial S}{\partial \rho}\right)_\theta\right)mE}{\left(\frac{\partial S}{\partial \theta}\right)_\rho\left[mE^2 - \frac{S\frac{\rho_s}{\rho_n} - \rho\left(\frac{\partial S}{\partial \rho}\right)_\theta}{\left(\frac{\partial S}{\partial \theta}\right)_\rho}\right]}\,(\eta^*_{-k} - \eta_k). \qquad (4.10)$$

We can likewise express $\delta\rho(k)$ in terms of $v_s(k)$. From equation (4.9) we have

$$\delta\rho(k) = \frac{\rho}{E}\sum_\beta k^{(\beta)} v_s^{(\beta)}(k) + \frac{1}{mE^2}\rho Sk^2 \delta\theta(k) - \frac{\sqrt{\rho_0}}{E}\,(\eta^*_{-k} - \eta_k).$$

Substituting (4.10) gives:

$$\delta\rho(k) = \frac{\rho}{E}\sum_\beta k^{(\beta)} v_s^{(\beta)}(k) - \frac{\rho^2 k^2 S\left(\frac{\partial S}{\partial \rho}\right)_\theta mE\sum_\beta k^{(\beta)} v_s^{(\beta)}}{mE^2\left(\frac{\partial S}{\partial \theta}\right)_\rho\left[mE^2 - Sk^2\frac{S\frac{\rho_s}{\rho_n} - \rho\left(\frac{\partial S}{\partial \rho}\right)_\theta}{\left(\frac{\partial S}{\partial \theta}\right)_\rho}\right]}$$

$$+ \frac{\sqrt{\rho_0}}{E}\left\{\frac{Sk^2\left(S + \rho\left(\frac{\partial S}{\partial \rho}\right)_\theta\right)mE}{\left(\frac{\partial S}{\partial \theta}\right)_\rho\left[mE^2 - Sk^2\frac{S\frac{\rho_s}{\rho_n} - \rho\left(\frac{\partial S}{\partial \rho}\right)_\theta}{\left(\frac{\partial S}{\partial \theta}\right)_\rho}\right]} - 1\right\}(\eta^*_{-k} - \eta_k).$$

$$(4.11)$$

Now, substituting the expression for $\delta\rho(k)$ and $\delta\theta(k)$ in the equation for $v^{(\alpha)}(k)$ [equation (4.5c)], multiplying by $k^{(\alpha)}$ and summing over α, we get:

$$mE\sum_\alpha k^{(\alpha)} v_s^{(\alpha)}(k) = -Sk^2 \delta\theta(k) + \frac{1}{\rho}\left(\frac{\partial \mathscr{P}}{\partial \rho}\right)_\theta k^2 \delta\rho(k) + \frac{1}{\rho}\left(\frac{\partial \mathscr{P}}{\partial \theta}\right)_\rho k^2 \delta\theta(k)$$

$$= \left(\frac{1}{\rho}\left(\frac{\partial \mathscr{P}}{\partial \theta}\right)_\rho - S\right)\frac{k^2\sqrt{\rho_0}\left(S + \rho\left(\frac{\partial S}{\partial \rho}\right)_\theta\right)mE}{\rho\left[mE^2\left(\frac{\partial S}{\partial \theta}\right)_\rho - Sk^2\left(S\frac{\rho_s}{\rho_n} - \rho\left(\frac{\partial S}{\partial \rho}\right)_\theta\right)\right]}(\eta^*_{-k} - \eta_k)$$

$$+ \frac{1}{\rho}\left(\frac{\partial \mathscr{P}}{\partial \rho}\right)_\theta k^2 \left\{ \frac{\rho}{E} - \frac{\rho^2 S k^2 \left(\frac{\partial S}{\partial \rho}\right)_\theta}{E\left[mE^2\left(\frac{\partial S}{\partial \theta}\right)_\rho - Sk^2\left(S\frac{\rho_s}{\rho_n} - \rho\left(\frac{\partial S}{\partial \rho}\right)_\theta\right)\right]} \right\}$$

$$\times \sum_\alpha k^{(\alpha)} v_s^{(\alpha)}(k) + \frac{1}{\rho}\left(\frac{\partial \mathscr{P}}{\partial \rho}\right)_\theta k^2 \frac{\sqrt{\rho_0}}{E}$$

$$- \left\{ \frac{Sk^2\left(S + \rho\left(\frac{\partial S}{\partial \rho}\right)_\theta\right)}{mE^2\left(\frac{\partial S}{\partial \theta}\right)_\rho - Sk^2\left(S\frac{\rho_s}{\rho_n} - \rho\left(\frac{\partial S}{\partial \rho}\right)_\theta\right)} - 1 \right\} (\eta_{-k}^* - \eta_k)$$

$$- \left(\frac{1}{\rho}\left(\frac{\partial \mathscr{P}}{\partial \theta}\right)_\rho - S\right) k^2 \frac{\rho\left(\frac{\partial S}{\partial \rho}\right)_\theta mE}{mE^2\left(\frac{\partial S}{\partial \theta}\right)_\rho - Sk^2\left(S\frac{\rho_s}{\rho_n} - \rho\left(\frac{\partial S}{\partial \rho}\right)_\theta\right)}$$

$$\times \sum_\alpha k^{(\alpha)} v_s^{(\alpha)}(k).$$

Solving this equation for $\sum_\alpha k^{(\alpha)} v_s^{(\alpha)}(k)$, we get:

$$\sum_\alpha k^{(\alpha)} v_s^{(\alpha)}(k) = \frac{k^2 \Delta(k,E)\sqrt{\rho_0}}{m\Omega(k,E)} (\eta_{-k}^* - \eta_k)$$

$$= \frac{\pi k^2}{m\sqrt{\rho_0}}\{\ll a_k - a_{-k}^\dagger, a_{-k} \gg_{E=0} \eta_{-k}^* + \ll a_k - a_{-k}^\dagger, a_k^\dagger \gg_{E=0} \eta_k\} \quad (4.12)$$

where

$$\Omega(k,E) = E^4 - \frac{E^2 k^2}{m}\left[S^2 \frac{\rho_s}{\rho_n\left(\frac{\partial S}{\partial \theta}\right)_\rho} - \left(\frac{\partial \mathscr{P}}{\partial \theta}\right)_\rho \left(\frac{\partial S}{\partial \rho}\right)_\theta \frac{1}{\left(\frac{\partial S}{\partial \theta}\right)_\rho} \right.$$

$$\left. + \left(\frac{\partial \mathscr{P}}{\partial \theta}\right)_\rho \right] + \frac{k^4}{m^2}\left(\frac{\partial \mathscr{P}}{\partial \rho}\right)_\theta S^2 \frac{\rho_s}{\rho_n\left(\frac{\partial S}{\partial \theta}\right)_\rho} = E^4 - E^2 k^2$$

$$\times \left[\frac{1}{m}\left(\frac{\partial \mathscr{P}}{\partial \rho}\right)_\theta + \frac{S^2 \rho_s \theta}{\rho_n m c_v}\right] + \frac{k^4}{m}\left(\frac{\partial \mathscr{P}}{\partial \rho}\right)_\theta \frac{S^2 \rho_s \theta}{\rho_n m c_v} = (E^2 - c_1^2 k^2)(E^2 - c_0^2 k^2).$$

$$(4.13)$$

Here $c_v = \theta \left(\dfrac{\partial S}{\partial \theta} \right)_\rho$, and the quantities c_0, c_1 are given by

$$c_{0,1}^2 = \frac{1}{2m} \left(\frac{\partial \mathscr{P}}{\partial \rho} \right)_\theta + \frac{1}{2} \frac{S^2 \rho_s \theta}{\rho_n m c_v}$$
$$\pm \sqrt{\frac{1}{4} \left[\frac{1}{m} \left(\frac{\partial \mathscr{P}}{\partial \rho} \right)_\theta + \frac{S^2 \rho_s \theta}{\rho_n m c_v} \right]^2 - \frac{1}{m} \left(\frac{\partial \mathscr{P}}{\partial \rho} \right)_\theta \frac{S^2 \rho_s \theta}{\rho_n m c_v}}, \qquad (4.14)$$

(the upper sign referring to c_0). The velocity c_0 tends to the normal speed of sound both for $\rho_s \to 0$ and for $\theta \to 0$. The velocity c_1 is the speed of the "second-sound" vibration peculiar to a superfluid liquid and tends to zero for $\rho_s \to 0$. Finally, in equation (4.12) the quantity $\Delta(E, k)$ has the form

$$\Delta(E, k) = \frac{1}{m} \left(\frac{\partial \mathscr{P}}{\partial \rho} \right)_\theta k^2 \frac{S^2 \theta}{\rho_n c_v} - E^2 \left[\frac{1}{\rho} \left(\frac{\partial \mathscr{P}}{\partial \rho} \right)_\theta - \left(\frac{\partial}{\partial \rho} \rho S \right)_\theta \frac{\theta}{\rho c_v} \right].$$

It is easy to show, as we did above for the case $\theta = 0$, that $\delta\rho(k)$ and $\delta\theta(k)$ contain extra small factor of order k compared to $v_s(k)$. Hence, once again, equation (4.7) is satisfied to lowest order. Equating the coefficients of η_k and η^*_{-k} on the right and left sides of (4.12), and using (4.7), we therefore finally obtain the expressions:

$$\ll a_k; a_{-k} \gg \approx \frac{\Delta(E, k)\rho_0}{2\pi\Omega(E, k)}, \qquad \ll a_k; a_k^\dagger \gg \approx -\frac{\Delta(E, k)\rho_0}{2\pi\Omega(E, k)}. \qquad (4.15)$$

Obviously in the limits $E \to 0$ or $\theta \to 0$ this reduces to the results obtained above.

From (4.15) and (4.13) we see that the Green's functions have poles corresponding to two types of elementary excitation:

$$E = c_0 k, \qquad E = c_1 k.$$

In equations (4.15) the effects of damping are not taken into account; this is due to the fact that we have been considering only the ideal-liquid approximation. It would be interesting to improve the asymptotic accuracy of (4.15) by making instead a "viscous-liquid approximation"; to do this we should have to take into account in the equations of section 3 the terms "of order μ" which we actually dropped. This problem is considerably simplified by the fact that to construct the Green's functions we do not need the complete hydrodynamic equations but only the linearized acoustic equations. The improved expressions for the Green's function obtained in this way would contain damping terms involving the viscosity, thermal conductivity and other kinetic coefficients.

In conclusion I should like to thank S. V. Iordanskii for considerable assistance in preparing these lectures for publication.

References

1. K. P. Gurov, "Quantum hydrodynamics," *Zh. Exper. i Theor. Fiz.*, **18**, 110 (1948); *Zh. Exper. i Theor. Fiz.*, **20**, 279 (1950).

2. S. V. Iordanskii, "Hydrodynamics of a rotating Bose system below the condensation point," *Docl. Akad. Naul SSSR* **153**, 1, 74-77 (1963); *Sov. Phys. "Doklady", English Transl.*

3. L. D. Landau, *Zh. Exper. i Theor. Fiz.*, **11**, 592, (1941); *Zh. Exper. i Theor. Fiz.*, **14**, 112, (1944); *English Transl.* see Collected Papers of L. D. Landau, ed. D. ter Haar, Gordon and Breach and Pergamon Press (1965).

4. D. N. Zubarev, "Double-time Green functions in statistical physics," *Uspekhi Fiz. Nauk*, **21**, 1, 71 (1960); *Sov. Phys.-Usp. English Transl.*, **3**, 320 (1960).

5. N. N. Bogoliubov, "Quasi-averages in statistical mechanics," JINR preprint, Dubna (1961).

CHAPTER 5

ON THE MODEL HAMILTONIAN OF SUPERCONDUCTIVITY

1. Statement of the Problem

The simplest model system considered in superconductivity theory is one described by a hamiltonian which retains only interactions between particles of opposite spin and momentum

$$H = \sum_f T(f) a_f^\dagger a_f - \frac{1}{2V} \sum_{f \cdot f'} \lambda(f) a_f^\dagger a_{-f}^\dagger a_{-f'} a_{f'} \qquad (1.1)$$

where $f = (p, s)$, $s \pm 1$ and p is the momentum vector. For a given volume $V = L^3$, allowed values of p are given by:

$$p_x = \frac{2\pi}{L} n_x, \quad p_y = \frac{2\pi}{L} n_y, \quad p_z = \frac{2\pi}{L} n_z$$

where n_x, n_y, n_z are integers. In (1.1) we also use the following notation:

$$-f = (-p, -s).$$

$T(f) = \frac{p^2}{2m} - \mu$, where μ is the chemical potential ($\mu > 0$)

$$\lambda(f) = \begin{cases} J, & \text{for } \left| \dfrac{p^2}{2m} - \mu \right| \le \Delta, \\[2mm] 0, & \text{for } \left| \dfrac{p^2}{2m} - \mu \right| > \Delta. \end{cases}$$

For such a system, the BCS method [1] and the method of compensation of dangerous diagrams lead to identical results. Moreover, as was shown in reference [2], the Hamiltonian (1.1) is of great methodological interest in its own right, since it provides on of the very few completely soluble problems of statistical mechanics.

In reference [2] we showed that for this purpose we can obtain an expression for the free energy which is exact in the limit $V \to \infty$. The proof given there was along the following lines: The hamiltonian (1.1) was divided in a particular way into two parts H_0 and H_1. The problem described by the Hamiltonian H_0 was solved exactly, and the effect of H_1 was calculated by perturbation theory. It was proved that any given term of the corresponding series expansion is asymptotically small in the limit $V \to \infty$., which led to the conclusion that it is always legitimate to neglect the effect of H_1 on passing to the limit of infinite volume. Needless to say, this kind of approach cannot pretend to mathematical rigour: however, it is worth pointing out that the problems of statistical physics are often handled by far cruder methods. For instance, a very commonly used device consists in the selective summation of various so-called "principal terms" of the perturbation series to the neglect of all the other terms, though the latter do not even tend to zero for $V \to \infty$.

Some doubt was cast on the results of reference [2] when various attempts to use the normal Feynman diagram technique (without taking into account the "anomalous contractions" $\overline{a_f a_{-f}}$, $\overline{a_f^\dagger a_{-f}^\dagger}$ generated by the canonical $u - v$ transformation (see below)) failed to give the results anticipated. Furthermore, in reference [3] the summation of a certain class of Feynman diagrams led to a solution which is fundamentally different from the one obtained in references [1] and [2], and it was concluded that the latter were invalid.

In view of this situation, we undertook an investigation [4] of the hierarchy of coupled equations for the Green's functions which did not involve recourse to perturbation theory. In this work it was shown that the Green's functions for the Hamiltonian H_0 satisfies the whole chain of equations for the exact Hamiltonian $H = H_0 + H_1$ to order $1/V$. This tends to confirm the results of reference [3] and reveal the "ineffectiveness" of the additional term H_1.

However, it is also possible to treat the problem form a purely mathematical standpoint. Once we have specified the Hamiltonian, say in the form (1.1), we have a perfectly well-defined mathematical problem, which we also may solve rigorously, without any "physical assumptions" whatsoever. We need not to content ourselves with the knowledge that the approximate expressions satisfy the exact equations to order $1/V$; on the contrary, we can actually evaluate the difference between the exact and approximate expressions.

With a view of the complete elucidation of the behavior of a system with the Hamiltonian (1.1), we shall adopt in this work just such a purely mathematical standpoint. We shall investigate the hamiltonian (1.1) at zero temperature and prove rigorously that the relative difference $(E - E_0)/E_0$ between the groundstate energies of H and H_0, and also the difference between the corresponding Green's functions, tends to zero for $V \to \infty$; we shall set a bound on the rate of decrease in each case.

For the methodology reasons we shall find in convenient to consider a rather more general Hamiltonian, which contains terms representing sources of creation and destruction of pairs;

$$\mathscr{H} = \sum_f T(f)a_f^\dagger a_f - \nu \sum_f \frac{\lambda(f)}{2}(a_{-f}a_f + a_f^\dagger a_{-f}^\dagger)$$
$$- \frac{1}{2V} \sum_{f,f'} \lambda(f)\lambda(f')a_f^\dagger a_{-f}^\dagger a_{-f'}a_{f'}, \tag{1.2}$$

where ν is a parameter on whose magnitude we place no restrictions; in particular, ν may be equal to zero. Notice that the case $\nu < 0$ need not be considered, since it can be reduced to the case $\nu > 0$ by the trivial gauge transformation

$$a_f \to ia_f, \quad a_f^\dagger \to -ia_f^\dagger.$$

We emphasize that the only motive for considering the case $\nu > 0$ is the light it sheds on the situation in the physical case $\nu = 0$.

For this investigation we shall not need all the specific properties of the functions $\lambda(f)$, $T(f)$ mentioned above; it will be quite sufficient for our purpose if they fulfil the following more general conditions:

1. The functions $\lambda(f)$ and $T(f)$ are real, piecewise continuous and obey the symmetry conditions

$$\lambda(-f) = -\lambda(f); \quad T(-f) = T(f)$$

2. $\lambda(f)$ is uniformly bounded in all space, while

$$T(f) \to \infty \quad \text{for} \quad |f| \to \infty$$

3. $\qquad\qquad \frac{1}{V}\sum_f |\lambda(f)| \leq \text{const} \quad \text{for} \quad V \to \infty$

4.
$$\lim_{V \to \infty} \frac{1}{2V} \sum_f \frac{\lambda^2(f)}{\sqrt{\lambda^2(f)x + T^2(f)}} > 1$$

for sufficiently small positive x.

We write the Hamiltonian (1.2) in the form

$$\mathscr{H} = \mathscr{H}_0 + \mathscr{H}_1, \tag{1.3}$$

where

$$\mathscr{H}_0 = \sum_f T(f)a^\dagger_f a_f - \frac{1}{2} \sum_f \lambda(f)\{(\nu + \sigma^*)a_{-f}a_f(\nu + \sigma)a^\dagger_f a^\dagger_{-f}\}$$
$$+ \frac{|\sigma|^2 V}{2} \tag{1.4}$$

$$\mathscr{H}_1 = -\frac{1}{2V}\left(\sum_f \lambda(f)a^\dagger_f a^\dagger_{-f} - V\sigma^*\right)\left(\sum_f \lambda(f)a_{-f}a_f - V\sigma\right) \tag{1.5}$$

where σ is a complex number.

Notice that if we fix σ by minimizing the ground-state energy of H_0, and then neglect H_1, we arrive at the well-known approximate solution considered in the work cited above [1,2,4]. Our present problem is to obtain bounds for the difference of the corresponding Green's functions. We shall prove that these differences vanish when we take the limit[x] $V \to \infty$.

2.   General Properties of the Hamiltonian

1.   In this section we shall establish certain general properties of the Hamiltonian H, (1.2). First, we consider the occupation number operator $n_f = a^\dagger_f a_f$. We shall prove that the quantities $n_f - n_{-f}$ are constant of motion.

---

[x]In recent years papers [7–12] have been published in which new methods were developed to find asymptotically exact expressions for many-time correlation functions and Green's functions at arbitrary temperature θ. In addition, bounds were obtained for free energy of system of BCS type. These bounds are exact in the limit $V \to \infty$. On the basis of analyzing and generalizing the results of [7–12], it became possible to formulate a new principle, the minimax principle [12], for a broad range of problems in statistical physics. (Remark added by the author in 1971.)

We have

$$a_{-f}a_f(n_f - n_{-f}) - (n_f - n_{-f})a_{-f}a_f = 0,$$

and also

$$a^\dagger_{-f}a^\dagger_f(n_f - n_{-f}) - (n_f - n_{-f})a^\dagger_{-f}a^\dagger_f = 0,$$

whence

$$\mathscr{H}(n_f - n_{-f}) - (n_f - n_{-f})\mathscr{H} = 0.$$

It follows that

$$\frac{d}{dt}(n_f(t) - n_{-f}(t)) = 0. \tag{2.1}$$

2. We next show that the wave function ϕ_H corresponding to the minimum eigenvalue of the Hamiltonian H satisfy the equation

$$(n_f - n_{-f})\phi_H = 0 \tag{2.2}$$

for arbitrary f.

To prove this, we assume the contrary. Since the operator $n_f - n_{-f}$ commute with H and with one another, we can always choose ϕ_H to be an eigenfunction of all this operators:

$$n_f - n_{-f} = \begin{cases} 1 \\ 0 \\ -1. \end{cases}$$

We denote by K_0, K_+, K_- respectively the classes of all those indices f for which

$$\begin{aligned} (n_f - n_{-f})\phi_H = 0 \qquad & f \in K_0 \\ (n_f - n_{-f} - 1)\phi_H = 0 \qquad & f \in K_+ \\ (n_f - n_{-f} + 1)\phi_H = 0 \qquad & f \in K_-. \end{aligned}$$

Our hypothesis then reduces to the statement that the classes K_+ and/or K_1 are not empty and that[y]

$$\langle \phi^*_H \mathscr{H} \phi_H \rangle \leq \langle \phi^* \mathscr{H} \phi \rangle \leq$$

---

[y]We use the notation $\langle \phi * \psi \rangle$ to denote the scalar product of the functions ϕ and ψ.

for arbitrary ϕ. We shall consider in particular functions ϕ satisfying the auxiliary conditions

$$(n_f - n_{-f})\phi = 0. \tag{2.3}$$

Now we notice that if $f \in K_+$, then

$$n_f = 1; \quad n_{-f} = 0,$$

while if $f \in K_-$, then

$$n_f = 0; \quad n_{-f} = 1.$$

Thus, we can write ϕ_H as a direct product

$$\phi_H = \phi_{K_0}\phi_{K_+}\phi_{K_-},$$

where

$$\phi_{K_+} = \prod_{f \in K_+} \delta(n_f - 1)\delta(n_{-f}); \quad \phi_{K_-} = \prod_{f \in K_-} \delta(n_f)\delta(n_{-f} - 1),$$

while ϕ_{K_0} is a function only of those n_f for which $f \in K_0$:

$$\phi_{K_0} = F(\ldots n_f \ldots); \quad \langle \phi_{K_0}^\dagger \phi_{K_0} \rangle = 1 \quad f \in K_0.$$

Further, we notice that

$$a_{-f}a_f\delta(n_f - 1)\delta(n_{-f}) = 0; \quad a_{-f}a_f\delta(n_f)\delta(n_{-f} - 1) = 0;$$
$$a_f^\dagger a_{-f}^\dagger\delta(n_f - 1)\delta(n_{-f}) = 0; \quad a_f^\dagger a_{-f}^\dagger\delta(n_f)\delta(n_{-f} - 1) = 0;$$

and hence

$$a_{-f}a_f\phi_{K_+}\phi_{K_-} = 0, \quad a_f^\dagger a_{-f}^\dagger\phi_{K_+}\phi_{K_-} = 0$$

if $f \in K_+$ or $f \in K_-$.

Accordingly

$$H\phi_H = \Big\{ \sum_{f \in K_+} T(f) + \sum_{f \in K_-} T(f) + \sum_{f \in K_0} T(f)n_f$$
$$- \frac{\nu}{2}\sum_{f \in K_0} \frac{\lambda(f)}{2}(a_{-f}a_f + a_f^\dagger a_{-f}^\dagger)$$
$$- \frac{1}{2V}\sum_{f \in K_0}\sum_{f' \in K_0} \lambda(f)\lambda(f')a_f^\dagger a_{-f}^\dagger a_{-f'}a_{f'} \Big\}\phi_H,$$

and so it follows that

$$\langle \phi_H^* H \varphi_H \rangle = \sum_{f \in K_+} T(f) + \sum_{f \in K_-} T(f)$$

$$+ \left\langle \phi_{K_0}^* \left\{ \sum_{f \in K_0} T(f) n_f - \frac{\nu}{2} \sum_{f \in K_0} \frac{\lambda(f)}{2} (a_{-f} a_f + a_f^\dagger a_{-f}^\dagger) \right. \right.$$

$$\left. \left. - \frac{1}{2V} \sum_{f \in K_0} \sum_{f' \in K_0} \lambda(f) \lambda(f') a_f^\dagger a_{-f}^\dagger a_{-f'} a_{f'} \right\} \phi_{K_0} \right\rangle.$$

Let us now divide the set $K_+ + K_-$ into two sets Q_+ and Q_-:

$$K_+ + K_- = Q_+ + Q_-$$

in such a way that Q_+ includes all indices f from the set $K_+ + K_-$ such that $T(f) \geq 0$, while Q_- includes all those for which $T(f) < 0$. Because of the symmetry property $T(f) = T(-f)$ the indices f and $-f$ always fall into Q_+ and Q_- as a pair. We can write

$$\langle \phi_H^* H \varphi_H \rangle = \sum_{f \in Q_+} |T(f)| + \sum_{f \in Q_-} |T(f)|$$

$$+ \left\langle \phi_{K_0}^* \left\{ \sum_{f \in K_0} T(f) n_f - \frac{\nu}{2} \sum_{f \in K_0} \frac{\lambda(f)}{2} (a_{-f} a_f + a_f^\dagger a_{-f}^\dagger) \right. \right.$$

$$\left. \left. - \frac{1}{2V} \sum_{f \in K_0} \sum_{f' \in K_0} \lambda(f) \lambda(f') a_f^\dagger a_{-f}^\dagger a_{-f'} a_{f'} \right\} \phi_{K_0} \right\rangle.$$

Let us now construct the function ϕ as a direct product by putting

$$\phi = \phi_{K_0} \phi_{Q_+} \phi_{Q_-},$$

where

$$\phi_{Q_+} = \prod_{f \in Q_+} \delta(n_f) \delta(n_{-f}); \quad \phi_{Q_-} = \prod_{f \in Q_-} \delta(n_f - 1) \delta(n_{-f} - 1).$$

(Here we the importance of the fact that f and $-f$ always belong to Q_+ or Q_- as a pair). For such function we have

$$\langle \phi^* H \varphi \rangle = -2 \sum_{f \in Q_-} |T(f)|$$

$$+ \left\langle \phi_{K_0}^* \left\{ \sum_{f \in K_0} T(f) n_f - \frac{\nu}{2} \sum_{f \in K_0} \frac{\lambda(f)}{2} (a_{-f} a_f + a_f^\dagger a_{-f}^\dagger) \right. \right.$$

$$\left. \left. - \frac{1}{2V} \sum_{f \in K_0} \sum_{f' \in K_0} \lambda(f) \lambda(f') a_f^\dagger a_{-f}^\dagger a_{-f'} a_{f'} \right\} \cdot \phi_{K_0} \right\rangle$$

$$- \frac{1}{2V} \sum_{f \in Q_-} \lambda^2(f).$$

It is obvious that

$$\langle \phi_H^* H \varphi_H \rangle > \langle \phi^* H \varphi \rangle.$$

On the other hand, ψ by construction satisfies all the auxiliary conditions (2.3), so that we arrive at a contradiction of our hypothesis. Thus, the statement (2.2) is proved.

A particular consequence of (2.2) is the fact that the total momentum of the state ψ_H is equal to zero:

$$\sum_{\vec{f}} \vec{f} n_{\vec{f}} \psi_H = \frac{1}{2} \sum_{\vec{f}} \vec{f} (n_f - n_{-f}) \phi_H = 0. \tag{2.4}$$

It is obvious from the above discussion that when attempting to find the eigenfunction ϕ_H for the minimum eigenvalue of H we may always restrict ourselves to the class of functions ϕ which satisfy the auxiliary conditions (2.3). For this special class of functions the hamiltonian H may be expressed in terms of Pauli operators. Consider the operators

$$b_f = a_{-f} a_f, \quad b_f^\dagger = a_f^\dagger a_{-f}^\dagger.$$

Independently of the auxiliary condition we have

$$b_f b_{f'} = b_{f'} b_f; \quad b_f^\dagger b\dagger_{f'} = b\dagger_{f'} b\dagger_f; \quad b_f^2 = 0; \quad (b_f^\dagger)^2 = 0;$$
$$b_f b_{-f}^\dagger - b_{-f}^\dagger b_f = 0; \quad f \neq f.$$

Moreover, the auxiliary conditions imply that

$$b_f^\dagger b_f + b_f b_f^\dagger = n_f n_{-f} + (1 - n_f)(1 - n_{-f}) = 1$$

since n_f and n_{-f} are either both equal to zero or both equal to one. It follows that, within the class of functions satisfying (2.3), the operators b_f, $b_f^\dagger$

constitute Pauli operators. When acting on this class of functions the Hamiltonian has the form

$$H = 2\left\{ \sum_{f>0} T(f) b_f^\dagger b_f - \frac{\nu}{2} \sum_{f>0} (b_f + b_f^\dagger) - \frac{1}{V} \sum_{f>0,f'>0} \lambda(f)\lambda(f') b_f^\dagger b_{f'} \right\}. \quad (2.5)$$

The restriction $f > 0$ is made in order to ensure that all the operators b_f shall be independent (since $b_f = -b_{-f}$). A Hamiltonian of this type was considered in a previous paper by the author [5].

3.   Upper Bound for the Minimum Eigenvalue of the Hamiltonian

We now consider the problem of finding an upper bound for the minimum eigenvalue of the Hamiltonian H (equation (1.2)). We start from the representation of H in the form (1.3), and denote by E_H the minimum eigenvalue of H and by $E_0(\sigma)$ the minimum eigenvalue of H_0 (equation (1.4)) Since the operator $H_1 \leq 0$, the minimum eigenvalue of H_0 cannot be smaller than the minimum eigenvalue of $H = H_0 + H_1$, i.e.

$$E_0(\sigma) \geq E_H \quad (3.1)$$

for arbitrary σ. Thus, the set of minimum eigenvalues of the Hamiltonians $H_0(\sigma)$ form an upper bound for E_H, and the optimum bound is obtained by minimizing $E_0(\sigma)$ with respect to σ.

We shall now proceed to calculate the eigenvalues of the Hamiltonian H_0. To perform the canonical transformation which diagonalizes the quadratic form H_0 (1.4), we write down the identity

$$H_0 = \sum_f \sqrt{\lambda^2(f)(\nu + \sigma^*)(\nu + \sigma) + T^2(f)} (u_f a_f^\dagger + v_{f a_{-f}}^*)(u_f a_f + v_f a_{-f}^\dagger)$$

$$+ \frac{1}{2} V \left\{ \sigma^* \sigma - \frac{1}{V} \sum_f \left[\sqrt{\lambda^2(f)(\nu + \sigma^*)(\nu + \sigma) + T^2(f)} - T(f) \right] \right\},$$

$$(3.2)$$

where

$$u_f = \frac{1}{\sqrt{2}} \sqrt{1 + \frac{T(f)}{\sqrt{\lambda^2(f)(\nu + \sigma^*)(\nu + \sigma) + T^2(f)}}},$$

$$v_f = \frac{-\epsilon(f)}{\sqrt{2}} \sqrt{1 - \frac{T(f)}{\sqrt{\lambda^2(f)(\nu + \sigma^*)(\nu + \sigma) + T^2(f)}}} \cdot \frac{\sigma + \nu}{|\sigma + \nu|},$$

$$(3.3)$$

and where we have put

$$\lambda(f) = \epsilon(f)|\lambda(f)|, \quad \epsilon(f) = \text{sign}\lambda(f).$$

(3.4)

Obviously:

$$u(-f) = u(f); \quad v(-f) = -v(f); \quad u_f^2 + v_f^2 = 1.$$

(3.5)

In general u is real and v complex. From (3.5) it follows that the operators

$$\alpha_f = u_f a_f + v_f a_{-f}^\dagger$$
$$\alpha_f^\dagger = u_f a_f^\dagger + v_f^* a_{-f}$$

(3.6)

are fermion operators. Accordingly, we can rewrite the expression (3.2) for H_0 as

$$H_0 = \sum_f \sqrt{\lambda^2(f)(\nu + \sigma^*)(\nu + \sigma) + T^2(f)} \alpha_f^\dagger \alpha_f$$
$$+ \frac{1}{2}V\left\{\sigma^*\sigma - \frac{1}{V}\sum_f \left[\sqrt{\lambda^2(f)(\nu + \sigma^*)(\nu + \sigma) + T^2(f)} - T(f)\right]\right\},$$

(3.7)

The minimum eigenvalue of H_0 is obviously obtained by putting the occupation number $\alpha_f^\dagger \alpha_f$ equal to zero. We then find the following expression for the ground state energy of the Hamiltonian

$$E_0(\sigma) = \frac{1}{2}V\left\{\sigma^*\sigma - \frac{1}{V}\sum_f \left[\sqrt{\lambda^2(f)(\nu + \sigma^*)(\nu + \sigma) + T^2(f)} - T(f)\right]\right\}. \quad (3.8)$$

To obtain the optimum upper bound for E_H we must minimize $E_0(\sigma)$ with respect to σ. For this purpose it is convenient to consider the cases $\nu = 0$ and $\nu > 0$ separately.

(i) *The case $\nu = 0$*

Writing $x = \sigma^*\sigma > 0$ we have

$$E_0(\sigma) = \frac{1}{2}VF(x),$$

where

$$F(x) = x - \frac{1}{V} \sum_f \left[\sqrt{\lambda^2(f)x + T^2(f)} - T(f) \right].$$

In this case the minimization condition obviously determines only the modulus of σ, not its phase. We have

$$F'(x) = 1 - \frac{1}{2V} \sum_f \frac{\lambda^2(f)}{\sqrt{\lambda^2(f)x + T^2(f)}};$$

$$F''(x) = \frac{1}{4V} \sum_f \frac{\lambda^4(f)}{(\sqrt{\lambda^2(f)x + T^2(f)})^3}.$$

Clearly $F''(x) > 0$ for $0 \le x \le \infty$, so that $F'(x)$ can have at most one zero in this interval. Taking into account the properties of the functions $\lambda(f)$ and $T(f)$ (see Section 1) we get

$$F'(0) < 0;, \quad F'(\infty) > 0$$

and so the interval $0 < x < \infty$ contains a single solution of the equation $F'(x) = 0$, and this solution defines the absolute minimum of the function $F(x)$. Our final result is therefore:

$$\frac{V}{2} \min F(x) \ge E_H (0 < x < \infty). \tag{3.9}$$

(ii) *The case $\nu > 0$*

We put $(\nu + \sigma^*)(\nu + \sigma) = x$ (so that, obviously, $x > 0$) and note the identity

$$\sigma^* \sigma = x + \nu^2 - \nu(\sigma + \nu + \sigma^* + \nu)$$
$$= (\sqrt{x} - \nu)^2 + 2\nu\{\sqrt{x} - (\sigma + \nu + \sigma^* + \nu)\}.$$

The root is to be taken here and subsequently, as the positive root. Then we can write

$$\sigma + \nu = \sqrt{x}\, e^{i\varphi}; \quad \sigma^* + \nu = \sqrt{x}\, e^{-i\varphi}$$

and

$$\sigma^* \sigma = (\sqrt{x} - \nu)^2 + 2\nu\sqrt{x}(1 - \cos\varphi).$$

Therefore, we can express $E_0(\sigma)$ in the form

$$E_0(\sigma) = \frac{V}{2}F(x) + V\nu\sqrt{x}(1 - \cos\varphi), \qquad (3.10)$$

where

$$F(x) = (\sqrt{x} - \nu)^2 - \frac{1}{V}\sum_f \left[\sqrt{\lambda^2(f)x + T^2(f)} - T(f)\right].$$

We then have

$$F'(x) = 1 - \frac{\nu}{\sqrt{x}} - \frac{1}{2V}\sum_f \frac{\lambda^2(f)}{\sqrt{\lambda^2(f)x + T^2(f)}};$$

$$F''(x) = \frac{\nu}{2x^{3/2}} + \frac{1}{4V}\sum_f \frac{\lambda^4(f)}{(\sqrt{\lambda^2(f)x + T^2(f)})^3}.$$

Since $F''(x) > 0$, clearly $F'(x)$ can have at most on e zero in the range $0 \le x \le \infty$. But,

$$F'(0) = -\infty, \quad F'(\infty) = 1,$$

and therefore there exists a value x_0 of x in the range $0 < x < \infty$ such that $F'(x_0) = 0$. This value defines the absolute minimum of the function $F(x)$.

It is clear from (3.10) that the unique choice of σ which corresponds to the absolute minimum of $E_0(\sigma)$ is

$$x = x_0, \quad \varphi = 0. \qquad (3.11)$$

Thus,

$$\sigma + \nu = \sqrt{x}, \quad \sigma = \sqrt{x} - \nu.$$

Hence, in the present case $(\nu > 0)$ both the amplitudes and the phase of σ are fixed; in fact, σ must be real. Our final result, therefore, is

$$\frac{V}{2}\min F(x) \ge E_H \quad (0 < x < \infty). \qquad (3.12)$$

It can be shown by the simple considerations of reference 2 that the complementary term $H - H_0 = H_1$ in equation (1.3) has no effect in the limit to infinite volume. To prove this rigorously, however, we need not only an upper bound for E_H but also a corresponding lower bound. To put in another way, we should like to be able to eliminate completely the term

$$\left(\sum_f \lambda(f)a_f^\dagger a_{-f}^\dagger - V\sigma^*\right)\left(\sum_f \lambda(f)a_{-f}a_f - V\sigma\right).$$

Formally this could be achieved by treating σ not as a c-number but as the operator

$$L = \frac{1}{V} \sum_f \lambda(f) a_{-f} a_f.$$

However, if σ is an operator we cannot carry out a canonical transformation from the fermions operators a to the new fermion operators σ. In spite of this, we shall now try to generalize the identity (3.2) to this case; all that is necessary is to determine the correct order of the operator. By this method we shall prove the theorem that the solution for H_0 is also asymptotically exact for H in the limit of infinite volume.

4.   Lower Bound for the Minimum Eigenvalue of the Hamiltonian

To derive the lower bound for the Hamiltonian (1.2) we first generalize the identity (1.3) in such a way as to reduce the term H_1 (equation (1.5)) identically to zero. The way to do this is to take σ not as a c-number but as an operator, which we shall call L:

$$L = \frac{1}{V} \sum_f \lambda(f) a_{-f} a_f. \tag{4.1}$$

Instead of the c-numbers $(\nu + \sigma^*)(\nu + \sigma)$ we introduce the operators:

$$K = (L + \nu)(L^\dagger + \nu) + \beta^2; \quad \widetilde{K} = (L^\dagger + \nu)(L + \nu) + \beta^2, \tag{4.2}$$

where β is a constant.

Consider now the following operators:

$$p_f = \frac{1}{\sqrt{2}} \sqrt{\sqrt{K\lambda^2(f) + T^2(f)} + T(f)}; \quad p_f = p_f^\dagger$$

$$q_f = -\frac{\epsilon(f)}{\sqrt{2}} \sqrt{\sqrt{K\lambda^2(f) + T^2(f)} - T(f)} \cdot \frac{1}{\sqrt{K}} (L + \nu) \tag{4.3}$$

Clearly

$$p_f q_f = -\frac{\lambda(f)}{2} (L + \nu) \tag{4.4}$$

$$p_f^2 = \frac{1}{2} \{ \sqrt{K\lambda^2(f) + T^2(f)} + T(f) \} \tag{4.5}$$

$$q_f^\dagger q_f = (L^\dagger + \nu)\frac{1}{2K}\{\sqrt{K\lambda^2(f) + T^2(f)} - T(f)\}(L + \nu). \tag{4.6}$$

Taking into account the identity

$$\xi^\dagger F(\xi\xi^\dagger)\xi = \xi^\dagger \xi F(\xi\xi^\dagger) \tag{4.7}$$

which holds for any arbitrary operator ξ, we can rewrite equation (4.6) in the form

$$q_f^\dagger q_f = (L^\dagger + \nu)(L + \nu)\frac{1}{2\widetilde{K}}\{\sqrt{\widetilde{K}\lambda^2(f) + T^2(f)} - T(f)\}$$

$$= \frac{1}{2}\{\sqrt{\widetilde{K}\lambda^2(f) + T^2(f)} - T(f)\} - \frac{\beta^2}{2\widetilde{K}}\{\sqrt{\widetilde{K}\lambda^2(f) + T^2(f)} - T(f)\}. \tag{4.8}$$

With a view to the subsequent application if lemma II of the Appendix (equations (A.9), (A.10)) we write (4.8) as follows

$$q_f^\dagger q_f = \frac{1}{2}\left\{\sqrt{\lambda^2(f)\left(K + \frac{2s}{V}\right) + T^2(f)} - T(f)\right\}$$

$$- \frac{1}{2}\left\{\sqrt{\lambda^2(f)\left(K + \frac{2s}{V}\right) + T^2(f)} - \sqrt{\lambda^2(f)\widetilde{K} + T^2(f)}\right\}$$

$$- \frac{\beta^2}{2\widetilde{K}}\{\sqrt{\widetilde{K}\lambda^2(f) + T^2(f)} - T(f)\}, \tag{4.9}$$

where s forms an upper bound for the expression $\frac{1}{V}\sum_f |\lambda(f)|^2$:

$$\frac{1}{V}\sum_f |\lambda(f)|^2 \le s. \tag{4.10}$$

Notice that the second term on the right-hand side of (4.9) is non-negative. In a similar way we can rewrite (4.5) as

$$p_f^2 = \frac{1}{2}\left\{\sqrt{\lambda^2(f)\left(K + \frac{2s}{V}\right) + T^2(f)} + T(f)\right\}$$

$$- \frac{1}{2}\left\{\sqrt{\lambda^2(f)\left(K + \frac{2s}{V}\right) + T^2(f)} - \sqrt{K\lambda^2(f) + T^2(f)}\right\}. \tag{4.11}$$

Now we consider the quantity

$$\Omega = \sum_f (a_f^\dagger p_f + a_{-f} q_f^\dagger)(p_f a_f + q_f a_{-f}^\dagger). \tag{4.12}$$

Using the fact that $q_f^\dagger q_f = q_{-f}^\dagger q_{-f}$ and equation (4.4), we obtain

$$\Omega = \sum_f a_f^\dagger p_f^2 a_f + \sum_f a_f q_f^\dagger q_f a_f^\dagger - \sum_f \frac{\lambda(f)}{2}$$
$$\times \left\{ (L^\dagger + \nu)a_{-f}a_f + a_f^\dagger a_{-f}^\dagger(L + \nu) \right\} + R_1 \tag{4.13}$$

where

$$R_1 = \sum_f \frac{\lambda(f)}{2} \left\{ (L^\dagger a_{-f} + a_{-f} L^\dagger)a_f + a_f^\dagger(a_{-f}^\dagger L + L a_{-f}^\dagger) \right\}. \tag{4.14}$$

Now observe that

$$\sum_f \frac{\lambda(f)}{2} \left\{ (L^\dagger + \nu)a_{-f}a_f + a_f^\dagger a_{-f}^\dagger(L + \nu) \right\} = V L^\dagger L + \frac{V}{2}(\nu L + \nu l^\dagger) \tag{4.15}$$

and consequently

$$\Omega + \frac{V}{2}L^\dagger L - \sum_f a_f^\dagger p_f^2 a_f + \sum_f a_f q_f^\dagger q_f a_f^\dagger$$
$$= -\frac{V}{2}\left\{ L^\dagger L + \nu(L + L^\dagger) \right\} + R_1. \tag{4.16}$$

In other words, by virtue of (4.9) and (4.11) we have

$$\sum_f \sum_f (a_f^\dagger p_f + a_{-f} q_f^\dagger)(p_f a_f + q_f a_{-f}^\dagger)$$

$$+ \frac{1}{2} \sum_f a_f^\dagger \left\{ \sqrt{\left(K + \frac{2s}{V}\right)\lambda^2(f) + T^2(f)} - \sqrt{K\lambda^2(f) + T^2(f)} \right\} a_f$$

$$+ \frac{1}{2} \sum_f a_f \left\{ \sqrt{\lambda^2(f)\left(K + \frac{2s}{V}\right) + T^2(f)} - \sqrt{\widetilde{K}\lambda^2(f) + T^2(f)} \right\} a_f^\dagger$$

$$+ \frac{1}{2} \sum_f a_f \frac{\beta^2}{2\widetilde{K}} \left\{ \sqrt{\widetilde{K}\lambda^2(f) + T^2(f)} - T(f) \right\} a_f^\dagger$$

$$-\frac{1}{2}\sum_f a_f^\dagger\left\{\sqrt{\lambda^2(f)\left(K+\frac{2s}{V}\right)+T^2(f)}+T(f)\right\}a_f$$

$$-\frac{1}{2}\sum_f a_f\left\{\sqrt{\lambda^2(f)\left(K+\frac{2s}{V}\right)+T^2(f)}-T(f)\right\}a_f^\dagger+\frac{V}{2}L^\dagger L$$

$$=-\frac{V}{2}\{L^\dagger L+\nu(L+L^\dagger)\}+R_1. \tag{4.17}$$

Let us introduce the notation:

$$\Delta_1=\frac{1}{2}\sum_f a_f^\dagger\left\{\sqrt{\left(K+\frac{2s}{V}\right)\lambda^2+T^2}-\sqrt{K\lambda^2+T^2}\right\}a_f \tag{4.18}$$

$$\Delta_2=\frac{1}{2}\sum_f a_f\left\{\sqrt{\left(K+\frac{2s}{V}\right)\lambda^2+T^2}-\sqrt{\widetilde{K}\lambda^2+T^2}\right\}a_f^\dagger \tag{4.19}$$

$$\Delta_3=\frac{1}{2}\sum_f a_f\frac{\beta^2}{2\widetilde{K}}\left\{\sqrt{\widetilde{K}\lambda^2+T^2}-T\right\}a_f^\dagger. \tag{4.20}$$

Then by virtue of lemma II (equations (A.9), (A.10)):

$$\Omega\geq 0,\quad \Delta_1\geq 0;\quad \Delta_2\geq 0;\quad \Delta_3\geq 0. \tag{4.21}$$

According to (4.17),

$$\Omega+\Delta_1+\Delta_2+\Delta_3-\frac{1}{2}\sum_f a_f^\dagger\left\{\sqrt{\left(K+\frac{2s}{V}\right)\lambda^2+T^2}+T\right\}a_f$$

$$-\frac{1}{2}\sum_f a_f\left\{\sqrt{\left(K+\frac{2s}{V}\right)\lambda^2+T^2}-T\right\}a_f^\dagger+\frac{V}{2}L^\dagger L$$

$$=-\frac{V}{2}\{L^\dagger L+\nu(L+L^\dagger)\}+R_1. \tag{4.22}$$

If we put

$$R_2=\frac{1}{2}\sum_f a_f^\dagger\left\{\left[\sqrt{\left(K+\frac{2s}{V}\right)\lambda^2+T^2}\right]a_f-a_f\sqrt{\left(K+\frac{2s}{V}\right)\lambda^2+T^2}\right\} \tag{4.23}$$

$$R_3=\frac{1}{2}\sum_f a_f\left\{\left[\sqrt{\left(K+\frac{2s}{V}\right)\lambda^2+T^2}\right]a_f^\dagger-a_f^\dagger\sqrt{\left(K+\frac{2s}{V}\right)\lambda^2+T^2}\right\} \tag{4.24}$$

then (4.22) becomes

$$\Omega + \Delta_1 + \Delta_2 + \Delta_3 - R_1 - R_2 - R_3 + \frac{V}{2}L^\dagger L$$

$$- \frac{1}{2}\sum_f a_f^\dagger a_f \left\{ \sqrt{\left(K + \frac{2s}{V}\right)\lambda^2 + T^2} + T \right\}$$

$$- \frac{1}{2}\sum_f a_f a_f^\dagger \left\{ \sqrt{\left(K + \frac{2s}{V}\right)\lambda^2 + T^2} - T \right\}$$

$$= -\frac{V}{2}\{L^\dagger L + \nu(L + L^\dagger)\} + R_1. \tag{4.25}$$

However

$$\frac{1}{2}\sum_f a_f^\dagger a_f \left\{ \sqrt{\left(K + \frac{2s}{V}\right)\lambda^2 + T^2} + T \right\}$$

$$+ \frac{1}{2}V\sum_f a_f a_f^\dagger \left\{ \sqrt{\left(K + \frac{2s}{V}\right)\lambda^2 + T^2} - T \right\}$$

$$= \frac{1}{2}\sum_f \left\{ \sqrt{\left(K + \frac{2s}{V}\right)\lambda^2 + T^2} - T \right\} + \sum_f T(f)a_f^\dagger a_f, \tag{4.26}$$

and so

$$\Omega + \Delta_1 + \Delta_2 + \Delta_3 - R_1 - R_2 - R_3 + \frac{V}{2}(L^\dagger L - LL^\dagger)$$

$$+ \frac{1}{2}\left[LL^\dagger - \frac{1}{V}\sum_f \left\{ \sqrt{\left(K + \frac{2s}{V}\right)\lambda^2 + T^2} - T \right\} \right]$$

$$= \sum_f T(f)a_f^\dagger a_f - \frac{V}{2}\{L^\dagger L + \nu(L^\dagger + L)\}$$

$$= \sum_f T(f)a_f^\dagger a_f - \nu \sum_f \frac{\lambda(f)}{2}(a_f a_f + a_f^\dagger a_{-f}^\dagger)$$

$$- \frac{1}{2V}\sum_{f,f'} \lambda(f)\lambda(f')a_f^\dagger a_{-f}^\dagger a_{-f'}a_{f'} = H \tag{4.27}$$

Thus, we finally obtain

$$H = \frac{1}{2}V\left\{ LL^\dagger - \frac{1}{V}\sum_f \left[\sqrt{\left(K + \frac{2s}{V}\right)\lambda^2(f) + T^2(f)} - T(f) \right] \right\}$$

$$+ \Omega + \Delta_1 + \Delta_2 + \Delta_3 - R_1 - R_2 - R_3 + \frac{V}{2}(L^\dagger L - LL^\dagger). \qquad (4.28)$$

The expression (4.28) for the Hamiltonian (1.2) is simply an identity. We shall treat the first term as a principal one; the terms R_1, R_2, R_3 we shall prove to be an asymptotically small, and the terms Ω, Δ_1, Δ_2, Δ_3 we shall drop. Since they are positive (equation (4.21)) we obtain in this way a lower bound for the eigenvalues of H.

It is easy to show that by virtue of (2.2)

$$-R_1 + \frac{V}{2}(L^\dagger L - LL^\dagger) = -\frac{1}{V}\sum_f \lambda^2(f), \qquad (4.29)$$

where, according to (4.10), $\frac{1}{V} sum_f |\lambda(f)|^2 \leq s$. Further, in view of lemma IV [(A.30) of the Appendix], we have

$$|R_2| + |R_3| \leq c \qquad (4.30)$$

where

$$c = \frac{4}{\pi}\frac{1}{V}\sum_f |\lambda(f)|^2 \left(1 + \frac{|\lambda(f)|\left(\frac{1}{V}\sum|\lambda(f)| + V\right)}{2\frac{1}{V}\sum|\lambda(f)|^2 + V\frac{T^2(f)}{\lambda^2(f)}}\right) \int_0^\infty \frac{\sqrt{t}}{(1+t^2)^2}\, dt. \quad (4.31)$$

Hence, from (4.21) we have the following inequality for any normalized function ϕ

$$\langle \phi^* H \phi \rangle \geq -(s+c)$$
$$+ \frac{1}{2}V\left\langle \phi^*\left(LL^\dagger - \frac{1}{V}\sum_f\left[\sqrt{\left\{(L+\nu)(L^\dagger+\nu) + \beta^2 + \frac{2s}{V}\right\}\lambda^2 + T^2(f)}\right.\right.\right.$$
$$\left.\left.\left. - T(f)\right]\right)\phi\right\rangle.$$

However, s and c are independent of β; hence, taking the limit $\beta \to 0$, we find

$$\langle \phi^* H \phi \rangle \geq -(s+c)$$
$$+ \frac{1}{2}V\left\langle \phi^*\left(LL^\dagger - \frac{1}{V}\sum_f\left[\sqrt{\left\{(L+\nu)(L^\dagger+\nu) + \frac{2s}{V}\right\}\lambda^2 + T^2(f)}\right.\right.\right.$$
$$\left.\left.\left. - T(f)\right]\right)\phi\right\rangle. \quad (4.32)$$

Now we have

$$LL^\dagger = (L + \nu)(L^\dagger + \nu) - \nu\{L + \nu + L^\dagger + \nu\} + \nu^2$$

$$LL^\dagger + \frac{2s}{V} = \left\{(L + \nu)(L^\dagger + \nu) + \frac{2s}{V}\right\} - \nu\{L + \nu + L^\dagger + \nu\} + \nu^2 \quad (4.33)$$

Setting

$$(L + \nu)(L^\dagger + \nu) + \frac{2s}{V} = X \quad (4.34)$$

we can write

$$LL^\dagger + \frac{2s}{V} = (\sqrt{X} - \nu)^2 + \nu(2\sqrt{X} - (L + \nu) - (L^\dagger + \nu)). \quad (4.35)$$

If, in the inequality of lemma I (A.1), (A.2), we put

$$\xi = L + \nu, \quad \xi^\dagger = L^\dagger + \nu$$

then we obtain

$$2\sqrt{(L + \nu)(L + \nu^\dagger) + \frac{s}{V}} - (L + \nu) - (L^\dagger + \nu) \geq 0, \quad (4.36)$$

$$2\sqrt{X} - (L + \nu) - (L^\dagger + \nu) \geq 0. \quad (4.37)$$

Therefore, defining a function $F(x)$ as in section 3, i.e.

$$F(x) = (\sqrt{x} - \nu)^2 - \frac{1}{V}\left[\sqrt{x\lambda^2(f) + T^2(f)} - T(f)\right],$$

we can write formula (4.32) in the form

$$\langle \phi^* H \phi \rangle \geq - (2s + c) + \frac{1}{V}\langle \phi^* F(X)\phi \rangle$$

$$+ V\frac{\nu}{2}\langle \phi^*\{2\sqrt{X} - (L + \nu + L^\dagger + \nu)\}\phi \rangle$$

$$\geq - (2s + c) + \frac{1}{2}\langle \phi^* F(X)\phi \rangle, \quad (4.38)$$

where X is now the operator defined by (4.34).

Let E_H be the lowest eigenvalue of H and ψ_H is the corresponding eigenfunction; let E_{H_0} be the lowest eigenvalue of H_0, which, by (3.11) is given by

$$E_{H_0} = \frac{V}{2} \min F(x).$$

Suppose the absolute minimum of $F(x)$ is obtained when

$$x = x_0 = C^2 \tag{4.39}$$

Then we have

$$\frac{V}{2} F(C^2) \geq E_H = \langle \phi_H^* H \phi_H \rangle \geq -(2s + c) + \frac{1}{2} V \langle \phi^* H \phi \rangle$$

$$\geq -(2s + c) + \frac{V}{2} F(C^2). \tag{4.40}$$

If we now take into account that the energy of the system must be proportional to the volume, we get as our final pair of bounds for the minimum eigenvalue of the Hamiltonian H (1.2):

$$0 \leq \frac{E_{H_0} - E_H}{V} \leq \frac{2s + c}{V}. \tag{4.41}$$

Now the quantities c [equation (4.31)] and s remain finite in the limit $V \to \infty$, according to the postulates of section 1. Hence the difference between the eigenvalues of the approximate Hamiltonian H_0 and the exact Hamiltonian H, divided by the volume of the system, decreases as $1/V$ in the limit $V \to \infty$. This result proves that the solution for the approximate Hamiltonian H_0(1.4) constitutes a solution for the exact Hamiltonian H which is asymptotically exact in the limit of infinite volume.

We shall now show that it is asymptotically correct (i.e., correct to order $1/V$) to treat the operator X defined by equation (4.34) as a c-number. Consider an arbitrary normalized function ϕ such that

$$\langle \phi^* H \phi \rangle - E_H \leq c_1 = \text{const.} \tag{4.42}$$

Thus, using (4.38), (4.40) and (4.42), we have

$$\langle \phi^* \left(F(X) - F(C^2) \right) \phi \rangle + \nu \langle \phi^* \left(2\sqrt{X} - (L + \nu + L^\dagger + \nu) \right) \phi \rangle \leq \frac{l}{V}$$

$$l = 2(2s + c + c_1). \tag{4.43}$$

To proceed, we notice that both terms on the left-hand side of (4.43) are positive. In fact, the second term is positive by virtue of lemma I of the Appendix (A.1), (A.2), and this implies that

$$\langle \phi^* \big(F(X) - F(C^2) \big) \phi \rangle \leq \frac{1}{V}, \tag{4.44}$$

while on the other hand, there exists a value of ξ such that

$$F(X) - F(C^2) = \frac{1}{2} F''(\xi)(X - C^2)^2. \tag{4.45}$$

However, $F''(x)$ is positive for all x

$$F''(x) = \frac{\nu}{2x^{3/2}} + \frac{1}{4} \frac{1}{V} \sum_f \frac{\lambda^4(f)}{(x\lambda^2(f) + T^2(f))^{3/2}};$$

$$\frac{1}{2} F''(\xi) \geq \alpha = \text{const} > 0. \tag{4.46}$$

From equations (4.44)-(4.46) it follows that

$$\langle \phi^* |X - C^2|^2 \phi \rangle \leq \frac{1}{\alpha V}. \tag{4.47}$$

Equation (4.47) shows that the operator X may with asymptotic accuracy be treated as a c-number.

Actually, in the case $\nu > 0$ we can obtain rather more complete information about the expectation values of the operators L, $L^\dagger$. In fact, we shall prove that the mean square deviation of the operator $\xi = L + \nu$ from the quantity C defined by (4.39) is asymptotically small in the limit of infinite volume.

We have the obvious inequality

$$(\sqrt{X} - C)^2 = \frac{(X - C^2)^2}{(\sqrt{X} + C)} \leq \frac{1}{C^2} (X - C^2)^2. \tag{4.48}$$

Hence, using (4.47), we get

$$\langle \phi^* (\sqrt{X} - C)^2 \phi \rangle \leq \frac{1}{\alpha C^2 V}. \tag{4.49}$$

Let us define the quantity C_0 by

$$\langle \phi^* \sqrt{X} \phi \rangle = C_0. \tag{4.50}$$

Then, from (4.49)

$$\langle \phi^*(\sqrt{X} - C_0)^2\phi \rangle \leq \langle \phi^*(\sqrt{X} - C)^2\phi \rangle \leq \frac{1}{\alpha C^2 V}. \qquad (4.51)$$

Here the first inequality follows from the identity

$$\langle \phi^*(\sqrt{X} - C)^2\phi \rangle = (C - C_0)^2 + \langle \phi^*(\sqrt{X} - C_0)^2\phi \rangle. \qquad (4.52)$$

From (4.51) we get the following inequality for the expectation value of the operator X:

$$\langle \phi^* X\phi \rangle - C_0^2 \leq \frac{1}{\alpha C^2 V}. \qquad (4.53)$$

Finally, from (4.51) and (4.52) follows an inequality for $(C - C_0)^2$, namely

$$(C - C_0)^2 \leq \frac{1}{\alpha C^2 V}. \qquad (4.54)$$

Now put

$$\xi = L + \nu, \quad \xi^\dagger = L^\dagger + \nu. \qquad (4.55)$$

Then, for the mean square deviation of σ from C_0 we have, by (4.34),

$$\langle \phi^*(C_0 - \xi)(C_0 - \xi^\dagger)\phi \rangle$$
$$\leq C_0^2 + \langle \phi^* X\phi \rangle - C_0 \langle \phi^*(\xi + \xi^\dagger)\phi \rangle. \qquad (4.56)$$

Using (4.53) and (4.54) we obtain

$$\langle \phi^*(C_0 - \xi)(C_0 - \xi^\dagger)\phi \rangle \leq 2C_0^2 + \frac{l}{\alpha C^2 V} - C_0 \langle \phi^*(\xi + \xi^\dagger)\phi \rangle$$
$$= \langle \phi^\dagger\{2\sqrt{X} - (\xi + \xi^\dagger)\}\phi \rangle C_0 + \frac{l}{\alpha C^2 V} \leq \frac{lC_0}{\nu V} + \frac{l}{\alpha C^2 V}. \qquad (4.57)$$

Thus, we finally obtain the following bound for the fluctuations of ξ:

$$\langle \phi^*(C - \xi)(C - \xi^\dagger)\phi \rangle = \langle \phi^*(C - C_0 + C_0 - \xi)(C - C_0 + C_0 - \xi^\dagger)\phi \rangle$$
$$\leq 2(C - C_0)^2 + 2\langle \phi^*(C_0 - \xi)(C_0 - \xi^\dagger)\phi \rangle$$
$$\leq \frac{2l}{\alpha C^2 V} + \frac{2lC_0}{\nu V} + \frac{2l}{\alpha C^2 V} \leq \frac{const}{V} = \frac{I}{V}, \quad (I = const). \qquad (4.58)$$

Note that this bound is applicable only for $\nu > 0$, since ν appears in the denominator of the right-hand side of the inequality (4.58).

A few comments on the results obtained here are appropriate at this stage. Suppose $\nu = 0$. Then, as we saw above, in any state with energy asymptotically close to the ground state E_H the operator $L^\dagger L$ is equal, with asymptotic accuracy, to the c-number C^2. However, these states possess no similar properties with respect to the operators $L, L^\dagger$ themselves, as we shall now see.

Consider a state ϕ_H with the ground state energy E_H; in general degeneracy can occur, so that we will have not just one ϕ_H but a linear manifold $\{\phi_H\}$ of possible states with the same (minimum) energy. Since in the case under consideration ($\nu = 0$) the operator $N = \sum_f a_f^\dagger a_f$, which represents the total number of particles in the system, commutes exactly with the Hamiltonian H, we can always choose from this manifold $\{\phi_H\}$ a function ϕ_H' for which N takes some definite value N_0. Then

$$\langle \phi_H^{*\prime} L \phi_H' \rangle = 0; \quad \langle \phi_H^{*\prime} L^\dagger \phi_H' \rangle = 0.$$

Consequently, L cannot have even an asymptotically well-defined value in the state ϕ_H', since if did $L^\dagger L$ would be (asymptotically) equal to zero in this state rather than to the finite quantity C^2.

Consider now the manifold $\{\phi\}$ of states with energies asymptotically close to E_H. Since L, $L^\dagger$ commute approximately with H, we might expect that we can choose from $\{\phi\}$ a function ϕ for which $L^\dagger$, L take asymptotically well-defined values.

Such is indeed the case. For instance, one function with the required properties is ϕ_{H_0}', the state corresponding to the minimum eigenvalue of H_0. Indeed, as we saw, ϕ_{H_0}' is defined by the relations

$$\alpha_f \phi = 0; \quad \text{where} \quad \alpha_f = u_f a_f - v_f \alpha_{-f}^\dagger$$

$$u_f = \frac{1}{\sqrt{2}} \sqrt{1 + \frac{T(f)}{\sqrt{\lambda^2(f)C^2 + T^2(f)}}}$$

$$u_f = \frac{\in(f)}{\sqrt{2}} \sqrt{1 - \frac{T(f)}{\sqrt{\lambda^2(f)C^2 + T^2(f)}}}.$$

We can express L in terms of the fermion operators α_f, $\alpha_f^\dagger$; the result is:

$$L = \frac{1}{V} \sum_f \lambda(f)\{u_f^2 \alpha_{-f} a_f - v_f^2 \alpha_f^\dagger \alpha_{-f}^\dagger - 2u_f v_f \alpha_f^\dagger a_f\} + \frac{1}{V} \sum_f u_f v_f \lambda(f).$$

However,

$$\frac{1}{V}\sum_f u_f v_f = \frac{1}{2V}\sum_f \frac{|\lambda(f)|^2 C}{\sqrt{\lambda^2(f)C^2 + T^2(f)}} = C,$$

and therefore,

$$\langle \psi^*_{H_0}(L^\dagger - C)(L - C)\phi_{H_0}\rangle \le \frac{\text{const}}{V};$$

$$\langle \psi^*_{H_0}(L - C)(L^\dagger - C)\phi_{H_0}\rangle \le \frac{\text{const}}{V}.$$

Thus, for ϕ_{H_0}, L and $L^\dagger$ are asymptotically equal to C. This is precisely the reason for the success of the approximation method which replaces Hm which conserves the particle number N exactly, by H_0, for which N is no longer an exact constant of motion.

We can now see that it is also possible to formulate the approximation method in such a way as not to break the law of conservation of particle number. To do so we introduce, instead of the Fermi operators α_f, the operators

$$\alpha_f = u_f a_f - v_f \frac{L}{|C|} a^\dagger_{-f},$$

which obey Fermi commutation relations with asymptotic accuracy. Then α_f decreases and $\alpha^\dagger_f$ increases the particle number N by unity. These operators are analogous to the operators

$$b_f = \frac{a^\dagger_f}{\sqrt{N_0}}\,\alpha_f,$$

which were introduced in the theory of superfluidity [6] to eliminate the condensate. Indeed, there is a strong analogy in general between the Bose-condensate operators a_0, $a^\dagger_0$ and the operators L, $L^\dagger$ in the present case.

As soon as we include in H a term containing sources of pairs (i.e., put $\nu > 0$) then L, $L^\dagger$ immediately take asymptotically well-defined values for eigenstates of H with energies near E_H. We can see here an analogy with the case of ferromagnetism in an isotropic medium; in the absence of the external magnetic field the direction of the axis of magnetization is not well-defined, but as soon as we introduce an arbitrary weak field acting in a given direction the magnetization vector immediately orient itself in that direction.

Finally, we point out that the relations

$$L^\dagger L \sim C^2 \quad (\nu = 0)$$

$$\nu + L \sim C; \quad \nu + L^\dagger \sim C \quad (\nu > 0)$$

enable us to prove that the correlation expectation values

$$\langle \phi_H^* \ldots a_{fl}(t_l) \ldots a_{fj}^\dagger(t_j) \ldots \phi_H \rangle$$

for the Hamiltonian H are also asymptotically equal to the corresponding expectation values for H_0. For $\nu > 0$ this is true for all averages of the type indicated; for $\nu = 0$ it is of course true only for those in which the numbers of creation and annihilation operators are equal, i.e. for averages of operators which conserve particle number. We shall now proceed to prove this statement.

5. Green's Functions (Case $\nu > 0$)

In this section we turn to the problem of finding asymptotic limits for the Green'd functions and correlations functions in the case $\nu > 0$. The existence of these limits implies that in the limit of infinite volume the solution of the equations of motion constructed for the Green's functions from the Hamiltonian H_0, (1.4), will differ by an asymptotically small amount from the corresponding solutions for the full model Hamiltonian H, (1.2).

Consider the equation of motion for the operators a_f, $a_f^\dagger$. From (1.2) and (4.1) we obtain

$$i\frac{da_f}{dt} = T(f)a_f - \lambda(f)a_{-f}^\dagger(\nu + L),$$

$$i\frac{da_f^\dagger}{dt} = -T(f)a_f^\dagger + \lambda(f)(\nu + L^\dagger)a_{-f}, \qquad (5.1)$$

and therefore

$$i\frac{da_{-f}}{dt} = T(f)a_{-f} - \lambda(f)a_f^\dagger(\nu + L),$$

$$i\frac{da_{-f}^\dagger}{dt} = -T(f)a_{-f}^\dagger + \lambda(f)(\nu + L^\dagger)a_f. \qquad (5.2)$$

We put [cf. (3.3) and (3.11)]

$$u_f = \frac{1}{\sqrt{2}} \sqrt{1 + \frac{T(f)}{\sqrt{C^2\lambda^2(f) + T^2(f)}}}$$

$$v_f = \frac{\in(f)}{\sqrt{2}} \sqrt{1 - \frac{T(f)}{\sqrt{C^2\lambda^2(f) + T^2(f)}}} \tag{5.3}$$

where C is a number given by (4.39), and introduce new fermion operators

$$\alpha_f^\dagger = u_f a_f^\dagger + v_f a_{-f}. \tag{5.4}$$

We then have

$$i\frac{d\alpha_f^\dagger}{dt} = u_f i\frac{da_f^\dagger}{dt} + v_f i\frac{da_{-f}}{dt}$$
$$= u_f\{-T(f)a_f^\dagger + \lambda(f)(\nu + L^\dagger)a_{-f}\} + v_f\{T(f)a_{-f} + \lambda(f)a_f^\dagger(\nu + L)\}$$
$$= -a_f^\dagger\{T(f)u_f - \lambda(f)v_f(\nu + L)\} + \{\lambda(f)(\nu + L^\dagger)a_f + T(f)v_f\}a_{-f}$$
$$= -a_f^\dagger\{T(f)u_f - \lambda(f)v_fC\} + \{\lambda(f)Ca_f + T(f)v_f\}a_{-f} + R_f,$$

where

$$R_f = R_f^{(1)} + R_f^{(2)}$$
$$R_f^{(1)} = u_f\lambda(f)(L^\dagger + \nu - C)a_{-f}$$
$$R_f^{(2)} = v_f\lambda(f)a_f^\dagger(L + \nu - C). \tag{5.5}$$

Now we introduce the following identities

$$T(f)u_f - \lambda(f)v_fC = \left[\sqrt{C^2\lambda^2(f) + T^2(f)}\right]u_f$$
$$T(f)v_f + \lambda(f)u_fC = -\left[\sqrt{C^2\lambda^2(f) + T^2(f)}\right]v_f. \tag{5.6}$$

It follows that

$$i\frac{d\alpha_f^\dagger}{dt} + \left[\sqrt{C^2\lambda^2(f) + T^2(f)}\right]\alpha_f^\dagger = R_f \tag{5.7}$$

and also

$$i\frac{d\alpha_f}{dt} - \left[\sqrt{C^2\lambda^2(f) + T^2(f)}\right]\alpha_f^\dagger = -R_f^\dagger. \tag{5.8}$$

The next step is to obtain bounds for various quantities connected with R and $R^\dagger$. We have

$$\langle\phi_H^* R_f R_f^\dagger \phi_H\rangle \le 2\langle\phi_H^* R_f^{(1)} R_f^{(1)\dagger}\phi_H\rangle + 2\langle\phi_H^* R_f^{(2)} R_f^{(2)\dagger}\phi_H\rangle$$
$$= 2u_f^2\lambda^2(f)\langle\phi_H^*(L^\dagger + \nu - C)a_{-f}a_{-f}^\dagger(L + \nu - C)\phi_H\rangle$$
$$+ 2v_f^2\lambda^2(f)\langle\phi_H^* a_f^\dagger(L + \nu - C)(L^\dagger + \nu - C)a_f\phi_H\rangle.$$

However, since $|a_{-f}a_{-f}^\dagger|^2 \le 1$, it follows that

$$\langle\phi_H^*(L^\dagger + \nu - C)a_{-f}a_{-f}^\dagger(L + \nu - C)\phi_H\rangle$$
$$\le \langle\phi_H^*(L^\dagger + \nu - C)(L + \nu - C)\phi_H\rangle$$

and also, using equation (A.18), that

$$\langle\phi_H^* a_f^\dagger(L + \nu - C)(L^\dagger + \nu - C)a_f\phi_H\rangle$$
$$\le \frac{2s}{V} + \langle\phi_H^* a_f^\dagger(L^\dagger + \nu - C)(L + \nu - C)a_f\phi_H\rangle$$
$$= \frac{2s}{V} + \langle\phi_H^*(L^\dagger + \nu - C)a_f^\dagger a_f(L + \nu - C)\phi_H\rangle$$
$$\le \frac{2s}{V} + \langle\phi_H^*(L^\dagger + \nu - C)a(L + \nu - C)\phi_H\rangle.$$

Thus,

$$\langle\phi_H^* R_f R_f^\dagger \phi_H\rangle \le 2\lambda^2(f)\langle\phi_H^*(L^\dagger + \nu - C)(L + \nu - C)\phi_H\rangle$$
$$+ 2\lambda^2(f)v_f^2\frac{2s}{V}. \qquad (5.9)$$

By an entirely analogous procedure we obtain

$$\langle\phi_H^* R_f^\dagger R_f \phi_H\rangle \le 2\lambda^2(f)\langle\phi_H^*(L + \nu - C)(L^\dagger + \nu - C)\phi_H\rangle$$
$$+ 2\lambda^2(f)u_f^2\frac{2s}{V}. \qquad (5.10)$$

Now, we proved above [cf. (4.58)] that

$$\langle\phi_H^*(L^\dagger + \nu - C)(L + \nu - C)\phi_H\rangle \le \frac{I}{V}$$
$$\langle\phi_H^*(L + \nu - C)(L^\dagger + \nu - C)\phi_H\rangle \le \frac{I}{V}$$

Hence, introducing the constant

$$\gamma = 2(I + 2s), \tag{5.11}$$

we can write

$$\langle \phi_H^* R_f R_f^\dagger \phi_H \rangle \leq \frac{\gamma}{V} |\lambda(f)|^2, \quad \langle \phi_H^* R_f^\dagger R_f \phi_H \rangle \leq \frac{\gamma}{V} |\lambda(f)|^2. \tag{5.12}$$

We can actually state a more general set of inequalities. Consider any set of operators A_f, each of which is a linear combination of operators a_f and $a_{-f}^\dagger$,

$$A_f = p_f a_f + q_f a_{-f}^\dagger \tag{5.13}$$

with bounded coefficients:

$$|p_f|^2 + |q_f|^2 \leq \text{const.} \tag{5.14}$$

We shall prove that

$$|\langle \phi_H^* A_{f_1} \dots A_{f_l} R_f A_{f_{l+1}} \dots A_{f_m} R_f^\dagger A_{f_{m+1}} \dots \phi_H \rangle| \leq \frac{\text{const}}{I}$$
$$|\langle \phi_H^* A_{f_1} \dots A_{f_l} R_f^\dagger A_{f_{l+1}} \dots A_{f_m} R_f A_{f_{m+1}} \dots \phi_H \rangle| \leq \frac{\text{const}}{I}. \tag{5.15}$$

Proof. We notice first of all that

$$L a_f - a_f L = 0; \quad L^\dagger a_f^\dagger - a_f^\dagger L^\dagger = 0;$$
$$|L a_f^\dagger - a_f^\dagger L| \leq \frac{2|\lambda(f)|}{V}; \quad |L^\dagger a_f - a_f L^\dagger| \leq \frac{2|\lambda(f)|}{V}.$$

Thus, for example,

$$\langle \phi_H^* A_{f_1} \dots (L + \nu - c) A_{f_j} \dots (L^\dagger + \nu - c) A_{f_i} \dots \phi_H \rangle$$
$$= Z + \langle \phi_H^* (L + \nu - c) A_{f_1} \dots A_{f_n} (L^\dagger + \nu - c) \phi_H \rangle,$$

where

$$|Z| \leq \frac{\text{const}}{V}.$$

Therefore,

$$\langle \phi_H^* A_{f_1} \dots (L + \nu - c) A_{f_j} \dots (L^\dagger + \nu - c) A_{f_i} \dots \phi_H \rangle$$

$$\leq \frac{\text{const}}{V} + |A_{f_1}| \dots |A_{f_n}| \langle \phi_H^*(L + \nu - c)(L^\dagger + \nu - c)\phi_H \rangle \leq \frac{\text{const}}{V}.$$
$$(5.16)$$

Similarly we can prove that

$$\langle \phi_H^* A_{f_1} \dots (L^\dagger + \nu - c)A_{f_j} \dots (L + \nu - c)A_{f_i} \dots \phi_H \rangle \leq \frac{\text{const}}{V}. \qquad (5.17)$$

Moreover, we have:

$$\langle \phi_H^* A_{f_1} \dots (L + \nu - c)A_{f_j} \dots (L + \nu - c)A_{f_i} \dots \phi_H \rangle$$
$$\leq \frac{\text{const}}{V} + \langle \phi_H^*(L + \nu - c)A_{f_1} \dots A_{f_n}(L + \nu - c)\phi_H \rangle$$
$$\leq \frac{\text{const}}{V} + \sqrt{\langle \phi_H^*(L + \nu - c)A_{f_1} \dots A_{f_n} A_{f_n}^\dagger A_{f_1}^\dagger (L^\dagger + \nu - c)\phi_H \rangle}$$
$$\times \sqrt{\langle \phi_H^*(L + \nu - c)(L + \nu - c)\phi_H \rangle}$$
$$\leq \frac{\text{const}}{V} + |A_{f_1}| \dots |A_{f_n}| \sqrt{\langle \phi_H^*(L + \nu - c)(L^\dagger + \nu - c)\phi_H \rangle}$$
$$\times \sqrt{\langle \phi_H^*(L + \nu - c)(L + \nu - c)\phi_H \rangle}$$
$$\leq \frac{\text{const}}{V} \qquad\qquad\qquad\qquad\qquad\qquad (5.18)$$

and it is easily shown that in a similar way

$$\langle \phi_H^* A_{f_1} \dots (L^\dagger + \nu - c)A_{f_j} \dots (L^\dagger + \nu - c)A_{f_i} \dots \phi_H \rangle \leq \frac{\text{const}}{V}. \qquad (5.19)$$

From (5.16)-(5.19) it follows that the inequalities (5.15) are satisfied.

We can now set about finding limits for the correlation functions. Using (5.7), we obtain

$$i\frac{d}{dt}\langle \phi_H^* \alpha_f^\dagger(t)\alpha_f \phi_H \rangle = -\sqrt{c^2\lambda^2(f) + T^2(f)}\langle \phi_H^* \alpha_f^\dagger(t)\alpha_f \phi_H \rangle$$
$$+ \langle \phi_H^* R_f(t)\alpha_f \phi_H \rangle. \qquad (5.20)$$

Here, and subsequently, we write

$$\alpha_f(0) = \alpha_f; \quad a_f^\dagger(0) = a_f^\dagger.$$

Since the equation

$$i\frac{dJ(t)}{dt} = -\Omega J(t) + R(t)$$

has the solution

$$J(t) = J(0)\, e^{i\Omega t} + e^{i\Omega t} \int_0^t e^{-i\Omega t}\, R(t)\, dt,$$

we may write

$$\langle \phi_H^* \alpha_f^\dagger(t)\alpha_f \phi_H \rangle = e^{i\sqrt{c^2\lambda^2(f)+T^2(f)}\,t}\langle \phi_H^* \alpha_f^\dagger \alpha_f \phi_H \rangle + e^{i\sqrt{c^2\lambda^2(f)+T^2(f)}\,t}$$

$$\times \int_0^t e^{i\sqrt{c^2\lambda^2(f)+T^2(f)}\,t}\,\langle \phi_H^* \alpha_f^\dagger(t)\alpha_f \phi_H \rangle\, dt. \qquad (5.21)$$

On the other hand, since ϕ_H is the eigenfunction of H corresponding to its minimum eigenvalue, the usual spectral representation gives

$$\langle \phi_H^* \alpha_f^\dagger(t)\alpha_f \phi_H \rangle = \int_0^t J_f(\nu)\, e^{-i\nu t}\, d\nu, \qquad (5.22)$$

where

$$J_f \geq 0; \quad \int_0^t J_f(\nu)\, d\nu \leq 1. \qquad (5.23)$$

Let us define a function

$$h(t) = \int_0^2 \omega^2 (2 - \omega)^2\, e^{-i\omega t}\, d\omega. \qquad (5.24)$$

Obviously this function is regular on the whole of the real axis. Integrating by parts, we easily see that for $|t| \to \infty$, $h(t)$ decreases in such a way that

$$|h(t)| \leq \frac{\text{const}}{|t|^3}. \qquad (5.25)$$

Thus, the integral

$$\int_{-\infty}^{\infty} |t h(t)|\, dt \qquad (5.26)$$

is finite.

We now put

$$\sqrt{c^2\lambda^2(f) + T^2(f)} = \Omega \tag{5.27}$$

and note that

$$h(\Omega t) = \frac{1}{\Omega} \int\limits_0^{2\Omega} \nu^2 (2\Omega - \nu)^2 \, e^{-i\nu t} \, d\nu. \tag{5.28}$$

It is clear form the above that

$$\int\limits_{-\infty}^{\infty} h(\Omega t) \, e^{-i\nu t} \, dt = 0, \quad \text{for} \quad \nu \geq 0 \tag{5.29}$$

and therefore that (5.22) implies

$$\int\limits_{-\infty}^{\infty} \langle \phi_H^* \alpha_f^\dagger(t) \alpha_f \phi_H \rangle h(\Omega t) \, dt = 0. \tag{5.30}$$

Thus, it follows from (5.21) that

$$\langle \phi_H^* \alpha_f^\dagger \alpha_f \phi_H \rangle \int\limits_{-\infty}^{\infty} e^{i\Omega t} h(\Omega t) \, dt$$

$$- \int\limits_{-\infty}^{\infty} h(\Omega t) e^{i\Omega t} \left(\int\limits_0^t e^{-i\Omega t'} \langle \phi_H^* R_f(t') \alpha_f \phi_H \rangle \, dt' \right) dt. \tag{5.31}$$

However,

$$\int\limits_{-\infty}^{\infty} e^{i\Omega t} h(\Omega t) \, dt = \frac{2\pi}{\Omega}. \tag{5.32}$$

Therefore

$$\langle \phi_H^* \alpha_f^\dagger \alpha_f \phi_H \rangle \leq \frac{\Omega}{2\pi} \int\limits_{-\infty}^{\infty} |h(\Omega t)| \left\{ \int\limits_0^t |\langle \phi_H^* R_f(t') \alpha_f \phi_H \rangle| \, dt' \right\} dt. \tag{5.33}$$

Using (5.12), we have

$$\langle \phi_H^* R_f \alpha_f \phi_H \rangle \leq \sqrt{|\langle \phi_H^* R_f^\dagger R_f \phi_H \rangle| \, |\langle \phi_H^* a_f^\dagger a_f \phi_H \rangle|}$$

$$\leq \left(\frac{\gamma}{V}\right)^{1/2} |\lambda(f)| \, |\langle \phi_H^* a_f^\dagger a_f \phi_H\rangle|^{1/2}. \qquad (5.34)$$

and consequently

$$\langle \phi_H^* \alpha_f^\dagger \alpha_f \phi_H\rangle$$

$$\leq \frac{\Omega}{2\pi} \int_{-\infty}^{\infty} |h(\Omega t)| \, |t| dt \left(\frac{\gamma}{V}\right)^{1/2} |\langle \phi_H^* a_f^\dagger a_f \phi_H\rangle|^{1/2}$$

$$\leq \frac{1}{2\pi\Omega} \int_{-\infty}^{\infty} |h(\tau)\tau| \, d\tau \left(\frac{\gamma}{V}\right)^{1/2} |\langle \phi_H^* a_f^\dagger a_f \phi_H\rangle|^{1/2} |\lambda(f)|.$$

Thus,

$$\langle \phi_H^* \alpha_f^\dagger \alpha_f \phi_H\rangle \leq \frac{|\lambda(f)|^2}{2\pi \left(C^2 |\lambda(f)|^2 + T^2(f)\right)} \frac{\gamma}{V} \left(\int_{-\infty}^{\infty} |h(\tau)\tau| \, d\tau\right)^2. \qquad (5.35)$$

From this formula we can immediately derive a number of inequalities. Using Schwartz's inequality and the fact $|a_f^\dagger a_f| \leq 1$, we obtain from (5.35)

$$|\langle \phi_H^* \alpha_{f_1}^\dagger \ldots \alpha_{f_s}^\dagger \alpha_{g_l} \ldots \alpha_{g_1} \phi_H\rangle|$$

$$\leq \sqrt{\langle \phi_H^* \alpha_{f_1}^\dagger \ldots \alpha_{f_s}^\dagger \alpha_{f_s} \ldots \alpha_{f_1} \phi_H\rangle \langle \phi_H^* \alpha_{g_1}^\dagger \ldots \alpha_{g_l}^\dagger \alpha_{g_l} \ldots \alpha_{g_1} \phi_H\rangle}$$

$$\leq \sqrt{\langle \phi_H^* \alpha_{f_1}^\dagger \alpha_{f_1} \phi\rangle \langle \phi_H^* \alpha_{g_1}^\dagger \alpha_{g_1} \phi\rangle} \leq \frac{\text{const}}{V}. \qquad (5.36)$$

We also have

$$|\langle \phi_H^* \alpha_{f_1} \ldots \alpha_{f_s} \phi_H\rangle|$$

$$\leq \sqrt{\langle \phi_H^* \alpha_{f_1} \ldots \alpha_{f_{s-1}} \alpha_{f_{s-1}}^\dagger \ldots \alpha_{f_1}^\dagger \phi_H\rangle \langle \phi_H^* \alpha_{f_s}^\dagger \alpha_{f_s} \phi_H\rangle}$$

$$\leq \sqrt{\langle \phi_H^* \alpha_{f_s}^\dagger \alpha_{f_s} \phi_H\rangle} \leq \frac{\text{const}}{\sqrt{V}} \qquad (5.37)$$

and

$$|\langle \phi_H^* \alpha_{f_1}^\dagger \ldots \alpha_{f_s}^\dagger \phi_H\rangle| \leq \sqrt{\langle \phi_H^* \alpha_{f_1}^\dagger \alpha_{f_1} \phi_H\rangle} \leq \frac{\text{const}}{\sqrt{V}}. \qquad (5.38)$$

We may now compare the expectation values

$$\langle \phi_H^* \mathscr{U}_{f_1} \ldots \mathscr{U}_{f_s} \phi_H\rangle$$

(where $\mathscr{U}_f$ may stand for a_f or $a_f^\dagger$) with the corresponding values calculated by using the Hamiltonian H_0 with $\nu + \sigma$ set equal to C. For convenience we shall denote the two kind of expectation values by

$$\langle \mathscr{U}_{f_1} \dots \mathscr{U}_{f_s} \rangle_H, \quad \text{and} \quad \langle \mathscr{U}_{f_1} \dots \mathscr{U}_{f_s} \rangle_{H_0}$$

respectively. We wish to establish bounds for the differences

$$\langle \mathscr{U}_{f_1} \dots \mathscr{U}_{f_s} \rangle_H - \langle \mathscr{U}_{f_1} \dots \mathscr{U}_{f_s} \rangle_{H_0}. \tag{5.39}$$

It is appropriate at this point to outline the calculation of the quantities $\langle \mathscr{U}_{f_1} \dots \mathscr{U}_{f_s} \rangle_{H_0}$. We use the transformation

$$a_f^\dagger = u_f \alpha_f^\dagger - v_f \alpha_{-f}$$
$$a_f = u_f \alpha_f - v_f \alpha_{-f}^\dagger$$

and then reduce the product $\mathscr{U}_{f_1} \dots \mathscr{U}_{f_s}$ to a sum of normal products (i.e., products in which all the $\alpha^\dagger$s precede all the αs). Since all terms of the type

$$\langle \alpha^\dagger \dots \alpha^\dagger \rangle_{H_0}; \quad \langle \alpha \dots \alpha \rangle_{H_0}; \quad \langle \alpha^\dagger \dots \alpha \rangle_{H_0} \tag{5.40}$$

are equal to zero, we obtain in this way an expression for $\langle \mathscr{U}_{f_1} \dots \mathscr{U}_{f_s} \rangle_{H_0}$.

We can apply the same procedure to calculate the quantities $\langle \mathscr{U}_{f_1} \dots \mathscr{U}_{f_s} \rangle_H$. Obviously the difference (5.39) is entirely due to terms proportional to expectation values of the form

$$\langle \alpha^\dagger \dots \alpha^\dagger \rangle_H; \quad \langle \alpha \dots \alpha \rangle_H; \quad \langle \alpha^\dagger \dots \alpha \rangle_H, \tag{5.41}$$

which, in general, unlike the terms (5.40), are not equal to zero. However, we have certain inequalities for the terms (5.41), namely formulae (5.36)-(5.38). We are therefore led to the result

$$|\langle \mathscr{U}_{f_1} \dots \mathscr{U}_{f_s} \rangle_H - \langle \mathscr{U}_{f_1} \dots \mathscr{U}_{f_s} \rangle_{H_0}| \leq \frac{\text{const}}{\sqrt{V}}. \tag{5.42}$$

We next turn to the double-time correlation functions. We shall prove that the absolute magnitudes of all differences of the type

$$\langle B_{f_1}(t) \dots B_{f_l} \mathscr{U}_{f_m}(\tau) \dots \mathscr{U}_{f_n}(\tau) \rangle_H - \langle B_{f_1}(t) \dots B_{f_l} \mathscr{U}_{f_m}(\tau) \dots \mathscr{U}_{f_n}(\tau) \rangle_{H_0}$$
$$\tag{5.43}$$

(where $\mathscr{U}_f$ and B_f may denote either a_f or $a_f^\dagger$) can be at most quantities of order $1/\sqrt{V}$.

We first notice that while

$$\langle \alpha_{f_1}^\dagger(t) \ldots \alpha_{f_j}(t) \mathscr{U}_{f_m}(\tau) \ldots \mathscr{U}_{f_n}(\tau) \rangle_{H_0} = 0, \tag{5.44}$$

we also have

$$|\langle \alpha_{f_1}^\dagger(t) \ldots \alpha_{f_j}(t) \mathscr{U}_{f_m}(\tau) \ldots \mathscr{U}_{f_n}(\tau) \rangle_H|$$
$$\leq \sqrt{\langle \alpha_{f_1}^\dagger(t) \alpha_{f_1}(t) \rangle_H \langle w^\dagger w \rangle_H} = \sqrt{\langle \alpha_{f_1}^\dagger \alpha_{f_1} \rangle_H \langle w^\dagger w \rangle_H}, \tag{5.45}$$

where

$$w = \ldots \alpha_{f_j}(t) \mathscr{U}_{f_m}(\tau) \ldots \mathscr{U}_{f_n}(\tau).$$

Therefore

$$|\langle \alpha_{f_1}^\dagger(t) \ldots \alpha_{f_j}(t) \mathscr{U}_{f_m}(\tau) \ldots \mathscr{U}_{f_n}(\tau) \rangle_H| \leq \sqrt{\langle \alpha_{f_1}^\dagger \alpha_{f_1} \rangle_H} \leq \frac{\text{const}}{\sqrt{V}}. \tag{5.46}$$

It follows that we need only prove that differences of the type

$$\langle a_{f_1}(t) \ldots a_{f_l}(t) \mathscr{U}_{f_m}(\tau) \ldots \mathscr{U}_{f_n}(\tau) \rangle_H - \langle a_{f_1}(t) \ldots \mathscr{U}_{f_n}(\tau) \rangle_{H_0}$$

have absolute magnitude of order equal to or less than $1/\sqrt{V}$.

Let us use the notation

$$\langle a_{f_1}(t) \ldots a_{f_l}(t) \mathscr{U}_{f_m}(\tau) \ldots \mathscr{U}_{f_n}(\tau) \rangle_H = \Gamma(t - \tau). \tag{5.47}$$

From (5.8) we have

$$i\frac{\Gamma(t - \tau)}{dt} - \{\Omega(f_1) + \ldots + \Omega(f_l)\}\Gamma(t - \tau) = \Delta(t - \tau), \tag{5.48}$$

where

$$\Omega(f) = \sqrt{C^2\lambda^2(f) + T^2(f)}$$

and

$$\Delta(t - \tau) = \Delta_1(t - \tau) + \ldots + \Delta_l(t - \tau)$$
$$\Delta_1(t - \tau) = -\langle R_{f_1}^\dagger(t) \alpha_{f_2}(t) \ldots \alpha_{f_l}(t) \mathscr{U}_{f_m}(\tau) \ldots \mathscr{U}_{f_n}(\tau) \rangle_H$$

$$\ldots$$

$$\Delta_l(t - \tau) = -\langle \alpha_{f_1}(t) \ldots \alpha_{f_{l-1}}(t) R^\dagger_{f_l}(t) \mathscr{U}_{f_m}(\tau) \ldots \mathscr{U}_{f_n}(\tau) \rangle_H.$$

However

$$|\Delta_s(t - \tau)| \leq \sqrt{\langle \alpha_{f_1}(t) \ldots \alpha_{f_{s-1}}(t) R^\dagger_{f_s}(t) \ldots \alpha_{f_l}(t) \alpha^\dagger_{f_l}(t) \ldots R_{f_s}(t) \ldots \alpha^\dagger_1(t) \rangle_H}$$

$$\times \sqrt{\langle 2 \mathscr{U}^\dagger_{f_n}(\tau) \ldots \mathscr{U}^\dagger_{f_m}(\tau) \mathscr{U}_{f_m}(\tau) \ldots \mathscr{U}_{f_n}(\tau) \rangle}$$

$$= \sqrt{\langle \alpha_{f_1} \ldots \alpha_{f_{s-1}} R^\dagger_{f_s} \ldots \alpha_{f_l} \alpha^\dagger_{f_l} \ldots R_{f_s} \ldots \alpha^\dagger_1 \rangle_H \langle \mathscr{U}^\dagger_{f_n} \ldots \mathscr{U}_{f_n} \rangle}$$

$$\leq \sqrt{\langle \alpha_{f_1} \ldots \alpha_{f_{s-1}} R^\dagger_{f_s} \ldots \alpha_{f_l} \alpha^\dagger_{f_l} \ldots R_{f_s} \ldots \alpha^\dagger_1 \rangle_H}$$

and therefore, using (5.15),

$$|\Delta_s(t - \tau)| \leq \frac{\text{const}}{\sqrt{V}}. \tag{5.49}$$

Consequently

$$|\Delta(t - \tau)| \leq \frac{s}{\sqrt{V}} \quad \text{where} \quad s = \text{const.} \tag{5.50}$$

From (5.48) we have [cf. (5.21)]

$$\Gamma(t - \tau) = \Gamma(0) \, e^{-i\{\Omega(f_1) + \ldots + \Omega(f_l)\}(t - \tau)}$$

$$+ \exp\left[-i\{\Omega(f_1) + \ldots + \Omega(f_l)\}(t - \tau)\right] \int_0^{t-\tau} e^{i\{\Omega(f_1) + \ldots + \Omega(f_l)\}\omega} \Delta(\omega) \, d\omega. \tag{5.51}$$

Hence, using (5.50),

$$\left| \Gamma(t - \tau) - \Gamma(0) \, e^{-i\{\Omega(f_1) + \ldots + \Omega(f_l)\}(t - \tau)} \right| \leq \frac{s}{\sqrt{V}} |t - \tau|. \tag{5.52}$$

We also have

$$\langle \alpha_{f_1}(t) \ldots \alpha_{f_l}(t) \ldots \mathscr{U}_{f_m}(\tau) \ldots \mathscr{U}_{f_n}(\tau) \rangle_{H_0}$$

$$= e^{-i\{\Omega(f_1) + \ldots + \Omega(f_l)\}(t - \tau)} \langle \alpha_{f_1} \ldots \alpha_{f_l} \ldots \mathscr{U}_{f_m} \ldots \mathscr{U}_{f_n} \rangle_{H_0}. \tag{5.53}$$

From (5.52) and (5.53) its follows that

$$D \equiv |\langle \alpha_{f_1}(t) \ldots \alpha_{f_l}(t) \mathscr{U}_{f_m}(\tau) \ldots \mathscr{U}_{f_n}(\tau) \rangle_H$$

$$-\langle\alpha_{f_1}(t)\ldots\alpha_{f_l}(t)\mathscr{U}_{f_m}(\tau)\ldots\mathscr{U}_{f_n}(\tau)\rangle_{H_0}|$$

$$\leq\frac{s}{\sqrt{V}}|t-\tau|+|\langle\alpha_{f_1}\ldots\alpha_{f_l}\mathscr{U}_{f_m}\ldots\mathscr{U}_{f_n}\rangle_H-\langle\alpha_{f_1}\ldots\alpha_{f_l}\mathscr{U}_{f_m}\ldots\mathscr{U}_{f_n}\rangle_{H_0}|.$$

The second term on the right-hand side is the difference of two equal-time expectation values. As we have proved above, [cf. (5.42)], such differences are all of order $1/\sqrt{V}$. Thus, we have succeeded in proving the following inequality for the double-time expectation value

$$|\langle B_{f_1}(t)\ldots B_{f_l}\mathscr{U}_{f_m}(\tau)\ldots\mathscr{U}_{f_n}(\tau)\rangle_H-\langle B_{f_1}(t)\ldots B_{f_l}\mathscr{U}_{f_m}(\tau)\ldots\mathscr{U}_{f_n}(\tau)\rangle_{H_0}|$$

$$\leq\frac{G_1}{\sqrt{V}}|t-\tau|+\frac{G_2}{\sqrt{V}};\quad G_1,\ G_2=\text{const.}\tag{5.54}$$

These inequalities can be generalized to the case of the multiple-time expectation values,

$$\langle P_s(t_s)P_{s-1}(t_{s-1})\ldots P_1(t_1)\rangle$$
$$P_j(t)=\mathscr{U}_1^{(j)}(t)\ldots\mathscr{U}_l^{(j)}(t)\tag{5.55}$$

where $\mathscr{U}_s^{(j)}(t)$ may, as usual, denote $\alpha_f(t)$ or $\alpha_f^\dagger(t)$. In fact, we shall prove that

$$|\langle P_s(t_s)\ldots P_1(t_1)\rangle_H-\langle P_s(t_s)\ldots P_1(t_1)\rangle_{H_0}|$$
$$\leq\frac{(K_s|t_s-t_{s-1}|+\ldots+K_2|t_2-t_1|+Q_s}{\sqrt{V}}\tag{5.56}$$

where

$$K_j=\text{const},\quad Q_s=\text{const.}\tag{5.57}$$

The proof is easily given by induction. We shall assume the relation (5.56) true for the $(s-1)$-time averages and prove it for the s-time ones. Reasoning as in the double-time case, we see that it will be sufficient to prove (5.56) for the average of the type

$$P_s(t)=\alpha_{f_1}(t)\ldots\alpha_{f_l}(t).$$

For such cases we have

$$\langle P_s(t_s)P_{s-1}(t_{s-1})\ldots P_1(t_1)\rangle_{H_0}$$

$$= \exp\{-i(\Omega_{f_1} + \ldots + \Omega_{f_l})(t_s - t_{s-1})\}\langle P_s(t_{s-1})P_{s-1}(t_{s-1})\ldots P_1(t_1)\rangle_{H_0}. \tag{5.58}$$

On the other hand, from (5.8) and (5.15), and argument leading to (5.52), we see that

$$|\langle P_s(t_s)P_{s-1}(t_{s-1})\ldots P_1(t_1)\rangle_H - \exp\{-i(\Omega_{f_1} + \ldots + \Omega_{f_l})(t_s - t_{s-1})\}$$

$$\times \langle P_s(t_{s-1})P_{s-1}(t_{s-1})\ldots P_1(t_1)\rangle_H| \le \frac{K_1^{(s)}|t_s - t_{s-1}|}{\sqrt{V}}, \tag{5.59}$$

where $K_1^{(s)} = \text{const.}$ Thus,

$$|\langle P_s(t_s)\ldots P_1(t_1)\rangle_H - \langle P_s(t_s)\ldots P_1(t_1)\rangle_H|$$

$$\le \frac{K_1^{(s)}|t_s - t_{s-1}|}{\sqrt{V}} + |\langle P_s(t_{s-1})P_{s-1}(t_{s-1})\ldots P_1(t_1)\rangle_H$$

$$- \langle P_s(t_{s-1})P_{s-1}(t_{s-1})\ldots P_1(t_1)\rangle_{H_0}| \tag{5.60}$$

But the second term on the right-hand side is the difference of two $(s-1)$-time correlation functions, for which, by hypothesis, the required inequalities has been established. Thus they are also true for the s-time expectation values.

Thus, the use of H_0 gives an asymptotically exact expression for all correlation functions of the type

$$\langle P_s(t_s)\ldots P_1(t_1)\rangle.$$

As a consequence, the same is true for the Green's functions constructed from operators of this type.

Note: We could have sharpened the above inequalities, replacing $\dfrac{\text{const}}{\sqrt{V}}$ by $\dfrac{\text{const}}{V}$ everywhere, had we chosen to replace C in the definition of u_f, v_f, and H_0 by the quantity

$$C_1 = \langle L + \nu\rangle_H = \langle L^\dagger + \nu\rangle_H. \tag{5.61}$$

Since we have [cf. (4.58)]

$$(C - C_1)^2 \le \frac{\text{const}}{V},$$

all inequalities of the type (5.12), (5.15), (5.35) remain valid; but we now also have the following useful relations:

$$|\langle A_{f_1} \ldots R_f \ldots A_{f_n} \rangle| \leq \frac{\text{const}}{V}$$

$$|\langle A_{f_1} \ldots R_f^\dagger \ldots A_{f_n} \rangle| \leq \frac{\text{const}}{V}. \tag{5.62}$$

To prove them it is sufficient to expand the expressions

$$\langle A_{f_1} \ldots R_f \ldots A_{f_n} \rangle; \quad \langle A_{f_1} \ldots R_f^\dagger \ldots A_{f_n} \rangle, \tag{5.63}$$

by expressing all the a's and $a^\dagger$'s in terms of α's and $\alpha^\dagger$'s. Then we can represent the expressions (5.63) as sums of terms of the type

$$\langle \alpha^\dagger \ldots \alpha \rangle_H$$
$$\langle (L + \nu - C_1) \ldots \alpha \rangle_H; \quad \langle \alpha^\dagger \ldots (L + \nu - C_1) \rangle_H$$
$$\langle (L^\dagger + \nu - C_1) \ldots \alpha \rangle_H; \quad \langle \alpha^\dagger \ldots (L^\dagger + \nu - C_1) \rangle_H$$
$$\text{const} \langle L + \nu - C_1 \rangle_H \equiv 0; \quad \text{const} \langle L^\dagger + \nu - C_1 \rangle_H \equiv 0 \tag{5.64}$$

and "commutation" terms of order $1/V$. (The last two terms in (5.64) are zero in virtue of (5.61).) Applying to (5.64) the inequality

$$|\langle AB \rangle| \leq \sqrt{|\langle AA^\dagger \rangle|} \sqrt{|\langle B^\dagger B \rangle|},$$

and also (5.35), we see that all these terms will be of order $1/V$, so that (5.62) is proved.

We shall now use these additional relations. Consider expression of the type

$$\langle \alpha_{f_1}^\dagger \ldots \alpha_{f_n}^\dagger \rangle,$$

which is obviously independent of t. For this reason we have

$$\frac{d}{dt} \langle \alpha_{f_1}^\dagger \ldots \alpha_{f_n}^\dagger \rangle_H = \left\langle \frac{d}{dt} \alpha_{f_1}^\dagger \ldots \alpha_{f_n}^\dagger \right\rangle_H + \ldots + \left\langle \alpha_{f_1}^\dagger \ldots \frac{d}{dt} \alpha_{f_n}^\dagger \right\rangle_H = 0. \tag{5.65}$$

Consequently, we get from (5.7)

$$(\Omega(f_1) + \ldots + \ldots \Omega(f_n)) \langle \alpha_{f_1}^\dagger \ldots \alpha_{f_n}^\dagger \rangle_H = \langle R_{f_1} \ldots \alpha_{f_n}^\dagger \rangle_H + \ldots + \langle \alpha_{f_1}^\dagger \ldots R_{f_n} \rangle_H. \tag{5.66}$$

According to (5.62), we have

$$|\langle R_{f_1} \ldots \alpha^\dagger_{f_n} \rangle_H + \ldots + \langle \alpha^\dagger_{f_1} \ldots R_{f_n} \rangle_H| \le \frac{D}{V}; \quad D = \text{const.} \tag{5.67}$$

Therefore

$$|\langle \alpha^\dagger_{f_1} \ldots \alpha^\dagger_{f_n} \rangle| \le \frac{D}{V(\Omega(f_1) + \ldots + \ldots \Omega(f_n))}. \tag{5.68}$$

Hence, by taking the complex conjugate, we also have

$$|\langle \alpha_{f_1} \ldots \alpha_{f_n} \rangle| \le \frac{D}{V(\Omega(f_1) + \ldots + \ldots \Omega(f_n))}. \tag{5.69}$$

Using our new inequalities (5.68) and (5.69) in place of the old ones (5.37) and (5.38) [but keeping (5.36)] we can prove the following relation,

$$|\langle \mathcal{U}_{f_1} \ldots \mathcal{U}_{f_s} \rangle_H - \langle \mathcal{U}_{f_1} \ldots \mathcal{U}_{f_s} \rangle_{H_0}| \le \frac{\text{const}}{V} \tag{5.70}$$

which replaces (5.42).

In a similar way we can sharpen the inequalities for all the correlation functions of the type considered above. Rather than give a general proof, we shall merely find a bound for the difference

$$\langle \alpha_{f_1}(t) \ldots \alpha_{f_l}(t) \alpha^\dagger_{g_1}(\tau) \ldots \alpha^\dagger_{g_r}(\tau) \rangle_H - \langle \alpha_{f_1}(t) \ldots \alpha^\dagger_{g_r}(\tau) \rangle_{H_0}. \tag{5.71}$$

Defining

$$\Gamma_{H,H_0}(t - \tau) = \langle \alpha_{f_1}(t) \ldots \alpha^\dagger_{g_r}(\tau) \rangle_{H,H_0}, \tag{5.72}$$

we have

$$i\frac{d\Gamma_H(t - \tau)}{dt} = (\Omega(f_1) + \ldots + \Omega(f_l))\Gamma_H(t - \tau) + \Delta_H(t - \tau), \tag{5.73}$$

where

$$\Delta_H(t - \tau) = -\sum_j \langle \alpha_{f_1}(t) \ldots R^\dagger_{f_j}(t) \ldots \alpha_{f_l}(t)\alpha^\dagger_{g_1}(\tau) \ldots \alpha^\dagger_{g_r}(\tau) \rangle_H. \tag{5.74}$$

Differentiating (5.74) with respect to t, we find

$$i\frac{d\Delta_H(t - \tau)}{dt} = -(\Omega(g_1) + \ldots + \Omega(g_r))\Delta_H(t - \tau) + \zeta_H(t - \tau), \tag{5.75}$$

where

$$\zeta(t - \tau) = -\sum_{j,s} \langle \alpha_{f_1}(t) \ldots R^{\dagger}_{f_j}(t) \ldots \alpha_{f_l}(t) \alpha^{\dagger}_{g_1}(\tau) \ldots R_{f_s}(\tau) \ldots \alpha^{\dagger}_{g_r}(\tau) \rangle_H.$$
(5.76)

However, from (5.15), we have

$$|\zeta(t - \tau)| \leq \frac{Q}{V}, \quad \text{where} \quad Q = \text{const.}$$
(5.77)

Thus, from (5.75) we obtain in the usual way [cf., e.g., (5.48)-(5.52)]

$$\left| \Delta_H(t - \tau) - \Delta_H(0) \exp\{i[\Omega(g_1) + \ldots + \Omega(g_r)](t - \tau)\} \right| \leq \frac{Q}{V}|t - \tau|. \quad (5.78)$$

From (5.62) and (5.74), we obtain

$$|\Delta_H(0)| \leq \frac{Q_1}{V}; \quad Q_1 = \text{const.}$$
(5.79)

Therefore,

$$|\Delta_H(t - \tau)| \leq \frac{Q_1 + Q|t - \tau|}{V}.$$
(5.80)

Substituting this inequality in (5.73), we find

$$\left| \Gamma_H(t - \tau) - \Gamma_H(0) \exp\{i[\Omega(f_1) + \ldots + \Omega(F_l)](t - \tau)\} \right|$$
$$\leq \frac{Q_1|t - \tau| + Q|t - \tau|^2 \frac{1}{2}}{V}$$
(5.81)

On the other hand we have

$$\Gamma_{H_0}(t - \tau) = \Gamma_{H_0}(0) \exp\{i[\Omega(f_1) + \ldots + \Omega(F_l)](t - \tau)\},$$
(5.82)

so that

$$|\Gamma_H(t - \tau) - \Gamma_{H_0}(t - \tau)| \leq |\Gamma_H(0) - \Gamma_{H_0}(0)|$$
$$+ \frac{Q_1|t - \tau| + Q|t - \tau|^2 \frac{1}{2}}{V}.$$
(5.83)

By (5.70),

$$|\Gamma_H(0) - \Gamma_{H_0}(0)| = \langle \alpha_{f_1} \ldots \alpha_{f_l} \alpha^{\dagger}_{g_1} \ldots \alpha^{\dagger}_{g_r} \rangle_H$$

$$- \langle \alpha_{f_1} \ldots \alpha_{f_l} \alpha^\dagger_{g_1} \ldots \alpha^\dagger_{g_r} \rangle_{H_0} \leq \frac{Q_2}{V}; \quad Q_2 = \text{const.}, \quad (5.84)$$

and so, finally

$$|\langle \alpha_{f_1}(t) \ldots \alpha_{f_l}(t) \alpha^\dagger_{g_1}(\tau) \ldots \alpha^\dagger_{g_r}(\tau) \rangle_H - \langle \alpha_{f_1}(t) \ldots \alpha^\dagger_{g_r}(\tau) \rangle_{H_0}|$$
$$\leq \frac{Q_2 + Q_1|t - \tau| + Q|t - \tau|^2 \frac{1}{2}}{V}. \quad (5.85)$$

By proceeding further along these lines we could easily replace quantities of order $1/\sqrt{V}$ by quantities of order $1/V$ in all inequalities obtained in this section.

6.   Green's Function (Case $\nu = 0$)

In the preceding section we derived all the necessary limits for the Green's functions in the case $\nu > 0$. Since some of the inequalities used in that section [e.g., (4.58)] are meaningless when $\nu = 0$, the results cannot be carried over directly to this case, which therefore requires special consideration.

Since now the operators L and $L^\dagger$ do not take asymptotically well-defined values in the lowest energy eigenstate ϕ_H, we shall work with fermion operators defined rather differently from those used earlier,

$$\alpha_f = u_f a_f + v_f a^\dagger_{-f} \frac{L}{C}, \quad (6.1)$$

where

$$u_f = \frac{1}{\sqrt{2}} \sqrt{1 + \frac{T(f)}{\sqrt{C^2 \lambda^2(f) + T^2(f)}}}$$

$$v_f = -\frac{\in(f)}{\sqrt{2}} \sqrt{1 - \frac{T(f)}{\sqrt{C^2 \lambda^2(f) + T^2(f)}}}. \quad (6.2)$$

These operators do not obey Fermi commutation relation exactly; however, they do obey them with asymptotic accuracy.

To obtain the required limits we shall first have to establish a number of inequalities. Consider first of all the expressions

$$\sum \Omega(f) \alpha^\dagger_f \alpha_f,$$

where

$$\Omega(f) = \sqrt{C^2\lambda^2(f) + T^2(f)}. \tag{6.2'}$$

Substituting (6.1), we have

$$\sum \Omega(f)\alpha_f^\dagger \alpha_f = \sum \Omega(f)\left\{u_f a_f^\dagger + v_f \frac{L^\dagger}{C}a_{-f}\right\}\left\{u_f a_f + v_f a_{-f}^\dagger \frac{L}{C}\right\}$$

$$= \sum \Omega(f)\left\{u_f^2 a_f^\dagger a_f + v_f^2 \frac{L^\dagger}{C}a_{-f} a_{-f}^\dagger \frac{L^\dagger}{C} + u_f v_f \frac{L^\dagger}{C}a_{-f} a_f + u_f v_f a_f^\dagger a_{-f}^\dagger \frac{L^\dagger}{C}\right\}.$$

However, we have

$$\sum \Omega(f) v_f^2 \frac{L^\dagger}{C}a_{-f} a_{-f}^\dagger \frac{L^\dagger}{C} = -\sum \Omega(f) v_f^2 \frac{L^\dagger}{C}a_f^\dagger a_f \frac{L^\dagger}{C} + \sum \Omega(f) v_f^2 \frac{L^\dagger L}{C^2}$$

$$= -\sum \Omega(f) v_f^2 a_f^\dagger \frac{L^\dagger}{C}a_f \frac{L^\dagger}{C} + \sum \Omega(f) v_f^2 \frac{L^\dagger L}{C^2}.$$

Moreover, since

$$u_f v_f = -\frac{C\lambda(f)}{2\Omega(f)},$$

we also have

$$-\sum \Omega(f)\left\{u_f v_f \frac{L^\dagger}{C}a_{-f} a_f + u_f v_f a_f^\dagger a_{-f}^\dagger \frac{L^\dagger}{C}\right\} = VL^\dagger L.$$

Consequently,

$$\sum \Omega(f)\alpha_f^\dagger \alpha = \sum_f \Omega(f)\left\{u_f^2 a_f^\dagger a_f - v_f^2 a_f^\dagger \frac{L^\dagger L}{C^2}a_f\right\}$$

$$+ \sum \Omega(f) v_f^2 \frac{L^\dagger L}{C^2} - VL^\dagger L$$

$$= \sum_f \Omega(f)(u_f^2 - v_f^2)a_f^\dagger a_f - \sum_f \Omega(f) v_f^2 a_f^\dagger \frac{L^\dagger L - C^2}{C^2}a_f$$

$$+ \sum \Omega(f) v_f^2 \frac{L^\dagger L}{C^2} - VL^\dagger L$$

hold. Since,

$$\Omega(f)(u_f^2 - v_f^2) = T(f)$$

we have

$$H = \sum T(f)a_f^\dagger a_f - V\frac{L^\dagger L}{2}$$

$$= \sum \Omega(f) a_f^\dagger a_f + V \frac{L^\dagger L}{2} - \sum \Omega(f) v_f^2 \frac{L^\dagger L}{C^2}$$

$$+ \sum \Omega(f) v_f^2 a_f^\dagger \frac{L^\dagger L - C^2}{C^2} a_f.$$

As a result, we obtain

$$H = \sum \Omega(f) a_f^\dagger a_f + \frac{V}{2} \left\{ C^2 - \frac{2}{V} \sum \Omega(f) v_f^2 \right\} + \frac{V(L^\dagger L - C^2)}{2}$$

$$- \sum \Omega(f) v_f^2 \frac{L^\dagger L - C^2}{C^2} + \sum \Omega(f) v_f^4 \frac{L^\dagger L - C^2}{C^2}$$

$$+ \sum \Omega(f) v_f^2 \left\{ a_f^\dagger \frac{L^\dagger L - C^2}{C^2} a_f - v_f^2 \frac{L^\dagger L - C^2}{C^2} \right\}.$$

On the other hand

$$C^2 \frac{V}{2} - \sum \Omega(f) v_f^2 + \sum \Omega(f) v_f^4 = C^2 \frac{V}{2} - \sum \Omega(f) u_f^2 v_f^2$$

$$= \frac{V}{2} \left\{ C^2 - \frac{1}{2V} C^2 \sum \frac{\lambda^2(f)}{\sqrt{C^2 \lambda^2(f) + T^2(f)}} \right\} = \frac{V}{2} C^2 F'(C^2);$$

where

$$F(C^2) = C^2 - \frac{2}{V} \sum \Omega(f) v_f^2.$$

Hence

$$H = \sum \Omega(f) a_f^\dagger a_f - w + \frac{V}{2} (L^\dagger L - C^2) F'(C^2) + F(C^2), \qquad (6.3)$$

where

$$w = - \sum \Omega(f) v_f^2 \left\{ a_f^\dagger \frac{L^\dagger L - C^2}{C^2} a_f - v_f^2 \frac{L^\dagger L - C^2}{C^2} \right\}. \qquad (6.4)$$

By definition [cf. (3.8) and (4.39)], C^2 is a root of the equation

$$F'(x) = 0.$$

Moreover, we have

$$\langle \phi_H^* H \phi_H \rangle \leq F(C^2).$$

It follows that

$$\langle \phi_H^* \sum \Omega(f) a_f^\dagger a_f \phi_H \rangle \leq \langle \phi_H^* w \phi_H \rangle. \qquad (6.5)$$

We now set about finding an inequality for the expectation value of w. We recall the definition (6.1) of the operators α:

$$\alpha_f^\dagger = u_f a_f^\dagger + v_f \frac{L^\dagger}{C} a_{-f},$$

$$\alpha_{-f} = -v_f a_f^\dagger \frac{L}{C} + u_f a_{-f}.$$

From this results

$$u_f^2 a_f^\dagger - v_f \frac{L^\dagger}{C} \alpha_{-f} = u_f^2 a_f^\dagger + v_f^2 \frac{L^\dagger}{C} a_f^\dagger \frac{L}{C} = a_f^\dagger \left(u_f^2 + v_f^2 \frac{L^\dagger L}{C^2} \right).$$

We put

$$\eta_f^\dagger = a_f^\dagger v_f^2 \frac{C^2 - L^\dagger L}{C^2} = v_f^2 \frac{C^2 - L^\dagger L}{C^2} a_f^\dagger + \frac{2\lambda(f) L^\dagger}{VC^2} a_{-f}$$

$$\eta_f = v_f^2 \frac{C^2 - L^\dagger L}{C^2}{}_f = v_f^2 a_f \frac{C^2 - L^\dagger L}{C^2} + \frac{2\lambda(f)}{VC^2} a_{-f}^\dagger L; \qquad (6.6)$$

then

$$a_f^\dagger = u_f \alpha^\dagger - v_f \frac{L^\dagger}{C} \alpha_{-f} + \eta_f^\dagger$$

$$a_f = u_f \alpha - v_f \alpha_{-f}^\dagger \frac{L}{C} + \eta_f. \qquad (6.7)$$

Let us now go back to (6.4) and write

$$w = w_1 + w_2 + w_3;$$

$$w_1 = \sum \Omega(f) v_f^2 u_f \alpha_f^\dagger \frac{C^2 - L^\dagger L}{C^2} a_f$$

$$= \sum \Omega(f) v_f^2 u_f \alpha_f^\dagger a_f \frac{C^2 - L^\dagger L}{C^2} + \sum \Omega(f) v_f^2 u_f \alpha_f^\dagger \frac{2\lambda(f)}{VC^2} a_{-f}^\dagger L,$$

$$w_2 = \sum \Omega(f) v_f^2 \eta_f^\dagger \frac{C^2 - L^\dagger L}{C^2} a_f$$

$$= \sum \Omega(f) v_f^2 \eta_f^\dagger a_f \frac{C^2 - L^\dagger L}{C^2} + \sum \Omega(f) v_f^2 \eta_f^\dagger \frac{2\lambda(f)}{VC^2} a_{-f}^\dagger L,$$

$$w_3 = \sum \Omega(f) v_f^2 \left\{ -v_f \frac{L^\dagger}{C} \alpha_{-f} \frac{C^2 - L^\dagger L}{C^2} a_f - v_f^2 \frac{C^2 - L^\dagger L}{C^2} \right\}$$

$$= -\sum \Omega(f) v_f^3 \frac{L^\dagger}{C} \left\{ \frac{L^\dagger L}{C^2} \alpha_{-f} - \alpha_{-f} \frac{L^\dagger L}{C^2} \right\} a_f$$

$$+ \sum \Omega(f) v_f^2 \left\{ -v_f \frac{L^\dagger}{C} \left(\frac{C^2 - L^\dagger L}{C^2} \right) (\alpha_{-f} a_f + a_f \alpha_{-f}) \right.$$

$$\left. - v_f^2 \frac{C^2 - L^\dagger L}{C^2} \right\} + \sum \Omega(f) v_f^3 \frac{L^\dagger}{C} \left(\frac{C^2 - L^\dagger L}{C^2} \right) a_f \alpha_{-f}.$$

We can now find bounds for w_1, w_2, and w_3. For w_1 we use the inequality (4.47) proved above, which we write in the form

$$\left\langle \phi_H^* \left(\frac{C^2 - L^\dagger L}{C^2} \right) \phi_H \right\rangle \leq \frac{G}{V}$$

$$\left\langle \phi_H^* \left(\frac{C^2 - L L^\dagger}{C^2} \right) \phi_H \right\rangle \leq \frac{G}{V} \qquad (6.8)$$

where $G = $ const. We then have

$$\left| \langle \phi_H^* w_1 \phi_H \rangle \right| \leq \sum \Omega(f) v_f^2 u_f \left| \left\langle \phi_H^* \alpha_f^\dagger a_f \left(\frac{C^2 - L^\dagger L}{C^2} \right) \phi_H \right\rangle \right|$$

$$+ \sum \Omega(f) v_f^2 u_f \left| \langle \phi_H^* \alpha_f^\dagger a_f^\dagger L \phi_H \rangle \right| \frac{2|\lambda(f)|}{V C^2}$$

$$\leq \sum \Omega(f) v_f^2 u_f \sqrt{\langle \phi_H^* \alpha_f^\dagger a_f a_f^\dagger \alpha_f \rangle} \sqrt{\left\langle \phi_H^* \left(\frac{C^2 - L^\dagger L}{C^2} \right)^2 \phi_H \right\rangle}$$

$$+ \sum \Omega(f) v_f^2 u_f \sqrt{\langle \phi_H^* \alpha^\dagger \alpha_f \phi_H \rangle \langle \phi_H^* L^\dagger a_{-f} a_{-f}^\dagger L \phi_H \rangle} \frac{2|\lambda(f)|}{V C^2}$$

$$\leq \sum \Omega(f) v_f^2 u_f \left(\frac{G}{V} \right)^{1/2} \sqrt{\langle \phi_H^* \alpha^\dagger \alpha_f \phi_H \rangle}$$

$$+ \frac{1}{V} \sum \Omega(f) v_f^2 u_f \frac{2|\lambda(f)|}{V C^2} |L| \langle \phi_H^* \alpha^\dagger \alpha_f \phi_H \rangle$$

$$\leq \sqrt{\langle \phi_H^* \sum \Omega(f) \alpha^\dagger \alpha_f \phi_H \rangle} \left\{ \sqrt{\frac{G}{V} \sum_f \Omega(f) v_f^4 u_f^2} \right.$$

$$\left. + \frac{2|L|}{C^2 \sqrt{V}} \sqrt{\frac{1}{V} \sum_f \Omega(f) v_f^4 u_f^2 |\lambda^2(f)|} \right\}.$$

Consequently, we obtain

$$|\langle \phi_H^* w_1 \phi_H \rangle| \leq R_1 \sqrt{\langle \phi_H^* \sum \Omega(f) \alpha^\dagger \alpha_f \phi_H \rangle}; \quad R_1 = \text{conts}.$$

In an entirely analogous way we can show that

$$|\langle \phi_H^* w_2 \phi_H \rangle| \le R_2, \quad \text{where} \quad R_2 = \text{conts.}$$

We now come to w_3. We notice that

$$\alpha_{-f} a_f + a_f \alpha_{-f} = \left(-v_f a_f^\dagger \frac{L}{C} + u_f a_{-f} \right) a_f$$

$$+ a_f \left(-v_f a_f^\dagger \frac{L}{C} + u_f a_{-f} \right) = -\frac{v_f}{C} (a_f^\dagger L a_f + a_f a_f^\dagger L)$$

$$= -\frac{v_f}{C} (a_f^\dagger a_f + a_f a_f^\dagger) L = -\frac{v_f}{C} L.$$

Therefore [cf. the expression for w_3]:

$$\Delta \equiv \sum_f \Omega(f) v_f^2 \left\{ -v_f \frac{L^\dagger}{C} \left(\frac{C^2 - L^\dagger L}{C^2} \right) (\alpha_{-f} a_f + a_f \alpha_{-f}) - v_f^2 \frac{C^2 + L^\dagger L}{C^2} \right\}$$

$$= \sum_f \Omega(f) v_f^2 \left\{ v_f^2 \frac{L^\dagger}{C} \left(\frac{C^2 - L^\dagger L}{C^2} \right) \frac{L}{C} - v_f^2 \frac{C^2 + L^\dagger L}{C^2} \right\}$$

$$= \sum_f \Omega(f) v_f^4 \frac{L^\dagger}{C^2} \left(\frac{LL^\dagger + L^\dagger L}{C^2} \right) L - \sum_f \Omega(f) v_f^4 \left(\frac{C^2 - L^\dagger L}{C^2} \right)^2$$

and further [cf. (A.18)]

$$\langle \phi_H^\dagger \Delta \phi_H \rangle \le \sum_f \Omega(f) v_f^4 \left\langle \phi_H^* \frac{L^\dagger}{C} \left(\frac{LL^\dagger - L^\dagger L}{C^2} \right) \frac{L}{C} \phi_H \right\rangle$$

$$= \frac{2}{V^2 C^2} \sum_{f,f'} \Omega(f) v_f^4 \lambda^2(f') \left\langle \phi_H^* \frac{L^\dagger}{C} (1 - a_{f'}^\dagger a_{f'} - a_{-f'}^\dagger a_{-f'}) \frac{L}{C} \phi_H \right\rangle$$

$$\le 2 \frac{|L|^2}{C^4} \frac{1}{V} \sum_f \Omega(f) v_f^4 \frac{1}{V} \sum_{f'} \lambda^2(f') \le \text{const.}$$

We also have

$$\sum_f \Omega(f) |v_f|^3 \left\langle \phi_H^* \frac{L^\dagger}{C} \left(\frac{L^\dagger L}{C^2} \alpha_{-f} - \alpha_{-f} \frac{L^\dagger L}{C^2} \right) a_f \phi_H \right\rangle \le \text{const}$$

$$\sum_f \Omega(f) |v_f|^3 \left\langle \phi_H^* \frac{L^\dagger}{C} \left(\frac{C^2 - L^\dagger L}{C^2} \right) a_f \alpha_{-f} \phi_H \right\rangle$$

$$\leq R_3 \sqrt{\sum_f \Omega(f) \langle \phi_H^* \alpha_f^\dagger \alpha_f \phi_H \rangle}.$$

Thus, collecting the expressions for w_1, w_2, and w_3, we find:

$$\langle \phi_H^* w \phi_H \rangle \leq \gamma_1 \sqrt{\langle \phi_H^* \sum_f \Omega(f) \alpha_f^\dagger \alpha_f \phi_H \rangle} + \gamma_2,$$

where

$$\gamma_1 = \text{const}, \quad \gamma_2 = \text{const}.$$

Substituting this inequality in (6.5), we obtain

$$\langle \phi_H^* \sum_f \Omega(f) \alpha_f^\dagger \alpha_f \phi_H \rangle \leq \gamma_1 \sqrt{\langle \phi_H^* \sum_f \Omega(f) \alpha_f^\dagger \alpha_f \phi_H \rangle} + \gamma_2.$$

If we put

$$x = \langle \phi_H^* \sum_f \Omega(f) \alpha_f^\dagger \alpha_f \phi_H \rangle,$$

then

$$x^2 - \gamma_1 x \leq \gamma_2, \quad \left(x - \frac{\gamma_1}{2}\right)^2 \leq \gamma_2 + \frac{\gamma_1^2}{4},$$

and so

$$x \leq \frac{\gamma_1}{2} + \sqrt{\gamma_2 + \frac{\gamma_1^2}{4}}.$$

Thus,

$$\left\langle \phi_H \frac{1}{V} \sum_f \Omega(f) \alpha_f^\dagger \alpha_f \phi_H \right\rangle \leq \frac{R}{V}, \tag{6.9}$$

where

$$R = \left(\frac{\gamma_1}{2} + \sqrt{\gamma_2 + \frac{\gamma_1^2}{4}}\right)^2 = \text{const}.$$

We may now turn to the equation of motion. For $\nu = 0$ we obtain from equations (5.1), (5.2),

$$i\frac{da_f}{dt} = T(f)a_f - \lambda(f)a_{-f}^\dagger L$$

$$i\frac{da_{-f}}{dt} = T(f)a_{-f} - \lambda(f)a_f^\dagger L. \tag{6.10}$$

Therefore

$$i\frac{dL}{dt} = \frac{1}{V}\sum \lambda(f)\{T(f)a_{-f} + \lambda(f)a_f^\dagger L\}a_f$$

$$+ \frac{1}{V}\sum \lambda(f)a_{-f}\{T(f)a_f + \lambda(f)a_{-f}^\dagger L\}$$

$$= \frac{2}{V}\sum \lambda(f)T(f)a_{-f}a_f + \frac{1}{V}\sum \lambda^2(f)(a_f^\dagger a_f + a_{-f}a_{-f}^\dagger)L$$

$$= \frac{2}{V}\sum \lambda(f)T(f)a_{-f}a_f + \frac{1}{V}\sum \lambda^2(f)(a_f^\dagger a_f + a_f a_f^\dagger)L$$

$$= \frac{2}{V}\sum \lambda(f)T(f)a_{-f}a_f + \frac{1}{V}\sum \lambda^2(f)(2a_f^\dagger a_f - 1)L.$$

We now notice that

$$-2\lambda(f)T(f)u_f v_f\frac{L}{C} + \lambda^2(f)(2v_f^2 - 1)L$$

$$= \lambda^2(f)\frac{T(f)}{\Omega(f)}L + \lambda^2(f)\left(1 - \frac{T(f)}{\Omega(f)} - 1\right)L = 0.$$

Consequently,

$$i\frac{dL}{dt} = D_1 + D_2$$

$$D_1 = \frac{2}{V}\sum \lambda(f)T(f)\left\{a_{-f}a_f + u_f v_f\frac{L}{C}\right\}$$

$$D_2 = \frac{2}{V}\sum \lambda^2(f)(a_f^\dagger a_f - v_f^2)L. \qquad (6.11)$$

Now, from (6.7), we have

$$a_{-f}a_f + u_f v_f\frac{L}{C} = \left(u_f\alpha_{-f} + v_f\alpha_f^\dagger\frac{L}{C}\right)\left(u_f\alpha_f - v_f\alpha_{-f}^\dagger\frac{L}{C}\right)$$

$$+ \eta_{-f}a_f + a_{-f}\eta_f - \eta_{-f}\eta_f + u_f v_f\frac{L}{C}$$

$$= u_f^2\alpha_{-f}\alpha_f - v_f^2\alpha_f^\dagger\frac{L}{C}\alpha_{-f}^\dagger\frac{L}{C} - u_f v_f\alpha_{-f}\alpha_f^\dagger\frac{L}{C}$$

$$+ u_f v_f\alpha_f^\dagger\frac{L}{C}\alpha_f + \eta_{-f}a_f + a_{-f}\eta_f - \eta_{-f}\eta_f + u_f v_f\frac{L}{C}$$

$$= u_f^2\alpha_{-f}\alpha_f - u_f v_f(\alpha_{-f}\alpha_{-f}^\dagger + \alpha_{-f}^\dagger\alpha_{-f} - 1)\frac{L}{C}$$

$$- v_f^2 \alpha_f^\dagger \frac{L}{C} \alpha_{-f}^\dagger \frac{L}{C} + u_f v_f \left(\alpha_f^\dagger \frac{L}{C} \alpha_f + \alpha_{-f}^\dagger \alpha_{-f} \frac{L}{C} \right)$$
$$+ \eta_{-f} a_f + a_{-f} \eta_f - \eta_{-f} \eta_f. \tag{6.12}$$

We also have

$$\alpha_{-f} \alpha_{-f}^\dagger + \alpha_{-f}^\dagger \alpha_{-f} - 1 = \left(-v_f a_f^\dagger \frac{L}{C} + u_f a_{-f} \right) \left(-v_f \frac{L^\dagger}{C} a_f + u_f a_{-f}^\dagger \right)$$
$$+ \left(-v_f \frac{L^\dagger}{C} a_f + u_f a_{-f}^\dagger \right) \left(-v_f a_f^\dagger \frac{L}{C} + u_f a_{-f} \right) - 1$$
$$= v_f^2 a_f^\dagger \frac{LL^\dagger}{C^2} a_f + u_f^2 a_{-f} a_{-f}^\dagger - u_f v_f a_f^\dagger \frac{L}{C} a_{-f}^\dagger - u_f v_f a_{-f} \frac{L^\dagger}{C} a_f$$
$$+ v_f^2 \frac{L^\dagger}{C} a_f a_f^\dagger \frac{L}{C} + u_f^2 a_{-f}^\dagger a_{-f} - u_f v_f a_{-f}^\dagger a_f^\dagger \frac{L}{C} - u_f v_f \frac{L^\dagger}{C} a_f a_{-f} - 1$$
$$= v_f^2 a_f^\dagger \frac{LL^\dagger - L^\dagger L}{C^2} a_f + v_f^2 \frac{L^\dagger L}{C^2} + u_f^2 - 1 - u_f v_f a_f^\dagger \left(\frac{L}{C} a_{-f}^\dagger - a_{-f}^\dagger \frac{L}{C} \right)$$
$$- u_f v_f \left(a_{-f} \frac{L^\dagger}{C} - \frac{L^\dagger}{C} a_{-f} \right) a_f$$

and therefore

$$\alpha_{-f} \alpha_{-f}^\dagger + \alpha_{-f}^\dagger \alpha_{-f} - 1 = \frac{2}{V^2} \sum_g v_f^2 a_f^\dagger \frac{\lambda^2(g)}{C^2} (1 - a_g^\dagger a_g - a_{-g}^\dagger a_{-g}) a_f$$
$$+ v_f^2 \frac{L^\dagger L - C^2}{C^2} + u_f v_f a_f^\dagger a_f \frac{2\lambda(f)}{V} + u_f v_f a_f^\dagger a_f \frac{2\lambda(f)}{V}. \tag{6.13}$$

As a result we obtain

$$\langle \phi_H^* D_1 D_1^\dagger \phi_H \rangle = \frac{2}{V} \sum_f \lambda(f) T(f) u_f^2 \langle \phi_H^* \alpha_{-f} \alpha_f D_1^\dagger \phi_H \rangle$$

$$- \frac{2}{V} \sum_f \lambda(f) T(f) \left\langle \phi_H^* \alpha_f^\dagger \left\{ v_f^2 \frac{L}{C} \alpha_{-f}^\dagger \frac{L}{C} - u_f v_f \left(\frac{L}{C} \alpha_f + \alpha_f \frac{L}{C} \right) \right\} D_1^\dagger \phi_H \right\rangle$$

$$- \frac{4}{V^3} \sum_{f,g} \lambda(f) T(f) u_f v_f^3 \frac{\lambda^2(g)}{C^2} \left\langle \phi_H^* a_f^\dagger (1 - a_g^\dagger a_g - a_{-g}^\dagger a_{-g}) a_f \frac{L}{C} D_1^\dagger \phi_H \right\rangle$$

$$+ \frac{2}{V} \sum_f \lambda(f) T(f) \langle \phi_H^* (\eta_{-f} a_f - \eta_f a_{-f} + [a_{-f} \eta_f + \eta_f a_{-f}] - \eta_{-f} \eta_f) D_1^\dagger \phi_H \rangle$$

$$- \frac{2}{V} \sum_f \lambda(f) T(f) \left\langle \phi_H^* \left\{ v_f^2 \frac{L^\dagger L - C^2}{C^2} + 2 u_f v_f a_f^\dagger a_f \frac{2\lambda(f)}{V} \right\} \frac{L}{C} D_1^\dagger \phi_H \right\rangle.$$

Taking into account that

$$\langle \phi_H^* \alpha_{-f} \alpha_f D_1^\dagger \phi_H \rangle = \langle \phi_H^* \alpha_{-f} D_1^\dagger \alpha_f \phi_H \rangle$$
$$+ \langle \phi_H^* \alpha_{-f} (\alpha_f D_1^\dagger - D_1^\dagger \alpha_f) \phi_H \rangle,$$

we can now use (6.8) and (6.9) to establish that

$$\langle \phi_H^* D_1 D_1^\dagger \phi_H \rangle \le \frac{\Gamma_1}{V}, \quad \Gamma_1 = \text{const.} \tag{6.14}$$

In the same way we obtain

$$\langle \phi_H^* D_1^\dagger D_1 \phi_H \rangle \le \frac{\Gamma_2}{V}, \quad \Gamma_2 = \text{const.} \tag{6.15}$$

We now turn to the expression D_2. We have

$$a_f^\dagger a_f - v_f^2 = a_f^\dagger \eta_f + \eta_f^\dagger a_f - \eta_f^\dagger \eta_f$$
$$+ \left(u_f a_f^\dagger - v_f \frac{L^\dagger}{C} \alpha_{-f} \right) \left(u_f a_f - v_f \alpha_{-f}^\dagger \frac{L}{C} \right) - v_f^2$$
$$= u_f^2 a_f^\dagger a_f + v_f^2 \frac{L^\dagger}{C} \alpha_{-f} \alpha_{-f}^\dagger \frac{L}{C} - v_f^2 - u_f v_f \alpha_{-f}^\dagger \alpha_{-f}^\dagger \frac{L}{C}$$
$$- u_f v_f \frac{L^\dagger}{C} \alpha_{-f} \alpha_f + a_f^\dagger \eta_f + \eta_f^\dagger a_f + \eta_f^\dagger \eta_f$$
$$= u_f^2 a_f^\dagger a_f + v_f^2 \frac{L^\dagger}{C} (\alpha_{-f} \alpha_{-f}^\dagger + \alpha_{-f}^\dagger \alpha_{-f} - 1) \frac{L}{C} - v_f^2 \frac{C^2 - L^\dagger L}{C^2}$$
$$- v_f^2 \frac{L^\dagger}{C} \alpha_{-f} \alpha_{-f}^\dagger \frac{L}{C} - u_f v_f a_f^\dagger \alpha_{-f}^\dagger \frac{L}{C} - u_f v_f \frac{L}{C} \alpha_{-f} a_f$$
$$+ a_f^\dagger \eta_f + \eta_f^\dagger a_f - \eta_f^\dagger \eta_f. \tag{6.16}$$

From this relation and the inequalities (6.8) and (6.9), we can show that

$$\langle \phi_H^* D_2 D_2^\dagger \phi_H \rangle \le \frac{\Gamma_3}{V}, \quad \Gamma_3 = \text{const}$$
$$\langle \phi_H^* D_2^\dagger D_2 \phi_H \rangle \le \frac{\Gamma_3}{V}. \tag{6.17}$$

It follows from (6.11) that

$$\left\langle \phi_H^* \left(\frac{dL}{dt} \right)^\dagger \frac{dL}{dt} \phi_H \right\rangle \le \frac{\Gamma}{V}; \quad \Gamma = \text{const}$$

$$\left\langle \phi_H^* \frac{dL}{dt}\left(\frac{dL}{dt}\right)^\dagger \phi_H \right\rangle \leq \frac{\Gamma}{V}. \qquad (6.18)$$

Let us go back again to the equations of motion (6.10). Using (6.1) and (6.2), we have

$$
\begin{aligned}
i\frac{d\alpha_f^\dagger}{dt} &= i\frac{d}{dt}\left(u_f a_f^\dagger + v_f \frac{L^\dagger}{C} a_{-f}\right) = u_f\, i\frac{da_f^\dagger}{dt} + v_f \frac{L^\dagger}{C}\, i\frac{da_{-f}}{dt} \\
&\quad + v_f\, i\frac{dL^\dagger}{dt}\frac{a_{-f}}{C} = u_f\{-T(f)a_f^\dagger + \lambda(f)L^\dagger a_{-f}\} \\
&\quad + v_f \frac{L^\dagger}{C}\{T(f)a_{-f} + \lambda(f)a_f^\dagger L\} + v_f\, i\frac{dL^\dagger}{dt}\frac{a_{-f}}{C} \\
&= -a_f^\dagger\left\{T(f)u_f - \lambda(f)v_f\frac{L^\dagger L}{C}\right\} + \frac{L^\dagger}{C}\{u_f\lambda(f)C + T(f)v_f\}a_{-f} \\
&\quad + v_f\, i\frac{dL^\dagger}{dt}\frac{a_{-f}}{C} \\
&= -a_f^\dagger\{T(f)u_f - \lambda(f)v_f C\} + \frac{L^\dagger}{C}\{u_f\lambda(f)C + T(f)v_f\}a_{-f} \\
&\quad - a_f^\dagger \lambda(f)v_f\frac{C^2 - L^\dagger L}{C} + v_f\, i\frac{dL^\dagger}{dt}\frac{a_{-f}}{C}.
\end{aligned}
$$

However, [cf. (5.6)]

$$
\begin{aligned}
u_f\lambda(f)C + T(f)v_f &= \Omega(f)v_f \\
T(f)u_f - \lambda(f)v_f C &= \Omega(f)u_f
\end{aligned}
\qquad (6.19)
$$

and so

$$i\frac{d\alpha_f^\dagger}{dt} + \Omega(f)\alpha_f^\dagger = R_f, \qquad (6.20)$$

where

$$R_f = -a_f^\dagger\lambda(f)v_f\,\frac{C^2 - L^\dagger L}{C} + v_f(D_1^\dagger + D_2^\dagger)\frac{a_{-f}}{C}.$$

We now have

$$
\begin{aligned}
\langle \phi_H^\dagger R_f^\dagger R_f \phi \rangle &\leq 2\left\langle \phi_H^\dagger \frac{C^2 - L^\dagger L}{C} a_f a_f^\dagger \frac{C^2 - L^\dagger L}{C}\phi_H \right\rangle \lambda^2(f)v_f^2 \\
&\quad + 2\left\langle \phi_H^* \frac{a_{-f}^\dagger}{C}(D_1 + D_2)(D_1^\dagger + D_2^\dagger)\frac{a_{-f}}{C}\phi_H \right\rangle v_f^2 \\
&\leq 2\lambda^2(f)v_f^2\left\langle \phi_H^* \frac{(C^2 - L^\dagger L)^2}{C^2}\phi_H \right\rangle
\end{aligned}
$$

$$+ 2v_f^2 \Big\langle \phi_H^* \Big\{ \frac{a_{-f}^\dagger}{C}(D_1 + D_2)(D_1^\dagger + D_2^\dagger)\frac{a_{-f}}{C}$$

$$- (D_1 + D_2)\frac{a_{-f}^\dagger a_{-f}}{C^2}(D_1^\dagger + D_2^\dagger)\Big\}\phi_H \Big\rangle$$

$$+ 2\langle\phi_H^*(D_1 + D_2)(D_1^\dagger + D_2^\dagger)\phi_H\rangle\frac{v_f^2}{C^2}$$

and also

$$\langle\phi_H^\dagger R_f R_f^\dagger\phi\rangle \leq 2\Big\langle\phi_H^* a_f^\dagger\Big(\frac{C^2 - L^\dagger L}{C^2}\Big)^2 a_f\phi_H\Big\rangle\lambda^2(f)v_f^2 + \frac{2v_f^2}{C^2}$$

$$\times \langle\phi_H^*(D_1^\dagger + D_2^\dagger)(D_1 + D_2)\phi_H\rangle = 2\lambda^2(f)v_f^2\Big\langle\phi_H^*\Big\{a_f^\dagger\Big(\frac{C^2 - L^\dagger L}{C}\Big)$$

$$\times \Big(\frac{C^2 - L^\dagger L}{C}\Big)a_f - \Big(\frac{C^2 - L^\dagger L}{C}\Big)a_f^\dagger a_f\Big(\frac{C^2 - L^\dagger L}{C}\Big)\Big\}\phi_H\Big\rangle$$

$$+ 2\lambda^2(f)v_f^2\Big\langle\phi_H^*\Big(\frac{C^2 - L^\dagger L}{C^2}\Big)^2\phi_H\Big\rangle$$

$$+ \frac{2v_f^2}{C^2}\langle\phi_H^*(D_1^\dagger + D_2^\dagger)(D_1 + D_2)\phi_H\rangle.$$

Hence, it follows that

$$\langle\phi_H^\dagger R_f R_f^\dagger\phi\rangle \leq v_f^2\frac{S}{V}$$

$$\langle\phi_H^\dagger R_f^\dagger R_f\phi\rangle \leq v_f^2\frac{S}{V}, \quad \text{where} \quad S = \text{const.} \tag{6.21}$$

Once we have established equation (6.20) and the inequality (6.21), we can repeat word for word the arguments used in the preceding section in connection with the case $\nu > 0$. We now obtain

$$\langle\phi_H^*\alpha_H^\dagger\alpha_H\phi_H\rangle \leq \frac{S}{V}\frac{v_f^2}{2\pi\Omega^2(f)}\Big(\int_{-\infty}^{\infty}|h(\tau)\tau|\,d\tau\Big)^2. \tag{6.21'}$$

This inequality is a considerable improvement on (6.9); the latter merely showed that the average with respect to f of the sum of quantities $\langle\phi_H^*\alpha_H^\dagger\alpha_H\phi_H\rangle$ is of order $1/V$, whereas (6.21') shows this quantity is itself of order $1/V$ for each value of f separately.

From (6.21′) we can immediately obtain limits for the single-time averages. Let $\mathcal{U}_f$ represent either a_f or $a_f^\dagger$, and consider those operators of the form

$$\mathcal{U}_{f_1} \mathcal{U}_{f_2} \ldots \mathcal{U}_{f_K},$$

which conserve particle number. We shall prove that

$$\left| \langle \mathcal{U}_{f_1} \mathcal{U}_{f_2} \ldots \mathcal{U}_{f_K} \rangle_H - \langle \mathcal{U}_{f_1} \mathcal{U}_{f_2} \ldots \mathcal{U}_{f_K} \rangle_{H_0} \right| \leq \frac{\text{const}}{\sqrt{V}}. \qquad (6.22)$$

We begin by noticing that ϕ_H and ϕ_{H_0} satisfy the condition (2.3)

$$(a_f^\dagger a_f - a_{-f}^\dagger a_{-f})\phi = 0.$$

Therefore,

$$\langle \mathcal{U}_{f_1} \mathcal{U}_{f_2} \ldots \mathcal{U}_{f_K} \rangle$$

may be written as a sum of terms of the type

$$\langle \ldots a_f^\dagger a_f \ldots a_g^\dagger a_{-g}^\dagger \ldots a_{-h} a_h \ldots \rangle,$$

where the indices $\pm f$, $\pm g$, $\pm h, \ldots$ are all different. The number of indices g must of course be equal to the number of indices h. Now it is obvious that

$$\langle \ldots a_f^\dagger a_f \ldots a_g^\dagger a_{-g}^\dagger \ldots a_{-h} a_h \ldots \rangle_{H_0} = \prod_f v_f^2 \prod_g (-u_g v_g) \prod_h (-u_h v_h),$$

so that we need to only prove that

$$\left| \langle \ldots a_f^\dagger a_f \ldots a_g^\dagger a_{-g}^\dagger \ldots a_{-h} a_h \ldots \rangle_H \right| - \prod_f v_f^2 \prod_g (-u_g v_g) \prod_h (-u_h v_h) \leq \frac{\text{const}}{\sqrt{V}}. \qquad (6.23)$$

For the results proved above [see (6.12), (6.13), (6.16)] we have

$$a_{-h} a_h + u_h v_h \frac{L}{C} = u_h^2 \alpha_{-h} \alpha_h - u_h v_h \left\{ \frac{2}{V^2} \sum_f v_h^2 a_h^\dagger \frac{\lambda^2(f)}{C^2} (1 - a_f^\dagger a_f - a_{-f}^\dagger a_{-f}) \right.$$

$$\times a_h + v_h^2 \frac{L^\dagger L - C^2}{C^2} + u_h v_h a_h^\dagger a_h \frac{4\lambda(f)}{V} \bigg\} \frac{L}{C} - v_h^2 \alpha_h^\dagger \frac{L}{C} a_{-h}^\dagger \frac{L}{C}$$

$$+ u_h v_h \left(\alpha_h^\dagger \frac{L}{C} \alpha_h + a_{-h}^\dagger \alpha_{-h} \frac{L}{C} \right) + \eta_{-h} a_h + a_{-h} \eta_h - \eta_{-h} \eta_h \qquad (6.24)$$

and

$$a_f^\dagger a_f - v_f^2 = u_f^2 \alpha_f^\dagger \alpha_f - v_f^2 \frac{L^\dagger}{C} \alpha_{-f}^\dagger \alpha_{-f} \frac{L}{C} - u_f v_f \alpha_f^\dagger \alpha_{-f}^\dagger \frac{L}{C}$$

$$- u_f v_f \frac{L^\dagger}{C} \alpha_{-f} \alpha_f + v_f^2 \frac{L^\dagger}{C} \left\{ \frac{2}{V^2} \sum_g v_f^2 a_f^\dagger \frac{\lambda^2(g)}{C^2} (1 - a_g^\dagger a_g - a_{-g}^\dagger a_{-g}) a_f \right.$$

$$\left. + v_f^2 \frac{L^\dagger L - C^2}{C^2} + u_f v_f \frac{4\lambda(g)}{V} a_f^\dagger a_f \right\} + a_f^\dagger \eta_f + \eta_f^\dagger a_f - \eta_f^\dagger \eta_f. \qquad (6.25)$$

We now transfer each operator $\alpha^\dagger$ to the left-hand end of the term in which it occurs, each α to the right-hand end and each operator $(L^\dagger L - C^2)/C^2$ (occurring for instance, in η or $\eta^\dagger$) to the one end or the other - exactly which end is unimportant in this case. Since all the indices f, g are different the commutators arising from these permutations will be quantities of order $1/V$. Then, invoking once again the inequality

$$|\langle AB \rangle| \leq \sqrt{\langle AA^\dagger \rangle \langle B^\dagger B \rangle}$$

we see that as soon as an operator $\alpha^\dagger$ find itself adjacent to the left-hand end, an α to the right-hand end or an $(L^\dagger L - C^2)/C^2$ to either, we immediately have a quantity of order of magnitude $1/\sqrt{V}$ at most. Consequently

$$\left| \langle \ldots a_f^\dagger a_f \ldots a_g^\dagger a_{-g}^\dagger \ldots a_{-h} a_h \ldots \rangle_H \right.$$

$$\left. - \prod_f v_f^2 \left\langle \ldots (-u_g v_g) \frac{L^\dagger}{C} (-u_h v_h) \frac{L}{C} \ldots \right\rangle_H \right| \leq \frac{\text{const}}{\sqrt{V}}. \qquad (6.26)$$

However, the number of indices g is equal to number of indices h, and the quantities L and $L^\dagger$ may be permuted, to within an error of order $1/V$. Therefore

$$\left\langle \ldots (-u_g v_g) \frac{L^\dagger}{C} (-u_h v_h) \frac{L}{C} \ldots \right\rangle_H$$

differs from

$$\prod_g u_g v_g \prod_h u_h v_h \left\langle \left(\frac{L^\dagger L}{C^2} \right)^l \right\rangle_H$$

by terms of order $1/V$. On the other hand the expression

$$\left\langle \left(\frac{L^\dagger L}{C^2} \right)^l \right\rangle_H$$

differs form unity by terms of order $1/\sqrt{V}$ at most. This therefore concludes the proof of (6.3) and also of (6.22).

We now turn to the double-time correlation functions. We shall prove the following general inequality:

$$
\begin{aligned}
&\left| \langle \mathscr{B}_{f_1}(t) \ldots \mathscr{B}_{f_l}(t); \mathscr{U}_{g_1}(\tau) \ldots \mathscr{U}_{g_k}(\tau) \rangle_H \right. \\
&\left. - \langle \mathscr{B}_{f_1}(t) \ldots \mathscr{B}_{f_l}(t); \mathscr{U}_{g_1}(\tau) \ldots \mathscr{U}_{g_k}(\tau) \rangle_{H_0} \right| \leq \frac{K(t-\tau) + K_1}{\sqrt{V}}
\end{aligned}
$$

$$
K = \text{const}, \quad K_1 = \text{const} \tag{6.27}
$$

where the operators $\mathscr{B}_f$, $\mathscr{U}_g$ may represents either a or $a^\dagger$. As always in this context, we assume that the operator

$$
\mathscr{B}_{f_1} \ldots \mathscr{U}_{g_k}
$$

conserves particle number.

By virtue of the supplementary condition (2.3) obeyed by ϕ_H and ϕ_{H_0}, rearrangement of these operators in the "correct" order allows us to reduce the averages being considered to a sum of terms of the type

$$
\begin{aligned}
&\langle \ldots a_f^\dagger(t) a_f(t) \ldots a_g^\dagger(t) a_{-g}^\dagger(t) \ldots a_{-h}(t) a_h(t) \ldots \\
&\ldots a_k^\dagger(t) \ldots a_f(t) \ldots a_{g'}^\dagger(\tau) a_{f'}(\tau) \ldots a_{g'}^\dagger(\tau) a_{-g'}^\dagger(\tau) \ldots \\
&\ldots a_{-h'}(\tau) a_{h'}(\tau) \ldots a_{k'}(\tau) \ldots a_{q'}^\dagger(\tau) \ldots \rangle,
\end{aligned} \tag{6.28}
$$

where the number of operators α and $\alpha^\dagger$ is the same, and the indices $\pm f$, $\pm g$, $\pm h$, $\pm k$, $\pm q$ are all different form one another, as are the indices $\pm f'$, $\pm g'$, $\pm h'$, $\pm k'$, $\pm q'$. It follows that it is sufficient to prove (6.27) for averages of the type (6.28).

We start by using for the pairs $a^\dagger a$, $a^\dagger a^\dagger$, aa formulae (6.24) and (6.25), and for single operators a, $a^\dagger$ formulae (6.7). Next we transfer all operators $\alpha^\dagger(t)$ and $L^\dagger(t)L(t) - C^2$ to the left end and all operators $\alpha(\tau)$ and $L^\dagger(\tau)L(\tau) - C^2$ to the right end of terms in which they occur. Since, as we noted above, all the indices are different and since we permute in this way only operators with the same time argument, all the commutators arising from this process will be of order $1/V$. Again, as soon as either an $\alpha^\dagger(t)$ or an $L^\dagger(t)L(t) - C^2$ finds itself adjacent to the left-hand end, or an $\alpha(\tau)$ and $L^\dagger(\tau)L(\tau) - C^2$ to the right-hand end, we immediately get a quantity of order $1/V$ at worst. It

therefore only remains to prove that an inequality of the type (6.23) holds
for averages of the form

$$\Gamma(t-\tau)$$
$$= \langle \alpha_{f_1}(t) \ldots \alpha_{f_l}(t) L^k(t) L^{\dagger q}(t) L^{\dagger q_1}(\tau) L^{k_1}(\tau) \alpha_{g_1}^\dagger(\tau) \ldots \alpha_{g_r}^\dagger(\tau) \rangle. \qquad (6.29)$$

We now use the equation of motion (6.11) and the inequalities given
by (6.18)-(6.21). We find

$$i \frac{\partial \Gamma_H(t-\tau)}{\partial t} - \{\Omega(f_1) + \ldots + \Omega(f_l)\} \Gamma_H(t-\tau) = \Delta(t-\tau)$$

with

$$|\Delta(t-\tau)| \le \frac{G}{\sqrt{V}}, \quad \text{where} \quad G = \text{const.}$$

Hence, since

$$\Gamma_H(t-\tau) = e^{-i\{\Omega(f_1)+\ldots+\Omega(f_l)\}(t-\tau)} \Gamma_H(0)$$

$$+ e^{-i\{\Omega(f_1)+\ldots+\Omega(f_l)\}(t-\tau)} \int_0^{t-\tau} e^{-i\{\Omega(f_1)+\ldots+\Omega(f_l)\}z} \Delta(z)\, dz$$

we obtain

$$\left| \Gamma_H(t-\tau) - e^{-i\{\Omega(f_1)+\ldots+\Omega(f_l)\}(t-\tau)} \Gamma_H(0) \right| \le \frac{G|t-\tau|}{\sqrt{V}}. \qquad (6.30)$$

On the other hand

$$\Gamma_{H_0}(t-\tau) = e^{-i\{\Omega(f_1)+\ldots+\Omega(f_l)\}(t-\tau)} \Gamma_{H_0}(0) \qquad (6.31)$$

and therefore

$$\Gamma_{H_0}(t-\tau) = \langle \alpha_{f_1}(t) \ldots \alpha_{f_l}(t) \alpha_{g_1}^\dagger(\tau) \ldots \alpha_{g_r}^\dagger(\tau) \rangle_{H_0} C^{k+q+q_1+k_1}. \qquad (6.32)$$

Thus, we have

$$\left| \Gamma_H(t-\tau) - \Gamma_{H_0}(t-\tau) \right|$$
$$\le \left| \Gamma_H(0) - \Gamma_{H_0}(0) \right| + \frac{G|t-\tau|}{\sqrt{V}}$$

$$= \left| \langle \alpha_{f_1} \dots \alpha_{f_l} L^k (L^\dagger)^{q+q_1} L^{k_1} \alpha_{q_1}^\dagger \dots \alpha_{q_r}^\dagger \rangle_H \right.$$

$$\left. - C^{k+k_1+q+q_1} \langle \alpha_{f_1} \dots \alpha_{f_l} \alpha_{q_1}^\dagger \dots \alpha_{q_r}^\dagger \rangle_{H_0} \right| + \frac{G|t-\tau|}{\sqrt{V}}. \tag{6.33}$$

Suppose that two of indices, $f_1, \dots, f_l$, in (6.33) coincide. Then, that the expression

$$\alpha_f^2 = \left(u_f a_f + v_f a_{-f}^\dagger \frac{L}{C} \right) \left(u_f a_f + v_f a_{-f}^\dagger \frac{L}{C} \right)$$

$$= v_f^2 a_{-f}^\dagger \frac{L}{C} a_{-f}^\dagger \frac{L}{C} + u_f v_f \left\{ a_{-f}^\dagger \frac{L}{C} a_f + a_f a_{-f}^\dagger \frac{L}{C} \right.$$

$$= \frac{v_f^2 a_{-f}^\dagger}{C} (L a_{-f}^\dagger - a_{-f}^\dagger L) L - u_f v_f (a_{-f}^\dagger a_f + a_f a_{-f}^\dagger) \frac{L}{C}$$

$$= - 2 \frac{v_f^2 \lambda(f)}{C^2 V} a_{-f}^\dagger a_f L \tag{6.34}$$

is of order $1/V$, we see that $\langle \dots \rangle_H$ will be of the same order, while the corresponding average with respect to H_0 is simply equal to zero. The same result of course also applies in the case where one or more pairs of the indices $g_1, \dots, g_r$ are identical.

Next suppose that there is one (or more) index f_j among the f's which does not occur among the g's. Then we can transfer α_{f_j} to the right-hand end in the expression $\langle \dots \rangle_H$, obtaining in the process commutators of order $1/V$. We easily see that in this case the average with respect to H will be at most of order $1/\sqrt{V}$. Again, the corresponding average with respect to H_0 is rigorously equal to zero. The situation is the same if there is one (or more) index among g's which does not occur among the f's.

Thus, it only remains to consider the cases for which

1. All $f_1, \dots, f_l$ are different, and

2. The set $g_1, \dots, g_r$ is identical with the set $f_1, \dots, f_l$ (the order of enumeration being in general different).

Now we rearrange the operators on the right-hand side of (6.13) in the "correct" order; that is we replace $\alpha_{g_1}^\dagger \dots \alpha_{g_r}^\dagger$ by $\alpha_{f_1}^\dagger \dots \alpha_{f_l}^\dagger$. This rearrangement can of course be performed exactly for H_0; for H it introduces an error which, as usual, is asymptotically small. Next we notice that since the operators inside the pointed brackets conserve particle number, and the number of α's and $\alpha^\dagger$'s are equal, $k + k_1$ must be equal to $q + q_1$. We can therefore write our expression as

$$\langle \alpha_{f_1} \ldots \alpha_{f_l} L^k (L^\dagger)^{k+k_1} L^{k_1} \alpha_{f_l}^\dagger \ldots \alpha_{f_1}^\dagger \rangle$$

make the replacement $L^k (L^\dagger)^{k+k_1} L^{k_1} \to (L^\dagger L)^{k+k_1}$ and transfer this factor to the right-hand end; this process introduces an error of order $1/V$. Finally, we notice that

$$\Big| \langle \alpha_{f_1} \ldots \alpha_{f_l} \alpha_{f_l}^\dagger \ldots \alpha_{f_1}^\dagger (L^\dagger L)^{k+k_1} \rangle_H$$
$$- \langle \alpha_{f_1} \ldots \alpha_{f_l} \alpha_{f_l}^\dagger \ldots \alpha_{f_1}^\dagger \rangle_{H_0} C^{2(k+k_1)} \Big| \leq \frac{\text{const}}{\sqrt{V}}. \tag{6.35}$$

Thus, we obtain from (6.33)

$$\Big| \Gamma_H(t - \tau) - \Gamma_{H_0}(t - \tau) \Big| \leq \frac{G|t - \tau|}{\sqrt{V}} + \frac{K}{\sqrt{V}}$$
$$+ C^{2(k+k_1)} \Big| \langle \alpha_{f_1} \ldots \alpha_{f_l} \alpha_{f_l}^\dagger \ldots \alpha_{f_1}^\dagger \rangle_H - \langle \alpha_{f_1} \ldots \alpha_{f_l} \alpha_{f_l}^\dagger \ldots \alpha_{f_1}^\dagger \rangle_{H_0} \Big|. \tag{6.36}$$

However, since all the f's are different,

$$\langle \alpha_{f_1} \ldots \alpha_{f_l} \alpha_{f_l}^\dagger \ldots \alpha_{f_1}^\dagger \rangle_{H_0}$$
$$= \langle \alpha_{f_1} \alpha_{f_1}^\dagger \rangle_{H_0} \langle \alpha_{f_2} \alpha_{f_2}^\dagger \rangle_{H_0} \ldots \langle \alpha_{f_l} \alpha_{f_l}^\dagger \rangle_{H_0} = 1.$$

In the average with respect to H such a decomposition can also be made – not, of course, exactly in this case, but with an error of order $1/V$. This conclude our proof of (6.27).

As in the case $\nu > 0$, we could obtain analogous asymptotic limits for the multiple-time correlation function; we shall not do so here. The interested reader is now in a position to carry out all the relevant calculations himself, along the patterns developed above. As in the case $\nu > 0$ the inequalities can be sharpened from $\text{const}/\sqrt{V}$ to const/V by replacing the constant C in the Hamiltonian H_0 by the quantity

$$C_1 = \sqrt{\langle L^\dagger L \rangle_H}$$

which, generally speaking, differs from C by a term of order $1/\sqrt{V}$. We shall not give the proof of this statement here.

Appendix A.

In this appendix we prove various relations used in the text.[z] The operators considered are assumed to be completely continuous, since all operators occurring in the text are of this type.

Lemma I.

Let the operator ξ satisfies the condition

$$|\xi\xi^\dagger - \xi^\dagger\xi| \leq \frac{2s}{V}, \tag{A.1}$$

where s is a number, and let ε equal either $+1$ or -1. Then the following inequality holds,

$$2\sqrt{\xi^\dagger\xi + \frac{s}{V}} - \epsilon(\xi + \xi^\dagger) \geq 0. \tag{A.2}$$

Proof. Assume the contrary; then there exists a normed function φ such that

$$\left\{2\sqrt{\xi^\dagger\xi + \frac{s}{V}} - \epsilon(\xi + \xi^\dagger)\right\}\varphi = -\rho\varphi,$$

where $\rho > 0$. Then we have

$$\left(2\sqrt{\xi^\dagger\xi + \frac{s}{V}} + \rho\right)\varphi = \epsilon(\xi + \xi^\dagger)\varphi. \tag{A.3}$$

Now we use the fact that if $A\varphi = B\varphi$ and if the operators A, B are Hermitian, then

$$\langle\varphi^* A^2\varphi\rangle = \langle\varphi^* B^2\varphi\rangle. \tag{A.4}$$

From (A.4) and (A.1) we have

$$\left\langle\varphi^*\left(2\sqrt{\xi^\dagger\xi + \frac{s}{V}} + \rho\right)^2\varphi\right\rangle = \langle\varphi^*(\xi + \xi^\dagger)^2\varphi\rangle$$

$$= 2\langle\varphi^*(\xi\xi^\dagger + \xi^\dagger\xi)\varphi\rangle - \langle\varphi^*(\xi^\dagger - \xi)(\xi - \xi^\dagger)\varphi\rangle$$

$$\leq 2\langle\varphi^*(\xi\xi^\dagger + \xi^\dagger\xi)\varphi\rangle \leq 2\left\langle\varphi^*\left(\xi^\dagger\xi + \frac{2s}{V} + \xi^\dagger\xi\right)\varphi\right\rangle$$

$$\leq 4\left\langle\varphi^*\left(\xi^\dagger\xi + \frac{s}{V}\right)\varphi\right\rangle, \tag{A.5}$$

---

[z]We shall denote the norm of a functions as follows: $\|\varphi\| = \sqrt{\langle\varphi^*\varphi\rangle}$; and the norm of an operator $\mathcal{U}$ by $|\mathcal{U}| = \sup\|\mathcal{U}\varphi\|$, where $\|\varphi\| = 1$.

which cannot be satisfied for $\rho > 0$. Thus, (A.2) is proved.

Corollary. Interchanging ξ and $\xi^\dagger$, we also have

$$2\sqrt{\xi\xi^\dagger + \frac{s}{V}} - \epsilon(\xi + \xi^\dagger) \geq 0. \tag{A.6}$$

Similarly we can prove the inequalities:

$$2\sqrt{\xi\xi^\dagger + \frac{s}{V}} + i\epsilon(\xi - \xi^\dagger) \geq 0. \tag{A.7}$$

$$2\sqrt{\xi^\dagger\xi + \frac{s}{V}} + i\epsilon(\xi - \xi^\dagger) \geq 0. \tag{A.8}$$

Lemma II.

Let ξ satisfy the condition

$$|\xi\xi^\dagger - \xi^\dagger\xi| \leq \frac{2s}{V}. \tag{A.9}$$

Then

$$\sqrt{\xi\xi^\dagger + \frac{2s}{V} + A^2} - \sqrt{\xi^\dagger\xi + A^2} \geq 0, \tag{A.10}$$

where A is a real c-number.

Proof. Assume the contrary; then there exists a normed function φ such that

$$\left\{\sqrt{\xi\xi^\dagger + \frac{2s}{V} + A^2} - \sqrt{\xi^\dagger\xi + A^2}\right\}\varphi = -\rho\varphi. \tag{A.11}$$

Hence

$$\left\{\sqrt{\xi\xi^\dagger + \frac{2s}{V} + A^2} + \rho\right\}\varphi = \sqrt{\xi^\dagger\xi + A^2}\varphi. \tag{A.12}$$

Using (A.4) we therefore get

$$\left\langle \varphi^*\left(\sqrt{\xi\xi^\dagger + \frac{2s}{V} + A^2} + \rho\right)^2\varphi\right\rangle = \langle\varphi^*(\xi^\dagger\xi + A^2)\varphi\rangle$$

$$\leq \left\langle \phi^*\left(\xi\xi^\dagger + \frac{2s}{V} + A^2\right)\varphi\right\rangle, \tag{A.13}$$

which cannot be satisfied for $\rho > 0$.

Corollary. Interchanging the operators ξ and $\xi^\dagger$, we get

$$\sqrt{\xi^\dagger \xi + \frac{2s}{V} + A^2} - \sqrt{\xi\xi^\dagger + A^2} \geq 0. \tag{A.14}$$

Also, if α and λ are real c-numbers, we have

$$\sqrt{\lambda^2\left(\xi\xi^\dagger + \frac{2s}{V} + \alpha^2\right) + A^2} - \sqrt{\lambda(\xi^\dagger \xi + \alpha^2) + A^2} \geq 0, \tag{A.15}$$

$$\sqrt{\lambda^2\left(\xi^\dagger \xi + \frac{2s}{V} + \alpha^2\right) + A^2} - \sqrt{\lambda(\xi\xi^\dagger + \alpha^2) + A^2} \geq 0. \tag{A.16}$$

Note to Lemma II.
Put

$$\xi = \frac{1}{V}\sum_f \lambda(f)a_{-f}a_f + \nu \equiv L + \nu. \tag{A.17}$$

Then,

$$\xi\xi^\dagger - \xi^\dagger \xi = \frac{2}{V^2}\sum_f \lambda^2(f)(1 - a_f^\dagger a_f - a_{-f}^\dagger a_{-f}). \tag{A.18}$$

Suppose $\lambda(f)$ satisfies the condition

$$\frac{1}{V}\sum_f \lambda^2(f) \leq s.$$

Then

$$|\xi\xi^\dagger - \xi^\dagger \xi| \leq \frac{2s}{V},$$

and so

$$\sqrt{\lambda^2(f)\left\{(L+\nu)(L^\dagger+\nu) + \alpha^2 + \frac{2s}{V}\right\} + T^2(f)}$$
$$- \sqrt{\lambda^2(f)\{(L+\nu)(L^\dagger+\nu) + \alpha^2\} + T^2(f)} > 0. \tag{A.19}$$

Lemma III (generalization of lemma II).
Again let

$$|\xi\xi^\dagger - \xi^\dagger \xi| \leq \frac{2s}{V}.$$

Consider operators $\mathcal{U}$, $\mathcal{U}^\dagger$ with norm

$$|\mathcal{U}| \le 1, \quad |\mathcal{U}^\dagger| \le 1,$$

such that

$$|\mathcal{U}\xi^\dagger\xi\mathcal{U}^\dagger - \xi^\dagger\mathcal{U}\mathcal{U}^\dagger\xi| \le \frac{2l}{V}. \tag{A.20}$$

Then

$$2\sqrt{\xi\xi^\dagger + \frac{s+l}{V}} - \in(\xi\mathcal{U}^\dagger - \mathcal{U}\xi^d ag) \ge 0, \tag{A.21}$$

where $\in$ is equal either to $+1$ or to -1.

Proof. Assume the contrary. Then there exists a normed function φ such that

$$\left\{2\sqrt{\xi\xi^\dagger + \frac{s+l}{V}} - \in(\xi\mathcal{U}^\dagger - \mathcal{U}\xi^d ag)\right\}\varphi = -\rho\varphi, \quad \rho > 0. \tag{A.22}$$

Hence,

$$\left(2\sqrt{\xi\xi^\dagger + \frac{s+l}{V}} + \rho\right) = \in(\xi\mathcal{U}^\dagger - \mathcal{U}\xi^\dagger)\varphi. \tag{A.23}$$

According to 2-A.4, it follows that

$$\left\langle\varphi^*\left(2\sqrt{\xi\xi^\dagger + \frac{s+l}{V}} + \rho\right)^2\varphi\right\rangle = \langle\varphi^*(\xi\mathcal{U}^\dagger - \mathcal{U}\xi^\dagger)^2\varphi\rangle$$
$$= 2\langle\varphi^*\{\xi\mathcal{U}^\dagger\mathcal{U}\xi^\dagger + \mathcal{U}\xi^\dagger\xi\mathcal{U}\}\varphi\rangle - \langle\varphi^*(\xi\mathcal{U}^\dagger - \mathcal{U}\xi)(\mathcal{U}\xi^\dagger - \xi\mathcal{U}^\dagger)\varphi\rangle$$
$$\le 2\langle\varphi^*\{\xi\mathcal{U}^\dagger\mathcal{U}\xi^\dagger + \mathcal{U}\xi^\dagger\xi\mathcal{U}\}\varphi\rangle. \tag{A.24}$$

However, since by hypothesis $|\mathcal{U}| \le 1$, $|\mathcal{U}^\dagger| \le 1$, we have $|\mathcal{U}^\dagger\mathcal{U}| \le$, and consequently

$$\langle\varphi^*\xi\mathcal{U}^\dagger\mathcal{U}\xi^\dagger\rangle \le \langle\varphi^\dagger\xi\xi^\dagger\varphi\rangle. \tag{A.25}$$

From (A.20) and (A.25) we obtain

$$\langle\varphi^*\mathcal{U}\xi^\dagger\xi\mathcal{U}^\dagger\varphi\rangle = \langle\varphi^*\xi^\dagger\mathcal{U}\mathcal{U}^\dagger\xi\varphi\rangle + \langle\varphi^*\{\mathcal{U}\xi^\dagger\xi\mathcal{U}^\dagger - \xi^\dagger\mathcal{U}\mathcal{U}^\dagger\xi\}\varphi\rangle$$
$$\le \langle\varphi^*\xi^\dagger\mathcal{U}\mathcal{U}^\dagger\xi\varphi\rangle + \frac{2l}{V} \le \langle\varphi^*\xi^\dagger\xi\varphi\rangle + \frac{2l}{V}$$
$$\le \langle\varphi^*\xi\xi^\dagger\varphi\rangle + \frac{2(l+s)}{V} = \left\langle\varphi^*\left(\xi\xi^\dagger + \frac{2(l+s)}{V}\right)\varphi\right\rangle. \tag{A.26}$$

Thus, using (A.24), we can write

$$\left\langle \varphi^* \left(2\sqrt{\xi\xi^\dagger + \frac{s+l}{V}} + \rho\right)^2 \varphi\right\rangle \leq 4\left\langle \varphi^*\left(\xi\xi^\dagger + \frac{(l+s)}{V}\right)\varphi\right\rangle. \qquad (A.27)$$

However, such an inequality is impossible for $\rho > 0$, which proves the relation (A.21)

Note to Lemma III.

Put

$$\xi = L + \nu; \quad \mathscr{U} = a_g.$$

Then

$$\begin{aligned}
|\mathscr{U}\xi^\dagger\xi\mathscr{U}^\dagger - \xi^\dagger\mathscr{U}\mathscr{U}^\dagger\xi| &= |a_g(L^\dagger + \nu)(L + \nu)a_g^\dagger - (L^\dagger + \nu)a_g a_g^\dagger(L + \nu)| \\
&= |a_g(L^\dagger + \nu)(L + \nu)a_g^\dagger - (L^\dagger + \nu)a_g(L + \nu)a_g^\dagger \\
&\quad + (L^\dagger + \nu)a_g(L + \nu)a_g^\dagger - (L^\dagger + \nu)a_g a_g^\dagger(L + \nu)| \\
&\leq (|L| + \nu)\{|La_g^\dagger - a_g^\dagger L| + |a_g L^\dagger - L^\dagger a_g|\} \\
&\leq (|L| + \nu)\frac{4}{V}|\lambda(g)|,
\end{aligned}$$

where [cf. the definition (A.17)]

$$|L| \leq \frac{1}{V}\sum_f |\lambda(f)|,$$

since $|a_f| \geq 1$. Hence, by (A.21),

$$\begin{aligned}
2\sqrt{(L + \nu)(L^\dagger + \nu) + \frac{1}{V}\{s + (|L| + \nu)2|\lambda(g)|\}} \\
- \in\{(L + \nu)a_g^\dagger + a_g(L^\dagger + \nu)\} \geq 0.
\end{aligned} \qquad (A.28)$$

Putting $\mathscr{U} = ia_g$, we also obtain

$$\begin{aligned}
2\sqrt{(L + \nu)(L^\dagger + \nu) + \frac{1}{V}\{s + (|L| + \nu)2|\lambda(g)|\}} \\
+ i\in\{(L + \nu)a_g^\dagger - a_g(L^\dagger + \nu)\} \geq 0.
\end{aligned} \qquad (A.29)$$

Lemma IV.

Let β be a real c-number; let $\alpha^2 = \beta^2 + \dfrac{2s}{V}$ and let $\nu \geq 0$. Then

$$\sqrt{\{(L+\nu)(L^\dagger+\nu) + \alpha^2\}\lambda^2(f) + T^2(f)}\, a_f$$
$$- a_f\sqrt{\{(L+\nu)(L^\dagger+\nu) + \alpha^2\}\lambda^2(f) + T^2(f)} \leq \frac{\text{const}}{V}. \qquad (\text{A.30})$$

The same inequality holds with a_f replaced by $a_f^\dagger$.

Proof. Consider an arbitrary normed function φ and form the expression

$$\langle \varphi^* \{ \sqrt{(Q+\alpha^2)\lambda^2(f) + T^2(f)}(a_f + a_f^\dagger) - (a_f + a_f^\dagger)$$
$$\times \sqrt{(Q+\alpha^2)\lambda^2(f) + T^2(f)} \}\varphi \rangle = \mathscr{E}, \qquad (\text{A.31})$$

where

$$Q = (L+\nu)(L^\dagger+\nu).$$

To examine the expression (A.31) we use the following identity,

$$\sqrt{z} - \sqrt{z_0} = \frac{1}{\pi} \int\limits_0^\infty \Big\{ \frac{1}{z_0 + \omega} - \frac{1}{z + \omega} \Big\} \sqrt{\omega}\, d\omega,$$

where z_0 is an arbitrary positive number. We also observe that if A and B are operators,

$$-\frac{1}{A}B + B\frac{1}{A} = \frac{1}{A}(AB - BA)\frac{1}{A}.$$

Thus, we have

$$\mathscr{E} = \frac{1}{\pi} \int\limits_0^\infty \Big\langle \varphi^* \frac{\lambda^2(f)}{(Q+\alpha^2)\lambda^2(f) + T^2(f) + \omega} \{Q(a_f + a_f^\dagger) - (a_f + a_f^\dagger)Q\}$$
$$\times \frac{1}{(Q+\alpha^2)\lambda^2(f) + T^2(f) + \omega}\varphi \Big\rangle \sqrt{\omega}\, d\omega.$$

However,

$$Q a_f - a_f Q = (L+\nu)\{L^\dagger a_f - a_f L^\dagger\},$$
$$L^\dagger = \frac{1}{V}\sum_f \lambda(f)a_f^\dagger a_{-f}^\dagger, \quad L^\dagger a_f - a_f L^\dagger = -\frac{2}{V}\lambda(f)a_{-f}^\dagger,$$

and therefore

$$Q(a_f + a_f^\dagger) - (a_f + a_f^\dagger)Q$$
$$= -\frac{2}{V}\lambda(f)(L + \nu)a_{-f}^\dagger + \frac{2}{V}\lambda(f)a_{-f}(L^\dagger + \nu).$$

Thus, we find

$$|\mathscr{E}| = \left|\frac{\mathscr{E}}{i}\right| = \frac{2|\lambda(f)|^3}{\pi}$$

$$\times \left|\int_0^\infty \left\langle \varphi^* \frac{1}{(Q + \alpha^2)\lambda^2(f) + T^2(f) + \omega} \frac{(L + \nu)a_{-f}^\dagger - a_f(L^\dagger + \nu)}{i}\right.\right.$$

$$\times \left.\left. \frac{1}{(Q + \alpha^2)\lambda^2(f) + T^2(f) + \omega}\varphi\right\rangle \sqrt{\omega}\, d\omega\right|.$$

Using (A.29) and changing the variable of integration, we obtain

$$|\mathscr{E}| \leq \frac{4|\lambda(f)|^3}{\pi V}\int_0^\infty \left\langle \varphi^* \frac{\sqrt{Q + \dfrac{1}{V}(s + 2|\lambda(f)|)(|L| + \nu)}}{\left(Q + \alpha^2 + \dfrac{T^2(f)}{\lambda^2(f)} + \tau\right)^2}\varphi\right\rangle \sqrt{\tau}\, d\tau$$

However, by definition, $\alpha^2 = \beta^2 + \frac{2s}{V}$, therefore

$$\sqrt{Q + \frac{1}{V}(s + 2|\lambda(f)|)(|L| + \nu)}$$

$$\sqrt{Q + \alpha^2 + \frac{T^2(f)}{\lambda^2(f)} + \frac{2|\lambda(f)|(|L| + \nu)}{V}}$$

$$= \sqrt{Q + \alpha^2 + \frac{T^2(f)}{\lambda^2(f)}}\sqrt{1 + \frac{2|\lambda(f)|(|L| + \nu)}{VQ + V\alpha^2 + V\dfrac{T^2(f)}{\lambda^2(f)}}}$$

$$< \sqrt{Q + \alpha^2 + \frac{T^2(f)}{\lambda^2(f)}}\sqrt{1 + \frac{|\lambda(f)|(|L| + \nu)}{s + \dfrac{1}{2}V\dfrac{T^2(f)}{\lambda^2(f)}}}$$

$$< \left(1 + \frac{|\lambda(f)|(|L| + \nu)}{2s + V\dfrac{T^2(f)}{\lambda^2(f)}} \right) \sqrt{Q + \alpha^2 + \frac{T^2(f)}{\lambda^2(f)}}.$$

Put

$$\Lambda = Q + \alpha^2 + \frac{T^2(f)}{\lambda^2(f)} \geq \alpha^2.$$

Then

$$|\mathscr{E}| \leq \frac{4|\lambda(f)|^3}{\pi V} \left(1 + \frac{|\lambda(f)|(|L| + \nu)}{2s + V\dfrac{T^2(f)}{\lambda^2(f)}} \right) \int\limits_0^\infty \left\langle \varphi^* \frac{\sqrt{\Lambda}}{(\Lambda + \tau)^2} \varphi \right\rangle \sqrt{\tau}\, d\tau.$$

We now expand the function φ in terms of the eigenfunctions of the operator Λ:

$$\varphi = \sum C_\Lambda \varphi_\Lambda; \quad \sum |C_\Lambda|^2 = 1.$$

We then obtain

$$\int\limits_0^\infty \left\langle \varphi^* \frac{\sqrt{\Lambda}}{(\Lambda + \tau)^2} \varphi \right\rangle \sqrt{\tau}\, d\tau = \sum_\Lambda |C_\Lambda|^2 \int\limits_0^\infty \frac{\sqrt{\Lambda \tau}\, d\tau}{(\Lambda + \tau)^2}$$

$$= \sum_\Lambda |C_\Lambda|^2 \int\limits_0^\infty \frac{\sqrt{t}\, d\tau}{(1 + t)^2} = \int\limits_0^\infty \frac{\sqrt{t}\, d\tau}{(1 + t)^2}.$$

Thus, for an arbitrary normed function, φ,

$$|\mathscr{E}| = \left| \left\langle \varphi^* \left[\frac{\sqrt{(Q + \alpha^2)\lambda^2 + T^2}; a_f + a_f^\dagger}{i} \right] \varphi \right\rangle \right| \leq S_f,$$

where

$$S_f = \frac{4|\lambda(f)|^3}{\pi V} \left(1 + \frac{|\lambda(f)|\left(\sum |\lambda(f)|\dfrac{1}{V} + \nu\right)}{2\dfrac{1}{V}\sum |\lambda(f)|^2 + V\dfrac{T^2(f)}{\lambda^2(f)}} \right) \int\limits_0^\infty \frac{\sqrt{t}\, d\tau}{(1 + t)^2}.$$

However, the operator

$$\left[\frac{\sqrt{(Q+\alpha^2)\lambda^2+T^2};a_f+a_f^\dagger}{i}\right]$$

is Hermitian and consequently,

$$\left|[\sqrt{(Q+\alpha^2)\lambda^2+T^2};a_f+a_f^\dagger]\right| \leq S_f.$$

In an entirely analogous way we can show that

$$\left|[\sqrt{(Q+\alpha^2)\lambda^2+T^2};a_f-a_f^\dagger]\right| \leq S_f.$$

Since,

$$|\mathscr{U}|+|\mathscr{B}| \geq |\mathscr{U}+\mathscr{B}|,$$

it follows that

$$\left|[\sqrt{(Q+\alpha^2)\lambda^2+T^2};a_f]\right| \leq S_f.$$

Since we also have $S_f \leq \text{const}/V$, this proves the required lemma. (The second half of the lemma follows from the obvious fact that $|\mathscr{U}| \leq S_f$ then also $|\mathscr{U}^\dagger| \leq S_f$.)

Appendix B.

The Principle of Extinction of Correlations

In our lectures "The principle of extinction of correlations and the quasi-averages method" we formulated a principle of extinction of correlations between particles for system in a state of statistical equilibrium. The principle may be formulated as follows:

If $\mathscr{U}_s(x_s,t_s)$ represents either the field operator $\psi(x_s,t_s)$ or its adjoint $\psi^\dagger(x_s,t_s)$, then the correlation functions

$$\langle \mathscr{U}_1(x_1,t_1)\ldots\mathscr{U}_s(x_s,t_s)\ldots\mathscr{U}_n(x_n,t_n)\rangle, \tag{B.1}$$

may be decomposed into the product of correlation functions

$$\langle \mathscr{U}_1(x_1,t_1)\ldots\mathscr{U}_{s-1}(x_{s-1},t_{s-1}s)\rangle\langle\mathscr{U}_{s+1}(x_{s+1},t_{s+1})\ldots\mathscr{U}_n(x_n,t_n)\rangle \tag{B.2}$$

provided the set of points $x_1,\ldots,x_s$ is infinitely distant from the set of points $x_{s+1},\ldots,x_n$ (the times $t_1,\ldots,t_s,\ldots,t_n$ are assumed fixed). In cases where the number of creation and annihilation operators inside the brackets is not equal, the averages $\langle \ldots \rangle$ must be understood as quasi-averages.

The system described by our model Hamiltonian is one of the few for which the principle of extinction of correlation may be demonstrated by direct calculation. Below we prove this on the basis of the asymptotic limits derived in the text.

Consider the "vacuum expectation values" of the field operators in the coordinate representation,

$$\psi_-(t, x) = \frac{1}{\sqrt{V}} \sum_{(f<0)} a_f(t)\, e^{i(fx)}$$

$$\psi_+(t, x) = \frac{1}{\sqrt{V}} \sum_{(f>0)} a_f^\dagger(t)\, e^{-i(fx)}. \tag{B.3}$$

Here, f represents both momentum and spin indices (k, σ); the sums $f < 0$, $f > 0$ mean summation over all f with σ fixed ($\sigma = \pm$), and $(fx) = (\boldsymbol{k} \cdot \boldsymbol{r})$. For instance, we have,

$$\langle \psi_{\sigma_1}(t, x)\psi_{\sigma_2}^\dagger(t, x')\rangle_{H_0} = \frac{1}{V} \sum_{(f>0)} |u_f|^2\, e^{if(x-x')}\, \delta(\sigma_1 - \sigma_2)$$

$$= \left\{ \frac{1}{V} \sum_{(f>0)} e^{if(x-x')} - \frac{1}{V} \sum_{(f>0)} |v_f|^2\, e^{if(x-x')} \right\} \delta(\sigma_1 - \sigma_2) \tag{B.4}$$

where u_f and v_f are the coefficients of the canonical transformation. Obviously the term

$$\frac{1}{V} \sum_{(f>0)} |v_f|^2\, e^{if(x-x')}$$

goes over in the limit $V \to \infty$ to the integral

$$\frac{1}{(2\pi)^3} \int |v_f|^2\, e^{if(x-x')}\, d\vec{k}.$$

This integral is absolutely convergent, since

$$\int |v_f^2|^2\, d\vec{k} = \frac{1}{2} \int \frac{\{\sqrt{T^2(f) + \lambda^2(f)C^2} - T(f)\}^2}{T^2(f) + \lambda^2(f)C^2}\, d\vec{k} < \infty.$$

As for the expression

$$\frac{1}{V} \sum_{(f>0)} |u_f|^2 \, e^{if(x-x')}$$

we may similarly say that in the limit $V \to \infty$ it goes over to the delta-function

$$\frac{1}{(2\pi)^3} \int e^{if(x-x')} \, d\vec{k}.$$

However, we must of course understand the words "limit" and "convergence of a function" in a rather different sense; in fact, that appropriate to the theory of generalized functions. We shall digress for a moment to recall the meaning of the relation

$$f_V(x_1, \ldots, x_e) \xrightarrow[V \to \infty]{} f(x_1, \ldots, x_e) \tag{B.5}$$

or equivalently

$$f(x_1, \ldots, x_e) = \lim_{V \to \infty} f_V(x_1, \ldots, x_e)$$

in that theory.

Consider the class $C(q, r)$ (where q and r are positive integers) of continuous and infinitely differentiable functions $h(x_1, \ldots, x_e)$ such that for the entire space E_e of the point $(x_1, \ldots, x_e)$ the following relations are fulfilled

$$\{|x_1| + \ldots + |x_e|\}^\alpha |h(x_1, \ldots, x_e)| \le \text{conts}$$
$$\alpha = 0, 1, \ldots, r$$

$$\{|x_1| + \ldots + |x_e|\}^\alpha \left| \frac{\partial^{s_1 + \ldots + s_e} h}{\partial x_1^{s_1} \ldots \partial x_e^{s_e}} \right| \le \text{const}$$

$$\alpha = 0, 1, \ldots, r; \quad s_1 + \ldots + s_e = 0, 1, \ldots, q.$$

Then, if we can find positive numbers q, r such that for every function h of the class $C(q, r)$ we have

$$\int h(x_1, \ldots, x_e) f_V(x_1, \ldots, x_e) \, dx_1 \ldots dx_e \to$$

$$\int h(x_1, \ldots, x_e) f(x_1, \ldots, x_e) \, dx_1 \ldots dx_e$$

we shall say that the generalized limit relation (B.5) is fulfilled. As we saw above, the averages of products of $\psi(t, x)$ and $\psi^\dagger(t, x)$ may contain generalized functions; therefore we must understand the corresponding asymptotic relations (for the limit $V \to \infty$) in the sense described above.

Consider the expression

$$\langle \psi_{\sigma_1}(t_1, x_1) \psi^\dagger_{\sigma_2}(t_2, x_2) \rangle = \frac{1}{V} \sum_{(f>0)} \langle a_f(t_1) a^\dagger_f(t_2) \rangle \, e^{if(x-x')} \, \delta(\sigma_1 - \sigma_2).$$

We have

$$\int h(x_1 - x_2) \langle \psi_{\sigma_1}(t_1, x_1) \psi^\dagger_{\sigma_2}(t_2, x_2) \rangle \, dx_1$$

$$= \frac{1}{V} \sum_{(f>0)} \langle a_f(t_1) a^\dagger_f(t_2) \rangle \tilde{h}(f) \delta(\sigma_1 - \sigma_2),$$

where

$$\tilde{h}(f) = \int h(x) \, e^{i(f \cdot x)} \, dx.$$

By an appropriate choice of the indices q, r of the class $C(q,r)$ to which $h(x)$ belongs, we can arrange that $\tilde{h}(x)$ shall decrease faster than any desired power of $|f|^{-1}$ in the limit $|f| \to \infty$. For present purpose we need only ensure that

$$\frac{1}{V} \sum_f |\tilde{h}(f)| \leq K = \text{const}.$$

Then, noticing that according to (6.36)

$$\left| \langle a_f(t_1) a^\dagger_f(t_2) \rangle_H - \langle a_f(t_1) a^\dagger_f(t_2) \rangle_{H_0} \right| \leq \frac{s_1|t_1 - t_2| + s_2}{\sqrt{V}},$$

where s_1, $s_2 = \text{const}$, we obtain

$$\left| \int h(x_1 - x_2) \{ \langle \psi_{\sigma_1}(t_1, x_1) \psi^\dagger_{\sigma_2}(t_2, x_2) \rangle_H - \langle \psi_{\sigma_1}(t_1, x_1) \psi^\dagger_{\sigma_2}(t_2, x_2) \rangle_{H_0} \} dx_1 \right|$$

$$\leq \frac{1}{V} \sum_f \left| \langle a_f(t_1) a^\dagger_f(t_2) \rangle_H - \langle a_f(t_1) a^\dagger_f(t_2) \rangle_{H_0} \right| |\tilde{h}(f)|$$

$$\leq k \frac{s_1|t_1 - t_2| + s_2}{V} \xrightarrow[V \to \infty]{} 0.$$

Accordingly the following generalized limit relation holds,

$$\langle \psi_{\sigma_1}(t_1, x_1) \psi^\dagger_{\sigma_2}(t_2, x_2) \rangle_H - \langle \psi_{\sigma_1}(t_1, x_1) \psi^\dagger_{\sigma_2}(t_2, x_2) \rangle_{H_0} \to 0. \qquad (B.6)$$

We can see by direct calculations that

$$\langle\psi_{\sigma_1}(t_1,x_1)\psi_{\sigma_2}^\dagger(t_2,x_2)\rangle_{H_0} = \frac{1}{V}\sum_{(f>0)}|u_f|^2\, e^{-i\Omega(f)(t_1-t_2)+if(x_1-x_2)}\,\delta(\sigma_1-\sigma_2)$$

and hence, it is also true in the generalized sense that

$$\langle\psi_{\sigma_1}(t_1,x_1)\psi_{\sigma_2}^\dagger(t_2,x_2)\rangle_H$$
$$-\int|u_f|^2\exp\{-i\Omega(f)(t_1-t_2)+if(x_1-x_2)\}\,d\vec{k}\,\sigma(\sigma_1-\sigma_2)\xrightarrow[V\to\infty]{}0. \quad\text{(B.7)}$$

From (B.6) and (B.7) we finally get

$$\lim_{V\to\infty}\langle\psi_{\sigma_1}(t_1,x_1)\psi_{\sigma_2}^\dagger(t_2,x_2)\rangle_H$$
$$=\int|u_f|^2\exp\{-i\Omega(f)(t_1-t_2)+if(x_1-x_2)\}\,d\vec{k}\,\delta(\sigma_1-\sigma_2)$$
$$=\{\Delta(t_1-t_2,x_1-x_2)-F(t_1-t_2,x_1-x_2)\}\delta(\sigma_1-\sigma_2), \quad\text{(B.8)}$$

where

$$\Delta(t,x)=\int e^{-\Omega(f)t+ifx}\,d\vec{k}$$
$$F(t,x)=\int|v_f|^2\,e^{-\Omega(f)t+ifx}\,d\vec{k}. \quad\text{(B.9)}$$

In an entirely analogous way we obtain[aa]

$$\lim_{V\to\infty}\langle\psi_{\sigma_2}^\dagger(t_2,x_2)\psi_{\sigma_1}(t_1,x_1)\rangle_H = F(t_2-t_1,x_1-x_2)\delta(\sigma_1-\sigma_2). \quad\text{(B.10)}$$

Now consider the two-particle expressions

$$\langle\psi(t_1,x_1)\psi(t_2,x_2)\psi^\dagger(t_2',x_2')\psi^\dagger(t_1',x_1')\rangle.$$

We have

$$\langle\psi(t_1,x_1)\psi(t_2,x_2)\psi^\dagger(t_2',x_2')\psi^\dagger(t_1',x_1')\rangle$$
$$=\frac{1}{V^2}\sum\langle a_{f_1}(t_1)a_{f_2}(t_2)a_{g_2}^\dagger(t_2')a_{g_1}^\dagger(t_1')\rangle$$

[aa]This relation is also true in the generalized sense, owing to the absolute convergence of the integral defining $F(t,x)$.

$$\times \exp\{if_1x_1 + if_2x_2 - ig_2x_2' - ig_1x_1'\}. \tag{B.11}$$

Since the total momentum is conserved, and is equal to zero for ϕ_H (and ϕ_{H_0}) we see that the expressions

$$\langle a_{f_1}(t_1)a_{f_2}(t_2)a_{g_2}^\dagger(t_2')a_{g_1}^\dagger(t_1')\rangle \tag{B.12}$$

can be different form zero only if

$$f_1 + f_2 = g_2 + g_1. \tag{B.13}$$

We now recall that by (2.1) and (2.2) the quantity $n_f(t) - n_{-f}(t)$ (where $n_f = a_f^\dagger a_f$) is a constant of the motion and that ϕ_{H_0} (and ϕ_H) satisfy the additional condition

$$(n_f - n_{-f})\phi = 0.$$

Finally, we notice that

$$(n_f - n_{-f})a_h = a_h\{(n_f - n_{-f}) + \delta(f - h) + \delta(f + h)\}.$$

As a result we have, for arbitrary f,

$$\langle a_{f_1}(t_1)a_{f_2}(t_2)a_{g_2}^\dagger(t_2')a_{g_1}^\dagger(t_1')\rangle$$
$$= \langle\{1 + n_f - n_{-f}\}a_{f_1}(t_1)a_{f_2}(t_2)a_{g_2}^\dagger(t_2')a_{g_1}^\dagger(t_1')\rangle$$
$$= \langle\{1 + n_f(t_1) - n_{-f}(t_1)\}a_{f_1}(t_1)a_{f_2}(t_2)a_{g_2}^\dagger(t_2')a_{g_1}^\dagger(t_1')\rangle$$
$$= \langle a_{f_1}(t_1)\{1 + n_f(t_1) - n_{-f}(t_1) - \delta(f - f_1) + \delta(f + f_1)\}$$
$$\times\, a_{f_2}(t_2)a_{g_2}^\dagger(t_2')a_{g_1}^\dagger(t_1')\rangle$$
$$= \langle a_{f_1}(t_1)\{1 + n_f(t_2) - n_{-f}(t_1) - \delta(f - f_1) + \delta(f + f_1)\}$$
$$\times\, a_{f_2}(t_2)a_{g_2}^\dagger(t_2')a_{g_1}^\dagger(t_1')\rangle$$
$$= \langle a_{f_1}(t_1)a_{f_2}(t_2)\{1 + n_f(t_2) - n_{-f}(t_1) - \delta(f - f_1) + \delta(f + f_1)$$
$$-\, \delta(f - f_2) + \delta(f + f_2)\}a_{g_2}^\dagger(t_2')a_{g_1}^\dagger(t_1')\rangle = \ldots$$
$$= \langle a_{f_1}(t_1)a_{f_2}(t_2)a_{g_2}^\dagger(t_2')a_{g_1}^\dagger(t_1')$$
$$\times\, \{1 + n_f(t_2) - n_{-f}(t_1) - \delta(f - f_1) + \delta(f + f_1) + \delta(f + f_2)$$
$$-\, \delta(f - f_2) + \delta(f - g_2) - \delta(f + g_2) + \delta(f - g_1) - \delta(f + g_1)\}\rangle$$
$$= \{1 - \delta(f - f_1) + \delta(f + f_1) - \delta(f - f_2) + \delta(f + f_2)$$
$$+\, \delta(f - g_2) - \delta(f + g_2) + \delta(f - g_1) - \delta(f + g_1)\}$$

$$\times \langle a_{f_1}(t_1) a_{f_2}(t_2) a_{g_2}^\dagger(t_2') a_{g_1}^\dagger(t_1') \rangle.$$

This identity shows that the quantities (B.12) can be different from zero if, for arbitrary f, the following relation is satisfied,

$$-\delta(f - f_1) + \delta(f + f_1) - \delta(f - f_2) + \delta(f + f_2)$$
$$+ \delta(f - g_2) - \delta(f + g_2) + \delta(f - g_1) - \delta(f + g_1) = 0.$$

This relation can be fulfilled simultaneously with (B.13) only in the following cases:

$$f_1 + f_2 = 0; \quad g_1 + g_2 = 0 \tag{B.14}$$

$$f_1 = g_1; \quad f_2 = g_2 \tag{B.15}$$

$$f_1 = g_2; \quad f_2 = g_1. \tag{B.16}$$

Moreover, in the cases (B.15) and (B.16) we can always assume that $g_1 \neq g_2$, since

$$a_g^\dagger(t_2') a_g^\dagger(t_1') \phi_H = 0. \tag{B.17}$$

This last relation follows from the fact that

$$(n_g - n_{-g}) a_g^\dagger(t_2') a_g^\dagger(t_1') \phi_H = a_g^\dagger(t_2') a_g^\dagger(t_1')(n_g - n_{-g} + 2) \phi_H =$$
$$= 2 a_g^\dagger(t_2') a_g^\dagger(t_1') \phi_H.$$

Since the only possible eigenstates of $n_g - n_{-g}$ are ± 1 and 0, this relation can only be fulfilled by the satisfaction (B.17).

Thus, we can reduce (B.11) to the form

$$\langle \psi(t_1, x_1) \psi(t_2, x_2) \psi^\dagger(t_2', x_2') \psi^\dagger(t_1', x_1') \rangle$$

$$= \sum_{f, g} \frac{1}{V^2} \langle a_{-f}(t_1) a_f(t_2) a_g^\dagger(t_2') a_{-g}^\dagger(t_1') \rangle \exp\{if(x_2 - x_1) - ig(x_2' - x_1')\}$$

$$+ \sum_{\substack{f, g \\ \left(\substack{f \neq g \\ f+g \neq 0}\right)}} \frac{1}{V^2} \langle a_f(t_1) a_g(t_2) a_g^\dagger(t_2') a_f^\dagger(t_1') \rangle \exp\{if(x_1 - x_2') + ig(x_2 - x_2')\}$$

$$+ \sum_{\substack{f, g \\ \left(\substack{f \neq g \\ f+g \neq 0}\right)}} \frac{1}{V^2} \langle a_f(t_1) a_g(t_2) a_f^\dagger(t_2') a_g^\dagger(t_1') \rangle \exp\{if(x_1 - x_2') + ig(x_2 - x_1')\}.$$

$$\tag{B.18}$$

Now we turn to the limit $V \to \infty$. We consider the class $C(q, r)$ of functions $h(x, y)$ and fix q and r so that

$$\frac{1}{V^2} \sum_{f,g} |\tilde{h}(f, g)| \leq \text{const}$$

where

$$\tilde{h}(f, g) = \int h(x, y) \, e^{i(fx+gy)} \, dx \, dy.$$

Since for fixed t_1, t_2, t_2', t_1' we have [cf. (5.56)]

$$\langle a_f(t_1) a_g(t_2) a_f^\dagger(t_2') a_g^\dagger(t_1') \rangle_H$$

$$- \langle a_f(t_1) a_g(t_2) a_f^\dagger(t_2') a_g^\dagger(t_1') \rangle_{H_0} \leq \frac{\text{const}}{\sqrt{V}},$$

it follows that

$$\left| \int h(x, y) \{ \Gamma_H(t_1, t_2, t_2', t_1' | x, y) - \Gamma_{H_0}(t_1, t_2, t_2', t_1' | x, y) \} \right|$$

$$\leq \frac{\text{const}}{\sqrt{V}} \xrightarrow[V \to \infty]{} 0,$$

where

$$\Gamma_H(t_1, t_2, t_2', t_1' | x, y)$$

$$= \frac{1}{V^2} \sum_{\substack{f, g \\ \left(\substack{f \neq g \\ f+g \neq 0} \right)}} \langle a_f(t_1) a_g(t_2) a_f^\dagger(t_2') a_g^\dagger(t_1') \rangle \, e^{i(fx+gy)}.$$

Thus, we obtain the generalized limit relations

$$\Gamma_H(t_1, t_2, t_2', t_1', x_1 - x_2', x_2 - x_1')$$

$$- \Gamma_{H_0}(t_1, t_2, t_2', t_1', x_1 - x_2', x_2 - x_1') \xrightarrow[V \to \infty]{} 0.$$

However, a direct calculation, as in the case (B.4), shows that

$$\Gamma_{H_0}(t_1, t_2, t_2', t_1', x_1 - x_2', x_2 - x_1')$$

$$= -\frac{1}{V^2} \sum_{\substack{f, g \\ \left(\substack{f \neq g \\ f+g \neq 0} \right)}} |u_f|^2 |u_g|^2 e^{-i\Omega(f)(t_1-t_2') - i\Omega(g)(t_2-t_1')} \, e^{i(f(x_1-x_2')+g(x_2-x_1'))}$$

$$\rightarrow -\{\Delta(t_1 - t_2', x_1 - x_2') - F(t_1 - t_2', x_1 - x_2')\}$$
$$\times \{\Delta(t_2 - t_1', x_2 - x_1') - F(t_2 - t_1', x_2 - x_1')\}\delta(\sigma_1 - \sigma_2')\delta(\sigma_2 - \sigma_1') \tag{B.19}$$

where $\Omega(f)$ is defined by (6.2′) and $\Delta(t, x)$ and $F(t, x)$ by (B.9).
Consequently

$$\lim_{V \to \infty} \Gamma_H(t_1, t_2, t_2', t_1', x_1 - x_2', x_2 - x_1')$$
$$= -\{\Delta(t_1 - t_2', x_1 - x_2') - F(t_1 - t_2', x_1 - x_2')\}$$
$$\times \{\Delta(t_2 - t_1', x_2 - x_1') - F(t_2 - t_1', x_2 - x_1')\}\delta(\sigma_1 - \sigma_2')\delta(\sigma_2 - \sigma_1').$$

We can deal with the other terms on the right-hand side of (B.18) in an exactly similar way. Now let us put

$$\Phi_\sigma(t, x) = -\int u_f v_f \, e^{-i\Omega(f)t - ifx} \, d\vec{k}$$
$$= \int \frac{c\lambda(f)}{2\Omega(f)} e^{-i\Omega(f)t - ifx} \, d\vec{k}. \tag{B.20}$$

Then, we can write the generalized limit relation in the form

$$\lim_{V \to \infty} \langle \psi_{\sigma_1}(t_1, x_1)\psi_{\sigma_2}(t_2, x_2)\psi_{\sigma_2'}^\dagger(t_2', x_2')\psi_{\sigma_1'}^\dagger(t_1', x_1') \rangle$$
$$= \Phi_{\sigma_2}(t_1 - t_2, x_1 - x_2)\Phi_{\sigma_2'}(t_2' - t_1', x_2' - x_1')\delta(\sigma_1 + \sigma_2)\delta(\sigma_1' + \sigma_2')$$
$$+ \delta(\sigma_1 - \sigma_1')\delta(\sigma_2 - \sigma_2')\{\Delta(t_1 - t_1', x_1 - x_1') - F(t_1 - t_1', x_1 - x_1')\}$$
$$\times \{\Delta(t_2 - t_2', x_2 - x_2') - F(t_2 - t_2', x_2 - x_2')\}$$
$$- \delta(\sigma_1 - \sigma_2')\delta(\sigma_2 - \sigma_1')\{\Delta(t_1 - t_2', x_1 - x_2') - F(t_1 - t_2', x_1 - x_2')\}$$
$$\times \{\Delta(t_2 - t_1', x_2 - x_1') - F(t_2 - t_1', x_2 - x_1')\}. \tag{B.21}$$

By an entirely analogous procedure we can obtain formulae for other products of the field operators $\psi \cdot \psi^\dagger$.

We shall the example of (B.21) to illustrate the principle of extinction of correlations. We need only observe that

$$F(t, x) \to 0, \quad |x| \to \infty$$
$$\Phi(t, x) \to 0, \quad |x| \to \infty \tag{B.22}$$

and[bb]

$$\Delta(t, x) \to 0, \quad |x| \to \infty. \tag{B.23}$$

Let us fix the times t_1, t_2, t_2', t_1' and spatial differences

$$x_1 - x_1', \quad x_2 - x_2'$$

at some finite values. Now we let the remaining spatial differences

$$x_1 - x_2, \quad x_1' - x_2', \quad x_1 - x_2', \quad x_2 - x_1'$$

tend to infinity. Then the two-point function

$$\lim_{V \to \infty} \langle \psi_{\sigma_1}(t_1, x_1) \psi_{\sigma_2}(t_2, x_2) \psi_{\sigma_2'}^\dagger(t_2', x_2') \psi_{\sigma_1'}^\dagger(t_1', x_1') \rangle_H \tag{B.24}$$

will decompose into the product

$$\{\Delta(t_1 - t_1', x_1 - x_1') - F(t_1 - t_1', x_1 - x_1')\}$$
$$\times \{\Delta(t_2 - t_2', x_2 - x_2') - F(t_2 - t_2', x_2 - x_2')\}\delta(\sigma_1 - \sigma_1')\delta(\sigma_2 - \sigma_2')$$

which, by (B.8), is equal to

$$\lim_{V \to \infty} \langle \psi_{\sigma_1}(t_1, x_1) \psi_{\sigma_1'}^\dagger(t_1', x_1') \rangle_H \lim_{V \to \infty} \langle \psi_{\sigma_2}(t_2, x_2) \psi_{\sigma_2'}^\dagger(t_2', x_2') \rangle_H. \tag{B.25}$$

Now we consider a second aspect of the extinction of correlations. Again we fix the times t_1, t_2, t_2', t_1', and this time also the spatial differences

$$x_1 - x_2, \quad x_1' - x_2'.$$

Then we let the remaining spatial differences

$$x_1 - x_1', \quad x_2 - x_2', \quad x_1 - x_2', \quad x_2 - x_1'$$

tend to infinity. Then the function (B.24) decomposes into the product

$$\Phi(t_1 - t_2, x_1 - x_2)\Phi(t_2' - t_1', x_2' - x_1')\Phi_{\sigma_2}\Phi_{\sigma_2'}\delta(\sigma_1 + \sigma_2)\delta(\sigma_1' + \sigma_2'). \tag{B.26}$$

For $\nu > 0$,

$$\Phi_\sigma(t_1 - t_2, x_1 - x_2) = \lim_{V \to \infty} \langle \psi_{-\sigma}(t_1, x_1)\psi_\sigma(t_2, x_2) \rangle_H$$

[bb]The function $\Delta(t, x)$ is itself generalized; (B.23) is of course also true in the generalized sense.

$$\Phi_\sigma(t'_2 - t'_1, x'_2 - x'_1) = \lim_{V\to\infty} \langle \psi_{-\sigma}(t'_2, x'_2)\psi_\sigma(t'_1, x'_1)\rangle_H \qquad (B.27)$$

so that (B.24) decomposes into the product of averages

$$\lim_{V\to\infty} \langle \psi_{\sigma_1}(t_1, x_1)\psi^\dagger_{\sigma_2}(t_2, x_2)\rangle_H \lim_{V\to\infty} \langle \psi_{\sigma'_2}(t'_2, x'_2)\psi^\dagger_{\sigma'_1}(t'_1, x'_1)\rangle_H. \qquad (B.28)$$

The two relations (B.25) and (B.28) are the expressions of the principle of extinction of correlations for the two-particle average considered.

For the case $\nu = 0$,

$$\langle \psi(t_1, x_1)\psi(t_2, x_2)\rangle_H = 0,$$

and the relation (B.27) is no longer true. However, in this case we can introduce the "quasi-averages"

$$\langle \psi_{\sigma_1}(t_1, x_1)\psi_{\sigma_2}(t_2, x_2)\rangle_H = \lim_{\substack{\nu>0 \\ \nu\to 0}} \lim_{V\to\infty} \langle \psi_{\sigma_1}(t_1, x_1)\psi_{\sigma_2}(t_2, x_2)\rangle$$

$$= \Phi_{\sigma_2}(t_1 - t_2, x_1 - x_2)\delta(\sigma_1 + \sigma_2) \qquad (B.29)$$

and replace the product of averages in (B.28) by a product of quasi-averages.

Thus the relations obtained above illustrate the general principle of extinction of correlations.

References

1. J. Bardeen, L. Cooper and J. Schrieffer, *Phys. Rev.*, **108**, 1175 (1957).

2. N. N. Bogoliubov, D. N. Zubarev, and Yu. A. Tserkovnikov, *Docl. Akad. Nauk SSSR*, **117**, 788 (1957); *Sov. Phys. "Doklady", English Transl.*, **2**, 535, (1957).

3. R. E. Prange, *Bull. Am. Phys. Soc.*, **4**, 225 (1959).

4. N. N. Bogoliubov, D. N. Zubarev, and Yu. A. Tserkovnikov, *Zh. Exper. i Teor. Fiz.*, **39**, 120 (1960); *Sov. Phys. JETP, English Transl.*, **12**, 88 (1961).

5. N. N. Bogoliubov, *Zh. Exper. i Teor. Fiz.*, **34**, 73 (1958); *Sov. Phys. JETF, English Transl.*, **7**, 51 (1958).

6. N. N. Bogoliubov, *Izv. Akad. Nauk SSSR Ser. Fiz.* 11, 77 (1947); translation in D. Pines, *The Many-Body Problem*, Benjamin, New Work (1961).

7. N. N. Bogoliubov, Jr. Preprint JINR P4-4184. Dubna, 1968.

8. N. N. Bogoliubov, Jr. Preprint JINR P4-4175. Dubna, 1968.

9. N. N. Bogoliubov, Jr. Preprint ITPh-67-1. Kiev, 1967.

10. N. N. Bogoliubov, Jr. Preprint ITPh-68-65. Kiev, 1968.

11. N. N. Bogoliubov, Jr. Preprint ITPh-68-67. Kive, 1968.

12. N. N. Bogoliubov, Jr. *Yad. Phys.* **10**, 425 (1969) [*Sov. J. Nucl. Phys.* **10**, *243 (1970)*.

******

In connection with this work I should like to express sincere gratitude to D. N. Zubarev, S. V. Tyablikov, Yu. A. Tserkovnikov, and E. N. Yakovlev.

PART II

CHAPTER 6

MODEL HAMILTONIANS WITH FERMION INTERACTION

In many-body theory most problems of physical interest are rather complicated and usually insoluble. Model systems permitting a mathematical treatment of these problems are therefore acquiring considerable interest.

Unfortunately, however, in concrete problems in many-body theory there is usually no adequate correspondence between a real system and its mathematical models; one must be content with a model whose properties differ substantially from those of the real system and in solving problems one must use approximate methods lacking the necessary mathematical rigour.

Of considerable interest in this connection is the study of those few models which have some resemblance to real systems yet admit exact solution; fundamental properties of many-body can be established in this way. Systems of non-interacting particles can be taken as examples of systems which can be solved exactly. Although, of course, this model seems rather trivial, it is used as a starting point in most problems in many-body theory. In the theory of metals one can often leave the mutual interaction of the valence electron out of consideration. In the shell model of nuclei in its simplest form one can explain many general properties of nuclear spectra without introducing interaction between the particles into the treatment.

One of the most important problems in statistical physics is the study of exactly soluble cases. The fact is that this study makes an essential contribution to our understanding of the extremely complex problems of statistical physics and, in particular, serves as a basis for the approximate methods used in this field. Up to the present time, the class exactly soluble dynamical model systems has consisted mainly of one- and two-dimensional systems.

In this study we shall concentrate on the treatment of certain model systems of a general type which can be solved exactly, e.g.   models

249

with four-fermion pair interaction which have as their origin the BCS model problems and are applicable in the theory of superconductivity; the determination of asymptotically exact solutions for these models has been investigated by Bogolyubov, Zubarev and Tserkovnikov [1,2]. In these papers an approximation procedure was formulated in which ideas of a method based on the introduction of "approximating (trial) Hamiltonians" were propounded and reasons were given for believing that the solution obtained was asymptotically exact on passing to the usual statistical mechanical limit $V \to \infty$.

1.   General Treatment of the Problem. Some Preliminary Results

We begin here with the problem of the asymptotic calculation of quasi-averages based on works [13, 14, 17].

For a natural approach to the proper formulations of these we recall a series of results which we established earlier for a hamiltonian of the form

$$H = T - 2VgJJ^\dagger \tag{1}$$

in which

$$T = \sum_f T(f)a_f^\dagger a_f; \quad T(f) = \frac{p^2}{2m} - \mu$$

$$J = \frac{1}{2V} \sum_f \lambda(f)a_f^\dagger a_{-f}^\dagger \tag{2}$$

and a_f and $a_f^\dagger$ are Fermi amplitudes, m, μ and g are positive constants, and

$$f = (\boldsymbol{p}, \sigma)$$

where σ is the spin index, which takes the values $\pm\frac{1}{2}$. The function

$$\lambda(f) = \lambda(\boldsymbol{p}, \sigma)$$

occurring in (2) is a real and continuous in the spherical layer

$$\left| \frac{p^2}{2m} - \mu \right| \le \Delta$$

(where Δ is a certain positive constant) and equal to zero outside it; it also possesses the property of antisymmetry

$$\lambda(-f) = -\lambda(f); \quad -f = (-\boldsymbol{p}, -\sigma).$$

We note finally that in summation "over f" the components p_α ($\alpha = 1$, 2, 3) of the vector $\boldsymbol{p}$ take the values $2\pi n_\alpha/L$, while n_α runs over all integers $(-\infty, \infty)$. $L^3 = V$, where V is the volume of the system, which below will tend to ∞.

To introduce the quasi-averages we add to the Hamiltonian H terms with "pair sources", e.g.

$$-\nu V(J + J^\dagger)$$

where ν is a positive constant.

Thus, the Hamiltonian under consideration will be

$$\Gamma = T - 2VgJJ^\dagger - -\nu V(J + J^\dagger). \tag{3}$$

The quasi-averages for the Hamiltonian H are introduced as the limits, as $V \to \infty$, of the usual averages for the Hamiltonian Γ, with the *sequential* passage to the limit

$$\prec \ldots \succ_H = \lim_{\substack{\nu \to 0 \\ \nu > 0}} \lim_{V \to \infty} \langle \ldots \rangle_\Gamma.$$

We have shown [7][cc] that the simplest binary-type correlation averages

$$\langle a_f^\dagger(t) a_f(tau) \rangle_\Gamma, \quad \langle a_f^\dagger(t) a_{-f}^\dagger(\tau) \rangle_\Gamma, \quad \langle a_{-f}(t) a_f(tau) \rangle_\Gamma$$

are asymptotically ($V \to \infty$) close to the corresponding averages taken for the "trial Hamiltonian":[dd]

$$\Gamma_a(C) = T - 2Vg(CJ^\dagger + C^*J - C^*C) - \nu V(J + J^\dagger)$$
$$= T - 2Vg\left\{ \left(C + \frac{\nu}{2g}\right)J^\dagger + \left(C^* + \frac{\nu}{2g}\right)J \right\} + 2VgC^*C.$$

(There is, however, another approach, viz. to deal with infinite volume form the start. Such a situation was studied in a paper by Petrina (1970).) The quantity C accuring in $\Gamma_a(C)$ is determined from the condition that the free energy has an absolute minimum

$$f\{\Gamma_a(C)\} = \min$$

[cc]We note that this techniques has also been found to be useful in the study of exactly solvable quasi-spin models (see, for example, [9, 33] and [34]).

[dd]In this formulas we should have written $2VgC^*C \cdot \hat{1}$, where $\hat{1}$ is the unit operator; however, since this will not lead to misunderstandings anywhere, below we shall not write out the unit operator explicitly.

in the whole complex C-plane.

Since $\Gamma_a(C)$ is a quadratic form in the Fermi amplitudes, this Hamiltonian can be diagonalized by means of a $u - v$ transformation:

$$a_f = u(f)\alpha_f - v(f)\alpha_f^\dagger, \tag{4}$$

where α_f and $\alpha_f^\dagger$ are new Fermi amplitudes, and

$$u(f) = \frac{1}{2}\sqrt{1 + \frac{T(f)}{E(f)}},$$

$$v(f) = -\frac{\lambda(f)\left(C + \dfrac{\nu}{2g}\right)}{\sqrt{2}\left|\lambda(f)\left(C + \dfrac{\nu}{2g}\right)\right|}\sqrt{1 - \frac{T(f)}{E(f)}},$$

$$E(f) = \sqrt{T^2(f) + 4\lambda^2(f)g^2\left|C + \frac{\nu}{2g}\right|^2}.$$

In the new Fermi amplitudes the Hamiltonian will take the form

$$\Gamma_a(C) = \left\{2gC^*C - \frac{1}{2V}\sum_f (E(f) - T(f))\right\}V + \sum_f E(f)\alpha_f^\dagger\alpha_f \tag{5}$$

so that the free energy per unit volume calculated on the basis of this Hamiltonian will be

$$f\{\Gamma_a(C)\} = 2gC^*C - \frac{1}{2V}\sum_f \{E(f) - T(f)\}$$

$$- \frac{\theta}{V}\sum_f \ln(1 + e^{-E(f)/\theta}). \tag{6}$$

It is clear form this that the absolute minimum of $f\{\Gamma_a(C)\}$ as a function of the complex variable is found at a real value of C, and at a value such that

$$C + \frac{\nu}{2g} > 0.$$

This minimazing value of C depends general on V and ν:

$$C = C(V, \nu).$$

In our above-mentioned paper (Bogolyubov, Jr., 1967) it was shown that

$$C(V, \nu) \rightarrow C(\nu) \tag{7}$$

as

$$V \rightarrow \infty \quad (\nu(\text{fixed}) > 0)$$

and

$$C(\nu) \xrightarrow[\substack{\nu \to 0 \\ (\nu > 0)}]{} C(0).$$

Here $C = C(\nu)$ realizes the absolute minimum of the asymptotic expression

$$f_\infty\{\Gamma_a(C)\} = \lim_{V \to \infty} \{\Gamma_a(C)\}$$

$$= 2gC^*C - \frac{1}{2(2\pi)^3} \int df\{E(f) - T(f)\}$$

$$- \frac{\theta}{(2\pi)^3} \int df \, \ln(1 + e^{-E(f)/\theta}).$$

As throughout the book, the "integral over f" denotes integration $\boldsymbol{p}$ and summation over σ:

$$\int df(\ldots) = \sum_\sigma \int d\boldsymbol{p}(\ldots).$$

The value $C(0) \geq 0$ is chosen as the number giving the absolute minimum of the function

$$f_\infty\{H_a(C)\} = 2gC^2 - \frac{1}{2(2\pi)^3} \int df\{E(f) - T(f)\}$$

$$- \frac{\theta}{(2\pi)^3} \int df \, \ln(1 + e^{-E(f)/\theta}),$$

in which

$$E(f) = \sqrt{T^2(f) + 4g^2C^2\lambda^2(f)}.$$

As we have noted already, we have proved in our work that the difference of the binary averages constructed on the basis of the model (Γ) and trial (Γ_a) Hamiltonians tends to zero as $V \rightarrow \infty$ for any fixed value of $\nu > 0$.

On the other hand, in view of (4) and (5), these averages are calculated easily for Γ_a. For example, we have

$$\langle a_f^\dagger(t) a_f(\tau) \rangle_{\Gamma_a} = u^2(f) e^{iE(f)(t-\tau)} \frac{e^{-E(f)/\theta}}{1 + e^{-E(f)/\theta}}$$

$$+ v^2(f)e^{-iE(f)(t-\tau)}\frac{1}{1+e^{-E(f)/\theta}},$$

$$\langle a_f^\dagger(t)a_{-f}^\dagger(\tau)\rangle_{\Gamma_a} = u(f)v(f)\left\{ e^{iE(f)(t-\tau)}\frac{e^{-E(f)/\theta}}{1+e^{-E(f)/\theta}}\right.$$

$$\left.+ e^{-iE(f)(t-\tau)}\frac{1}{1+e^{-E(f)/\theta}}\right\}, \tag{8}$$

As can be seen, the right-hand sides here are defined for *all* $\boldsymbol{p}$, and not only for the quasi-discrete values

$$\boldsymbol{p} = \left(\frac{2\pi n_1}{L},\ \frac{2\pi n_2}{L},\ \frac{2\pi n_3}{L}\right),$$

which are the only values for which the amplitudes a_f and $a_f^\dagger$, and thereby the left-hand sides of the expressions (8), are defined.

Moreover, the right-hand sides of (8) as functions of f depend on V only through the quantity $C = C(V,\nu)$. The passage to the limit $V \to \infty$ therefore reduces, because of (7), to replacing $C(V,\nu)$ by $C(\nu)$ in these functions. The subsequent passage to the limit $\nu \to 0$ $(\nu > 0)$ corresponds to replacement of $C(\nu)$ by $C(0)$.

Thus, if we put $\nu = 0$, $C = C(0)$ in the functions of f occurring in the right-hand sides of (8), these functions will represent the corresponding quasi-averages for the Hamiltonian H.

We note that the most complicated of our proofs was that establishing the relation

$$\langle\ldots\rangle_\Gamma - \langle\ldots\rangle_{\Gamma_a} \to 0 \tag{9}$$

as $V \to \infty$ for binary expressions of the type indicated above. To prove them, we had to show first that

$$\langle (J - C(V,\nu))(J^\dagger - C(V,\nu))\rangle_\Gamma \to 0 \tag{10}$$

as $V \to \infty$ and then establish the asymptotic relations (9). We must emphasize that the problem of investigating situations with more complicated averages was not solved in the papers cited [4, 6, 7]. Moreover, the proof of the properties (10) and (7) was based on the specific features of the Hamiltonian (1) and could not be extended to model Hamiltonians of more general form.

In our work [4, 6–8], we have constructed a new method which enables us to extend the above-mentioned results to the case of many-time averages of

Fermi amplitudes or field functions, and, therefore, for model Hamiltonians of more complicated structure. If we wished to apply this method to the investigation of the Hamiltonian Γ it would be appropriate to start from the representation:

$$\Gamma = \Gamma_a(C(\nu)) - 2gV(J - C(V, \nu))(J^\dagger - C(V, \nu))$$
$$= T - \frac{1}{2}\sum_f \Lambda(f)\{a_f^\dagger a_{-f}^\dagger + a_{-f}a_f\} + 2gVC^2$$
$$- 2gV(J - C(V, \nu))(J^\dagger - C(V, \nu)), \tag{11}$$

where

$$\Lambda(f) = 2g\lambda(f)\left(C(\nu) + \frac{\nu}{2g}\right). \tag{12}$$

In this case, in view of (7) and (10)

$$\langle(J - C(V, \nu))(J^\dagger - C(V, \nu))\rangle_\Gamma \leq \varepsilon_V \to 0. \tag{13}$$

However, we shall not study only this Hamiltonian here.

As will be shown under much wider conditions in the following chapters, model Hamiltonians with properly chosen source terms can also be reduced to a form similar to (11).

We turn now to consider the situation when

$$\Gamma = \Gamma_a + H_1, \tag{14}$$

where

$$\Gamma_a = \sum_f T(f)a_f^\dagger a_f - \frac{1}{2}\sum_f \{\Lambda^*(f)a_{-f}a_f + \Lambda(f)a_f^\dagger a_{-f}^\dagger\} + \mathscr{K},$$

where

$$\mathscr{K} = \text{const}, \quad T(f) = \frac{p^2}{2m} - \mu,$$

and

$$H_1 = -V\sum_\alpha G_\alpha(J_\alpha - C_\alpha)(J_\alpha^\dagger - C_\alpha^*), \tag{15}$$

where

$$J_\alpha = \frac{1}{2V}\sum_f \lambda_\alpha(f)a_f^\dagger a_{-f}^\dagger.$$

Here the summation over f runs over the above-mentioned quasi-discrete set, which we shall call the set ϕ_V. We shall examine this model system under the following conditions, which we shall call conditions **I** (which were formulated in $[13, 14]$).

1. The functions $\lambda_\alpha(f)$ and $\Lambda(f)$ are defined and bounded in the whole space ϕ of the points $f = (\boldsymbol{p}, \sigma)$.

2. The series

$$\sum_\alpha |G_\alpha| |\lambda_\alpha(f)|^2 = P(f)$$

converges uniformly in ϕ and the function $P(f)$ represented by it satisfies the inequalities:

$$P(f) \leq M_1 = \text{const},$$

$$\frac{1}{V} \sum_f P(f) \leq M_2 = \text{const}.$$

3. The inequality

$$\left\langle \sum_\alpha |G_\alpha| (J_\alpha - C_\alpha)(J_\alpha^\dagger - C_\alpha^*) \right\rangle_\Gamma \leq \varepsilon_V$$

is satisfied, where

$$\varepsilon_V \to 0 \quad \text{as} \quad V \to \infty.$$

4. The function $\Lambda(f)$ and the constants C_α satisfy the inequalities

$$\frac{1}{V} \sum_f |\Lambda(f)|^2 \leq M_\Lambda,$$

$$\sum_\alpha |G_\alpha| |C_\alpha|^2 \leq M_C.$$

5. The functions $\lambda_\alpha(f)$ and $\Lambda(f)$ are antisymmetric with respect to reflection:[ee]

$$\lambda_\alpha(-f) = -\lambda_\alpha(f), \quad \Lambda(-f) = -\Lambda(f).$$

[ee]This last condition is not, essentially, a restrictive one. Any sum of the form

$$\sum_f F(f) a_f^\dagger a_{-f}^\dagger; \quad \sum_f F(f) a_{-f} a_f$$

Here α takes integer values. (In general, the sums over α imply an infinite number of terms; in the next sections, however, we consider problems in which α takes a finite number of values.)

With these conditions we prove a series of theorems on the asymptotic closeness of averages taken over the Hamiltonians Γ and Γ_a respectively.

Apart from conditions **I** when considering field functions

$$\psi_\sigma(r,t) = \frac{1}{\sqrt{V}} \sum_p a_{p\sigma}(t) e^{i(p \cdot r)},$$

$$\psi_\sigma^\dagger(r,t) = \frac{1}{\sqrt{V}} \sum_p a_{p\sigma}^\dagger(t) e^{-i(p \cdot r)}$$

(here the sum runs over the set of quasi-discrete p) we shall now have to impose the following additional conditions, which we will call conditions **I'**:

1. The functions
$$\Lambda(f) = \Lambda(\boldsymbol{p}, \sigma) \quad (\sigma = \pm\tfrac{1}{2})$$
   are defined and bounded in the whole space E of points $\boldsymbol{p}$ and are independent of V.

2. The discontinuities of these functions form a set of measure zero in the space E (see [31]).

These conditions **I'** mean that as a result of the passage to the limit $(V \to \infty)$, we go over form the sums to Riemann integrals. In fact, the integrability in the Riemann sense of some bounded functions is ensured by the fact that the set of its discontinuities is of a measure zero.

2. Calculation of the Free Energy for Model System with Attraction

In the following we shall study dynamical systems which correspond to attraction of fermions. We shall begin by calculating the free energy for

---

can always be reduced to the form

$$\sum_f \frac{F(f) - F(-f)}{2} a_f^\dagger a_{-f}^\dagger; \quad \sum_f \frac{F(f) - F(-f)}{2} a_{-f} a_f,$$

in which the coefficient function $[F(f) - F(-f)]/2$ is already antisymmetric with respect to the reflection: $f \to -f$.

model Hamiltonians with four-fermion interactions. This problem, as we have shown [5], is of great interest in the study of model problems in the problem of superconductivity and serves as an example of an exact calculation of the free energy for model systems of the BCS type [34, 35].[ff] The results and upper bounds obtained here also constitute a proof of the Theorem 1 formulated in this section.

We shall start from the Hamiltonian

$$H = T - 2V \sum_{1 \leq \alpha s} J_\alpha J_\alpha^\dagger. \tag{16}$$

If we take the following Fermi-operator expressions for the operators T and J_α:

$$T = \sum_f T(f) a_f^\dagger a_f, \qquad J_\alpha = \frac{1}{2V} \sum_f \lambda_\alpha(f) a_f^\dagger a_{-f}^\dagger, \tag{17}$$

we obtain the usual BCS Hamiltonian

$$H = \sum_f T(f) a_f^\dagger a_f - \frac{1}{2V} \sum_{f,f'} \mathscr{J}(f, f') a_f^\dagger a_{-f}^\dagger a_{-f'} a_{f'}. \tag{18}$$

In fact it is not necessary for our discussion that the operators T and J_α have the explicit form (17)

It is sufficient to impose the following general conditions:

$$|J_\alpha| \leq M_1, \qquad |TJ_\alpha - J_\alpha T| \leq M_2,$$
$$|J_\alpha^\dagger J_\beta - J_\beta J_\alpha^\dagger| \leq \frac{M_3}{V}, \qquad |J_\alpha J_\beta - J_\beta J_\alpha| \leq \frac{M_3}{V}, \tag{19}$$

where M_1, M_2 and M_3 are constants as $V \to \infty$ and the symbol $|\ldots|$ denotes the norm of the indicated operators. We assume also that the free energy per unit volume for the Hamiltonian $H = T$ is bounded by a constant and that the number of terms s in the sum (16) is fixed.

We thus start from the Hamiltonian (16) with the condition (19). We take the trial Hamiltonian to have the usual form

$$H^0 = T - 2V \sum_{1 \leq \alpha \leq s} (C_\alpha J_\alpha^\dagger + C_\alpha^* J_\alpha) + 2V \sum_{1 \leq \alpha \leq s} |C_\alpha|^2. \tag{20}$$

[ff]This treatment also solves a number of problems raised by Wentzel [35] concerning the asymptotically exact calculation of the free energy and can also be applied to certain quasi-spin Hamiltonians (cf. [12, 15] and [32]).

Here the C_α are complex constants determined from the condition that the function

$$f_{H^0} = -\frac{1}{V}\,\theta \ln \mathrm{Tr}\, e^{-H^0/\theta} \tag{21}$$

have its absolute minimum value in the domain of all the complex variables $(C_1, \ldots, C_s)$. We shall denote this complex set of points $(C_1, \ldots, C_s)$ by $\{E^s\}$. By making use of the minimizing values of C, we calculate the free energy per unit volume for the trail Hamiltonian:

$$f_{H^0}(C) = \min_{\{E^s\}} f_{H^0}(C). \tag{22}$$

We also take the corresponding free energy for the Hamiltonian (16):

$$f_H = -\frac{1}{V}\,\theta \ln \mathrm{Tr}\, e^{-H/\theta}. \tag{23}$$

We shall prove that the difference $F_{H^0} - f_h$ tends to zero as $V \to \infty$. For this it is convenient to consider fist the auxiliary problem with the Hamiltonian

$$\Gamma = H - V \sum_{1 \le \alpha \le s} (\nu_\alpha J_\alpha + \nu_\alpha^* J_\alpha^\dagger), \tag{24}$$

where $\nu_1, \ldots, \nu_s$ are arbitrary non-zero complex parameters. In this problem, the corresponding trial Hamiltonian has the form

$$\Gamma^0 = H^0 - V \sum_{1 \le \alpha \le s} (\nu_\alpha J_\alpha + \nu_\alpha^* J_\alpha^\dagger). \tag{25}$$

The complex quantities $C = (C_1, \ldots, C_s)$ occurring here are also determined from the condition for the absolute minimum of the function

$$f_{\Gamma^0}(C) = -\frac{1}{V}\,\theta \ln \mathrm{Tr}\, e^{-\Gamma^0/\theta}. \tag{26}$$

We shall obtain an upper bound for the difference $f_{\Gamma^0} - f_\gamma$, and show this difference to be asymptotically small as $V \to \infty$. Here

$$f_{\Gamma^0} = \min_{\{E^s\}} f_{\Gamma^0}(C)$$

and f_Γ is the free energy per unit volume for the Hamiltonian Γ. Although this bound for $f_{\Gamma^0} - f_\gamma$ will be found for $|\nu_\alpha| 0$, it turns out that it is uniform with respect to $\nu_\alpha \to 0$, so that we can then pass to the limit $\nu_\alpha = 0$ $(1 \le \alpha \le s)$. We thereby obtain a bound for $f_{H^0} - f_H$, proving it to be asymptotically small as $V \to \infty$.

We therefore begin by treating the trial Hamiltonian Γ^0. It is not difficult to show that the problem of the absolute minimum of the function (26) has a solution and that this absolute minimum is realized for finite values $C_k - C_k^0$ $(1 \le k \le s)$. This can be seen by using the inequalities

$$\sum_\alpha \{|C_\alpha|^2 + (|C_\alpha| + 2M_1)^2\} - 4M_1^2 s + \gamma + f_T \ge f_{\Gamma^0}(C)$$

$$\ge \sum_\alpha \{|C_\alpha|^2 + (|C_\alpha| - 2M_1)^2\} - 4M_1^2 s - \gamma + f_T$$

$$\ge \sum_\alpha |C_\alpha|^2 - 4M_1^2 s + \gamma + f_T,$$

$$C = (C_1, \ldots, C_s), \quad \gamma = 2M_1 \sum_\alpha |\nu_\alpha| \quad (1 \le \alpha \le s). \tag{27}$$

Thus, the function $f_{\Gamma^0}(C)$ has an absolute minimum at some point $C^0 = (C_1^0, \ldots, C_s^0)$.

Because the function $f_{\Gamma^0}(C)$ is continuously differentiable, at the point $C = C^0$ we have

$$\frac{\partial f_{\Gamma^0}(C)}{\partial C_\alpha} = 0 \quad (1 \le \alpha \le s),$$

i.e. an equation for C_α:

$$C_\alpha = \langle J_\alpha \rangle_{\Gamma^0} = \frac{\mathrm{Tr}\, J_\alpha\, e^{-\Gamma^0/\theta}}{\mathrm{Tr}\, e^{-\Gamma^0/\theta}}.$$

Taking into account the conditions (19) on the operators J_α, we obtain $|C_\alpha| \le M_1 = \mathrm{const.}$

We turn now to the derivation of inequalities limiting the difference in the free energies per unit volume $f_{\Gamma^0} - f_\Gamma$ in terms of averages of

$$\frac{\mathscr{A}}{V} = -2 \sum_{1 \le \alpha \le s} (J_\alpha - C_\alpha)(J_\alpha^\dagger - C_\alpha^*). \tag{28}$$

For this we note that $\Gamma = \Gamma^0 + \mathscr{A}$ and introduce the intermediate Hamiltonian

$$\Gamma^t = \Gamma^0 + t\mathscr{A},$$

which for $t = 0$ coincides with the trial Hamiltonian (25) and for $t = 1$ coincides with the original Γ (24). The constants $C = (C_1, \ldots, C_s)$ occurring in Γ^t are assumed to be fixed and to be independent of the parameter t.

Let us consider the partition function[gg] and free energy for the intermediate Hamiltonian Γ^t:

$$Q_t = \operatorname{Tr} e^{-\Gamma^t/\theta}, \quad f_t(C_1, \ldots, C_s) = -\frac{\theta}{V} \ln Q_t, \quad Q_t = e^{-V f_t/\theta}. \tag{29}$$

Differentiating the equality (29) twice with respect to t using operator differentiation, we arrive at the formula

$$-\frac{V}{\theta} \frac{\partial^2 f_t}{\partial t^2} + \frac{V^2}{\theta^2} \left(\frac{\partial f_t}{\partial t} \right) = \frac{1}{\theta^2 Q_t} \int_0^1 \operatorname{Tr}\{\mathscr{A}\, e^{-(\Gamma^t/\theta)\tau}\mathscr{A}\, e^{-(\Gamma^t/\theta)(1-\tau)}\}\, d\tau.$$

Taking into account that

$$\frac{\partial f_t}{\partial t} = \frac{1}{V} \frac{\operatorname{Tr}\mathscr{A}\, e^{-\Gamma^t/\theta}}{\operatorname{Tr} e^{-\Gamma^t/\theta}} = \frac{1}{V} \langle \mathscr{A} \rangle_t,$$

we find

$$-\frac{\partial^2 f_t}{\partial t^2} = \frac{1}{\theta V} \left\{ \frac{1}{Q_t} \int_0^1 \operatorname{Tr}\{\mathscr{A}\, e^{-(\Gamma^t/\theta)\tau}\mathscr{A}\, e^{-(\Gamma^t/\theta)(1-\tau)}\}\, d\tau - \langle \mathscr{A} \rangle^2 \right\}$$

$$= \frac{1}{\theta V Q_t} \int_0^1 \operatorname{Tr}\{\mathscr{B}\, e^{-(\Gamma^t/\theta)\tau}\mathscr{B}\, e^{-(\Gamma^t/\theta)(1-\tau)}\}\, d\tau;$$

$$\mathscr{B} = \mathscr{A} - \langle \mathscr{A} \rangle.$$

Going over to a matrix representation in which the Hamiltonian Γ^t is diagonal, we have

$$-\frac{\partial^2 f_t}{\partial t^2} = \frac{1}{\theta V Q_t} \int_0^1 d\tau \sum_{n,m} \mathscr{B}_{nm}\mathscr{B}_{mn}\, e^{-\{(E_m^t - E_n^t)/\theta\}\tau - E_n^t/\theta}$$

[gg]Mathematical questions concerning the existence and analytical properties of partition functions have been considered by the author in [11]. The theorem proved there has been further generalized by H. D. Maison [29].

$$= \frac{1}{\theta V Q_t} \int\limits_0^1 d\tau \sum_{n,m} |\mathscr{B}_{nm}|^2 \, e^{-\{(E_m^t - E_n^t)/\theta\}\tau - E_n^t/\theta} \geq 0.$$

Hence it follows, in particular, that $\partial_t^2/\partial t^2 \leq 0$, and therefore $\partial f_t/\partial t = \langle \mathscr{A} \rangle_t/V$ decreases with increase of the parameter t. Furthermore, taking into account that f_Γ does not depend on C, we have

$$f_{\Gamma^0}(C) - f_\Gamma = - \int\limits_0^1 \frac{\partial f_t}{\partial t} \, dt = \int\limits_0^1 \frac{\langle \mathscr{A} \rangle_t}{V} \, dt \geq 0.$$

Since this relation is true for all $C = (C_1, \ldots, C_s)$, we also have

$$\min_{\{E^s\}} f_{\Gamma^0} \geq f_\Gamma, \quad f_{\Gamma^0} \geq f_\Gamma.$$

We shall integrate the inequality $\langle \mathscr{A} \rangle_{\Gamma^t} \geq \langle \mathscr{A} \rangle_\Gamma$ $(0 \leq t \leq 1)$. Substituting the expression (28) for $\mathscr{A}$, we convince ourselves that the inequality

$$f_{\Gamma^0}(C) - f_\Gamma \leq 2 \sum_{1 \leq \alpha \leq s} \langle (J_\alpha - C_\alpha)(J_\alpha^\dagger - C^*\alpha) \rangle_\Gamma$$

is true for any $C = (C_1, \ldots, C_s)$. We put here $C_\alpha = \langle J_\alpha \rangle_\Gamma$ $(0 \leq t \leq 1)$ and note that

$$f_{\Gamma^0} = \min_{\{E^s\}} f_{\Gamma^0}(C) \leq f_{\Gamma^0}(\langle J \rangle_\Gamma).$$

Thus,

$$f_{\Gamma^0} - f_\Gamma \leq f_{\Gamma^0}(\langle J \rangle_\Gamma) - f_\Gamma \leq 2 \sum_{1 \leq \alpha \leq s} \langle (J_\alpha - \langle J_\alpha \rangle_\Gamma)(J_\alpha^\dagger - \langle J_\alpha^\dagger \rangle_\Gamma) \rangle_\Gamma$$

and, finally,

$$0 \leq f_{\Gamma^0} - f_\Gamma \leq 2 \sum_{1 \leq \alpha \leq s} \langle (J_\alpha - \langle J_\alpha \rangle_\Gamma)(J_\alpha^\dagger - \langle J_\alpha^\dagger \rangle_\Gamma) \rangle_\Gamma, {}^{\text{hh}} \tag{30}$$

where, as always,

$$f_{\Gamma^0} = \min_{\{E^s\}} f_{\Gamma^0}(C). \tag{31}$$

---

hh For convenience, in the reminder of this section we shall omit the subscript in statistical averages over the Hamiltonian Γ, i.e. we shall write $\langle \ldots \rangle \equiv \langle \ldots \rangle_\Gamma$.

Let us recall our main problem. We want to show that the difference $f_{\Gamma^0} - f_\Gamma$ is asymptotically small as $V \to \infty$. It follows from (30) that we shall have solved our problem if we can demonstrate the asymptotic smallness of the average on the right-hand side of (30).

Taking into account the main idea of a paper by the author (Bogolyubov, Jr., 1966a), we express this right-hand side in terms of $\partial^2 f / \partial \nu_\alpha^* \partial_\alpha$. Differentiating, we have

$$-\frac{1}{\theta}\frac{\partial^2 f}{\partial \nu_\alpha^* \partial_\alpha} = \frac{V}{\theta^2}\int_0^1 \frac{\mathrm{Tr}\left(D^{(\alpha)}\,\mathrm{e}^{-(\tau/\theta)\Gamma}\,(D^{(\alpha)})^\dagger\,\mathrm{e}^{-\{(1-\tau)/\theta\}\Gamma}\right) d\tau}{\mathrm{Tr}\,\mathrm{e}^{-\Gamma/\theta}},$$

where

$$D^{(\alpha)} = J_\alpha - \langle J_\alpha \rangle, \quad (1 \le \alpha \le s).$$

Going over to the matrix representation in which Γ is diagonal, we find

$$-\frac{1}{\theta}\frac{\partial^2 f}{\partial \nu_\alpha^* \partial_\alpha} = \frac{V}{Q\theta^2}\sum_{n,m}\int_0^1 D_{nm}^{(\alpha)}\,\mathrm{e}^{-(\tau/\theta)E_m}\,(D_{mn}^{(\alpha)})^\dagger\,\mathrm{e}^{-\{(1-\tau)/\theta\}E_m}\,d\tau$$

$$= \frac{V}{Q\theta^2}\sum_{n,m}\left|D_{nm}^{(\alpha)}\right|^2\int_0^1 \mathrm{e}^{-(\tau/\theta)E_m-\{(1-\tau)/\theta\}E_m}\,d\tau$$

$$= \frac{V}{Q\theta^2}\sum_{n,m}\left|D_{nm}^{(\alpha)}\right|^2\frac{\mathrm{e}^{-E_m/\theta} - \mathrm{e}^{-E_n/\theta}}{E_n - E_m} \ge 0.$$

Using Hölders inequality, we have the following bound:

$$\frac{V}{Q}\sum_{n,m}\left|D_{nm}^{(\alpha)}\right|^2\left|\mathrm{e}^{-E_m/\theta} - \mathrm{e}^{-E_n/\theta}\right|$$

$$\le \left(-\frac{\partial^2 f}{\partial \nu_\alpha^* \partial_\alpha}\right)^{3/2}\left(\frac{V}{Q}\sum_{n,m}\left|D_{nm}^{(\alpha)}\right|^2\left|E_n - E_m\right|^2\left(\mathrm{e}^{-E_m/\theta} - \mathrm{e}^{-E_n/\theta}\right)\right)^{1/3}.$$

We carry out the simple transformations:

$$\frac{V}{Q}\sum_{n,m}\left|D_{nm}^{(\alpha)}\right|^2\left|E_n - E_m\right|^2\left(\mathrm{e}^{-E_m/\theta} - \mathrm{e}^{-E_n/\theta}\right)$$

$$= \frac{V}{Q}\,\mathrm{Tr}\,\mathrm{e}^{-\Gamma/\theta}\left\{\left(\Gamma D^{(\alpha)} - D^{(\alpha)}\Gamma\right)\left((D^{(\alpha)})^\dagger\Gamma - \Gamma(D^{(\alpha)})^\dagger\right)\right.$$

$$+ \left((D^{(\alpha)})^\dagger \Gamma - \Gamma (D^{(\alpha)})^\dagger \right) \left((D^{(\alpha)})^\dagger \Gamma - \Gamma (D^{(\alpha)})^\dagger \right) \}$$
$$= V \left\langle (\Gamma J_\alpha - J_\alpha \Gamma)(\Gamma J_\alpha - J_\alpha \Gamma)^\dagger + (\Gamma J_\alpha - J_\alpha \Gamma)^\dagger (\Gamma J_\alpha - J_\alpha \Gamma) \right\rangle$$
$$\leq 2V \mathscr{M}^2,$$

where

$$\mathscr{M} = M_2 + 4 M_1 M_3 s + 2 M_3 \sum_{1 \leq \alpha \leq s} |\nu_\alpha|.$$

Hence we obtain

$$\frac{V}{Q} \sum_{n,m} |D_{nm}^{(\alpha)}|^2 |e^{-E_m/\theta} - e^{-E_n/\theta}| \leq \left(-\frac{\partial^2 f}{\partial \nu_\alpha^* \partial_\alpha} \right)^{2/3} (2V \mathscr{M}^2)^{1/3}.$$

Furthermore,

$$\frac{V}{Q} \sum_{n,m} |D_{nm}^{(\alpha)}|^2 e^{-E_n/\theta} \leq \theta \frac{V}{Q} \sum_{n,m} \frac{|D_{nm}^{(\alpha)}|^2}{(E_n - E_m)} \left(e^{-E_m/\theta} - e^{-E_n/\theta} \right)$$
$$+ \frac{V}{Q} \sum_{n,m} |D_{nm}^{(\alpha)}|^2 |e^{-E_m/\theta} - e^{-E_n/\theta}|,$$

where

$$\frac{V}{Q} \sum_{n,m} |D_{nm}^{(\alpha)}|^2 e^{-E_n/\theta} = \frac{V}{Q} \operatorname{Tr} D^{(\alpha)} (D^{(\alpha)})^\dagger e^{-\Gamma/\theta} = V \langle D^{(\alpha)} (D^{(\alpha)})^\dagger \rangle$$
$$= V \left\langle (J_\alpha - \langle J_\alpha \rangle)(J_\alpha^\dagger - \langle J_\alpha^\dagger \rangle) \right\rangle.$$

Thus, we finally obtain

$$\langle (J_\alpha - \langle J_\alpha \rangle)(J_\alpha^\dagger - \langle J_\alpha^\dagger \rangle) \rangle$$
$$\leq \left(-\frac{\partial^2 f}{\partial \nu_\alpha^* \partial_\alpha} \right) \frac{\theta}{V} + \frac{(2 \mathscr{M}^2)^{1/3}}{V^{2/3}} \left(-\frac{\partial^2 f}{\partial \nu_\alpha^* \partial_\alpha} \right)^{2/3}.$$

Substituting this inequality into (30), we find

$$0 \leq f_{\Gamma^0} - f_\Gamma \leq 2 \frac{\theta}{V} \sum_{1 \leq \alpha \leq s} \left(-\frac{\partial^2 f}{\partial \nu_\alpha^* \partial_\alpha} \right)$$
$$+ \frac{2}{V^{2/3}} (2 \mathscr{M}^2)^{1/3} \sum_{1 \leq \alpha \leq s} \left(-\frac{\partial^2 f}{\partial \nu_\alpha^* \partial_\alpha} \right)^{2/3}. \tag{32}$$

Hence we can see that our problem would be solved if we could show that the second derivatives $\left|\partial^2 f / \partial \nu_\alpha^* \partial_\alpha\right|$ are bounded by a constant as $V \to \infty$. Unfortunately, we are unable to prove such a statement. We must start from the boundedness of the first derivatives $\left|\partial f / \partial \nu_\alpha\right| \le M_1$ $(1 \le \alpha \le s)$.

Because of this, we develop a method in which it will not be necessary to use the boundedness of the second derivatives and by means of which we can demonstrate the asymptotic smallness of the difference

$$a = f_{\Gamma^0} - f_\Gamma.$$

For the following, it will be more convenient to transform to the polar variables r_α, φ_α:

$$r_\alpha = r_\alpha(\nu_\alpha, \nu_\alpha^*), \quad \varphi_\alpha = \varphi_\alpha(\nu_\alpha, \nu_\alpha^*), \quad (1 \le \alpha \le s)$$

in inequality (32). Accordingly,

$$f(\nu_1, \nu_1^*, \ldots, \nu_s, \nu_s^*) \to f(r_1, \varphi_1, \ldots, r_s, \varphi_s).$$

Then

$$\frac{\partial^2 f}{\partial \nu_\alpha^* \partial_\alpha} = \frac{1}{4}\left\{\frac{1}{r_\alpha}\frac{\partial}{\partial r_\alpha}\left(r_\alpha \frac{\partial f}{\partial r_\alpha}\right) + \frac{\partial^2 f}{\partial r_\alpha^2}\frac{1}{r_\alpha^2}\right\}. \tag{33}$$

We now make use of the inequality

$$\left|a(r_1, \ldots, r_s; \varphi_1, \ldots, \varphi_s)\right| \le \left|a(r_1, \ldots, r_s; \varphi_1, \ldots, \varphi_s)\right.$$
$$\left. - a(\xi_1, \ldots, \xi_s; \eta_1, \ldots, \eta_s)\right| + \left|a(\xi_1, \ldots, \xi_s; \eta_1, \ldots, \eta_s)\right|$$
$$\le \sum_{1 \le \alpha \le s}\left|\frac{\partial a}{\partial r_\alpha}\right|_{max}\left|r_\alpha - \xi_\alpha\right| + \sum_{1 \le \alpha \le s}\left|\frac{\partial a}{\partial \varphi_\alpha}\right|_{max}\left|\varphi_\alpha - \eta_\alpha\right|$$
$$+ \left|a(\xi_1, \ldots, \xi_s; \eta_1, \ldots, \eta_s)\right|, \tag{34}$$

we take

$$r_\alpha + l \le \xi_\alpha \le r_\alpha + 2l, \quad \varphi_\alpha \le \eta_\alpha \le \varphi_\alpha + \delta_\alpha, \quad \delta_\alpha = \frac{l}{r_\alpha}, \quad (1 \le \alpha \le s),$$

so that

$$a(\xi_1, \ldots, \xi_s; \eta_1, \ldots, \eta_s)$$

$$= \frac{\int\limits_{r_1+l}^{r_1+2l} \ldots \int\limits_{r_s+l}^{r_s+2l} dr_1 \ldots dr_s \int\limits_{\varphi_1}^{\varphi_1+\delta_1} \ldots \int\limits_{\varphi_s}^{\varphi_s+\delta_s} d\varphi_1 \ldots \varphi_s \, a(r_1,\ldots,r_s;\varphi_1,\ldots,\varphi_s) \prod\limits_{1\leq\alpha\leq s} r_\alpha}{(1/2)^3 \prod\limits_{1\leq\alpha\leq s}[(r_\alpha+2l)^2-(r_\alpha+l)^2]\delta_\alpha}$$

$$(35)$$

We note that

$$\left|\frac{\partial f}{\partial r_\alpha}\right| \leq 2M_1, \quad \left|\frac{\partial f}{\partial \varphi_\alpha}\right| \leq 2M_1 r_\alpha$$

and

$$\left|\frac{\partial a}{\partial r_\alpha}\right| \leq 4M_1, \quad \left|\frac{\partial a}{\partial \varphi_\alpha}\right| \leq 4M_1 r_\alpha. \tag{36}$$

Therefore, the first two terms in inequality (34) can be bounded as follows:

$$\sum_{1\leq\alpha\leq s}\left|\frac{\partial a}{\partial r_\alpha}\right|_{\max}|r_\alpha-\xi_\alpha| + \sum_{1\leq\alpha\leq s}\left|\frac{\partial a}{\partial \varphi_\alpha}\right|_{\max}|\varphi_\alpha-\eta_\alpha|$$
$$\leq 4M_1 s \cdot 2l + 4M_1 ls = 12M_1 sl. \tag{37}$$

Starting from formulae (32) and (35), we find a bound for the expression $a(\xi_1,\ldots,\xi_s;\eta_1,\ldots,\eta_s)$. We put (33) into the right-hand side of the inequality (32). We then multiply (32) by the product $r_1 r_2 \ldots r_s$ and integrate it over all values of the variables $r_1,\ldots,r_s,\varphi_1,\ldots,\varphi_s$ within the following limits:

$$r_\alpha + l \leq r_\alpha \leq r_\alpha + 2l, \quad \varphi_\alpha \leq \varphi_\alpha \leq \varphi_\alpha + \delta_\alpha, \quad \delta_\alpha = \frac{l}{r_\alpha}, \quad (1 \leq \alpha \leq s).$$

We then obtain

$$0 \leq \int\ldots\int a(r_1,\ldots,r_s,\varphi_1,\ldots,\varphi_s)r_1 r_2 \ldots r_s \, dr_1 \ldots dr_s d\varphi_1 \ldots \varphi_s$$
$$\leq \frac{\theta}{2V}\int\ldots\int\{F_1 r_2 r_3 \ldots r_s + F_2 r_1 r_3 \ldots r_s$$
$$+ \ldots + F_s r_1 r_2 \ldots r_{s-1}\}dr_1 \ldots dr_s d\varphi_1 \ldots \varphi_s$$
$$+ \frac{\mathscr{M}^{2/3}}{V^{2/3}}\int\ldots\int\{F_1^{2/3}r_1^{1/3}r_2 r_3 \ldots r_s + F_2^{2/3}r_1 r_2^{1/3}r_3 \ldots r_s$$

$$+ \ldots + F_s^{2/3} r_1 r_2 \ldots r_{s-1} r_s^{1/3}\} dr_1 \ldots dr_s d\varphi_1 \ldots \varphi_s. \tag{38}$$

Here

$$F_\alpha = \frac{\partial}{\partial r_\alpha}\left(r_\alpha \frac{\partial(-f)}{\partial r_\alpha}\right) + \frac{1}{r_\alpha}\frac{\partial}{\partial \varphi_\alpha}\left(\frac{\partial(-f)}{\partial \varphi_\alpha}\right) \geq 0, \quad (1 \leq \alpha \leq s).$$

By considering the separate terms of the part of (38) containing the factor $\theta/2V$, we see that bounds can be found for these by integrating successively over r_α and φ_α in each of them $(1 \leq \alpha \leq s)$ and using the inequalities (36) to find bounds for the resulting first derivatives $\partial f/\partial \varphi_\alpha$ and $\partial f/\partial r_\alpha$; for all the terms of this sum, we then find

$$\frac{\theta}{2V} \sum_{1 \leq \beta \leq s} \frac{2M_1(\delta_\beta + 2)}{l \cdot 2^{s-1}\delta_\beta} \prod_{1 \leq \alpha \leq s} \{(r_\alpha + 2l)^2 - (r_\alpha + l)^2\}\delta_\alpha. \tag{39}$$

By applying Hölder's inequality and using analogous reasoning for all terms in the sum containing the factor $\mathcal{M}^{2/3}/V^{2/3}$, we obtain

$$\frac{\mathcal{M}^{2/3}}{V^{2/3}2^{1/3}} \sum_{1 \leq \beta \leq s} \frac{\left(2M_1(\delta_\beta + 2)\right)^{2/3}}{l^{2/3} \cdot 2^{s-1}\delta_\beta^{2/3}} \prod_{1 \leq \alpha \leq s} \{(r_\alpha + 2l)^2 - (r_\alpha + l)^2\}\delta_\alpha. \tag{40}$$

Using the formulae (34)-(40), we now obtain a bound for $f_{\Gamma^0} - f_\Gamma = a(r_1, \ldots, r_s; \varphi_1, \ldots, \varphi_s)$:

$$0 \leq f_{\Gamma^0} - f_\Gamma \leq 12 M_1 l s + \frac{2\theta M_1}{V} \sum_{1 \leq \beta \leq s} \frac{(\delta_\beta + 2)}{l \delta_\beta}$$

$$+ \frac{\mathcal{M}^{2/3}}{V^{2/3}} (4M_1)^{2/3} \sum_{1 \leq \beta \leq s} \frac{(\delta_\beta + 2)^{2/3}}{l^{2/3}\delta_\beta^{2/3}}. \tag{41}$$

On the other hand, we note that $\delta_\beta = l/r_\beta$ and choose R in such a way,

$$R \geq |\nu_1|, \ldots, |\nu_s|,$$

that $R \geq r_\beta$ $(\beta = 1, \ldots, s)$ and $\delta_\beta \geq \delta = l/R$. Then, using the obvious inequality $(l + 2R)^{2/3} \leq l^{2/3} + (2R)^{2/3}$, we find from (41)

$$0 \leq f_{\Gamma^0} - f_\Gamma \leq 12 M_1 l s + \frac{2\theta M_1}{V l^2} s(l + 2R) + \frac{\mathcal{M}^{2/3}}{V^{2/3}} s \frac{(4M_1)^{2/3}}{l^{2/3}}$$

$$+ \frac{\mathcal{M}^{2/3}}{V^{2/3}} s \frac{(4M_1)^{2/3}}{l^{2/3}} 2^{2/3} R^{2/3}. \tag{42}$$

We now choose l, which is an arbitrary positive quantity, such that

$$12 M_1 l = \frac{\mathcal{M}^{2/3} (4M_1)^{2/3}}{V^{2/3} l^{2/3}}.$$

Then

$$l = \frac{P}{V^{2/5}}, \quad P = \frac{\mathcal{M}^{2/5}}{2^{2/5} 3^{3/5} M_1^{3/5}} = \text{const.}$$

Putting this expression for l into the inequality (42), we find

$$0 \le f_{\Gamma^0} - f_\Gamma \le 24 M_1 s \frac{P}{V^{2/5}} + \frac{2\theta M_1 s}{V^{3/5} P} + \frac{4\theta M_1 s R}{V^{1/5} P^2}$$

$$+ \frac{\mathcal{M}^{2/3}}{V^{2/15} P^{4/3}} s \, (4M_1)^{2/3} \, 2^{2/3} R^{2/3} \quad \text{for} \ |\nu_\alpha| < R \ \ (1 \le \alpha \le s).$$

Hence it is clear that the difference $f_{\Gamma^0} - f_\Gamma$ vanishes as $V \to \infty$.

We note that in the above bound, we can take the limit $\nu_\alpha = 0$ $(1 \le \alpha \le s)$ and finally prove the statement we made earlier about the asymptotic smallness of the difference $f_{\Gamma^0} - f_\Gamma$:

$$0 \le f_{\Gamma^0} - f_\Gamma \le 24 M_1 s \frac{\bar{P}}{V^{2/5}} + \frac{2\theta M_1 s}{V^{3/5} \bar{P}}, \tag{43}$$

where $\bar{P}$ is a simple combination of the original constants M_1, M_2, and M_3. It is also clear that the above bound is uniform as $\theta \to 0$, and, therefore, the inequality (43) is valid for $\theta \ge 0$.

We have thus proved the following theorem:

Theorem 1. *Let the Hamiltonian of the system be*

$$H = T - V \sum_{1 \le \alpha \le s} g_\alpha J_\alpha J_\alpha^\dagger \tag{44}$$

and let the operators T and J_α in (44) satisfy the following conditions:

$$T = T^\dagger, \quad |J_\alpha| \le M_1,$$

$$|T J_\alpha - J_\alpha T| \le M_2, \quad |J_\alpha J_\beta - J_\beta J_\alpha| \le \frac{M_3}{V},$$

$$|J_\alpha^\dagger J_\beta - J_\beta J_\alpha^\dagger| \leq \frac{M_3}{V}; \quad M_1,\ M_2,\ M_2 = \text{const.} \tag{45}$$

In addition, let the free energy per unit volume, calculated for the Hamiltonian T, be bounded by a constant:

$$|f(T)| \leq \mathcal{M}_0 = \text{const.} \tag{46}$$

We construct the trail Hamiltonian

$$H(C) = T - 2V \sum_\alpha g_\alpha (C_\alpha J_\alpha^\dagger + C_\alpha^* J_\alpha - C_\alpha C_\alpha^*), \tag{47}$$

where $C = (C_1, \ldots, C_s)$ and $C_1, \ldots, C_s$ are complex numbers. Then the following inequalities are valid:[ii]

$$0 \leq \min_{(C)} f\{H(C)\} - f(H) \leq \varepsilon\left(\frac{1}{V}\right), \tag{48}$$

where $\varepsilon(1/V) \to 0$ (as $V \to \infty$) uniformly with respect to θ in the interval $(0 \leq \theta \leq \theta_0)$ where θ_0 is an arbitrary fixed temperature.

3. Further Properties of the Expressions for the Free Energy

Having formulated Theorem 1, we shall now study the question of the existence of the limit

$$\lim_{V \to \infty} f(H).$$

We shall assume that, in addition to the conditions of Theorem 1, the following condition is fulfilled. For any complex $C_1, \ldots, C_s$ the limit

$$\lim_{V \to \infty} f\{H(C)\}$$

[ii](1) We shall denote the free energy per unit volume for some Hamiltonian A by $f(A)$, or, if we wish to emphasize its dependence on the volume, by $f_V(A)$.

(2) By $\min_{(C)}$ we shall always mean the *absolute minimum* of the function $f(C)$ in the space of all points C.

(3) $\varepsilon(1/V)$ is given by (cf.(43))

$$\varepsilon\left(\frac{1}{V}\right) = \frac{24 M_1 s \bar{P}}{\sqrt{q} V^{2/5}} + \frac{20 M_1 s}{\sqrt{g} V^{3/5} \bar{P}},$$

where $\bar{P}$ is a constant and $g > 0$ is the smallest of the g_α.

exists. We shall denote this limit by

$$f_\infty\{H(C)\}.$$

We put[jj]

$$F_V(C) = f\{H(C)\} - 2\sum_\alpha g_\alpha C_\alpha C_\alpha^*$$

and note that

$$\frac{\partial F_V(C)}{\partial C_\alpha} = -2g_\alpha \langle J_\alpha^\dagger \rangle_{H(C)}$$

$$\frac{\partial F_V(C)}{\partial C_\alpha^*} = -2g_\alpha \langle J_\alpha \rangle_{H(C)}.$$

Therefore, from (1) we have

$$\left|\frac{\partial F_V(C)}{\partial C_\alpha}\right| \le 2g_\alpha M_1, \quad \left|\frac{\partial F_V(C)}{\partial C_\alpha^*}\right| \le 2g_\alpha M_1,$$

whence

$$\left|F_V(C') - F_V(C'')\right| \le 4M_1 \sum_{1 \le \alpha \le s} g_\alpha |C_\alpha' - C_\alpha''|. \tag{49}$$

Thus, the set of functions

$$\{F_V(C)\} \quad (V \to \infty)$$

is uniformly continuous.

Since we have the convergence

$$F_V(C) \to F_\infty(C) = f_\infty\{H(C)\} - 2\sum_\alpha g_\alpha C_\alpha C_\alpha^* \quad (V \to \infty)$$

at each point C, we see that this convergence will be uniform on the set $\mathscr{M}(R)$ of points C defined by the inequalities

$$|C_1| \le R_1, \quad \ldots, \quad |C_s| \le R,$$

for any *fixed* value of R.

[jj]We stress that the function $F_V(C)$ need not be treated from the standpoint of the theory of functions of a complex variables. It is not difficult to see that, in essence, the function $F_V(C)$ is a function of a real variables and that these can be taken as the real and imaginary parts of the variables C_α.

Therefore,

$$\left| f_V\{H(C)\} - f_\infty\{H(C)\} \right| = \left| F_V(C) - F_\infty(C) \right|$$
$$\leq \eta_V(R) \to 0, \quad (V \to \infty) \qquad (50)$$

for $C \in \mathcal{M}(R)$.

On the other hand, it follows from (49) that

$$\left| F_V(C_1, C_2, \ldots, C_s) - F_V(0, C_2, \ldots, C_s) \right| \leq 4 M_1 g_1 |C_1|.$$

Hence, we have

$$f\{H(C_1, C_2, \ldots, C_s)\} - f\{H(0, C_2, \ldots, C_s)\}$$
$$= F_V(C_1, C_2, \ldots, C_s) - F_V(0, C_2, \ldots, C_s) + 2 g_1 |C_1|^2$$
$$\geq -4 g_1 M_1 M_1 |C_1| + 2 g_1 |C_1|^2. \qquad (51)$$

We denote the lower bound of $F\{H(C)\}$ in the space of the points C by

$$\inf_{(C)} f\{H(C)\}.$$

Obviously,

$$f\{H(C_1, C_2, \ldots, C_s)\} \geq \inf_{(C)} f\{H(C)\}$$

and, therefore, it follows from (51) that

$$f\{H(C)\} - \inf_{(C)} f\{H(C)\} \geq 2 g_1 |C_1|(|C_1| - 2 M_1).$$

Replacing C_1 by C_α ($\alpha = 1, 2, \ldots, s$) in the above discussion, we find also

$$f\{H(C)\} - \inf_{(C)} f\{H(C)\} \geq 2 g_\alpha |C_\alpha|(|C_\alpha| - 2 M_\alpha), \quad \alpha = 1, 2, \ldots, s.$$

Hence it is clear that if $|C_\alpha| > 2 M_1$ for at least one α, then

$$f\{H(C)\} > \inf_{(C)} f\{H(C)\}.$$

Therefore, the lower bound of $f\{H(C)\}$ on the set $\mathcal{M}(2M_1)$ is equal to the lower bound of this function on the whole space of points C. Since $F\{H(C)\}$ is continuous and the set $\mathcal{M}(2M_1)$ is bounded and closed, this lower bound is attained on $\mathcal{M}(2M_1)$, i.e. an absolute minimum of the function under consideration exists and is realized at certain points:[kk]

$$C = C^{(V)} \in \mathcal{M}(2M_1).$$

On the other hand, taking (49) into account and passing to the limit $V \to \infty$, we find

$$|F_\infty(C') - F_\infty(C'')| \le 4M_1 \sum_{1 \le \alpha \le s} g_\alpha |C'_\alpha - C''_\alpha|.$$

Hence, repeating exactly the above treatment, we see that the function

$$f_\infty\{H(C)\} = F_\infty(C) + 2\sum_\alpha g_\alpha C_\alpha C_\alpha^*$$

also has an absolute minimum in the space of all the points C, which is realized at certain points

$$C = \bar{C} \in \mathcal{M}(2M_1).$$

From (50), we have now:

$$f_\infty\{H(C^{(V)})\} - f\{H(C^{(V)})\} \le \eta_V(2M_1),$$
$$f\{H(\bar{C})\} - f_\infty\{H(\bar{C})\} \le \eta_V(2M_1).$$

But, by definition of the absolute minimum,

$$f_\infty\{H(C^{(V)})\} \ge f_\infty\{H(\bar{C})\},$$
$$f\{H(\bar{C})\} \ge f\{H(C^{(V)})\}.$$

Consequently

$$f_\infty\{H(\bar{C})\} - f\{H(C^{(V)})\} \le \eta_V(2M_1),$$
$$f\{H(C^{(V)})\} - f_\infty\{H(\bar{C})\} \le \eta_V(2M_1),$$

or

$$|f_\infty\{H(\bar{C})\} - \min_{(C)} f\{H(C)\}| \le \delta_V,$$

where

$$\delta_V = \eta_V(2M_1) \to 0, \quad (V \to \infty).$$

[kk]Generally speaking, the point $C^{(V)}$ of the absolute minimum is not unique.

Taking Theorem 1, we finally obtain

$$-\delta_V \leq f_\infty\{H(\bar{C})\} - f(H) \leq \varepsilon\left(\frac{1}{V}\right) + \delta_V.$$

Thus, we have now proved the following theorem:

Theorem 2. *If the conditions of Theorem 1 are fulfilled, and if for any complex values of $C_1, \ldots, C_s$ the limit*

$$f_\infty\{H(C)\} = \lim_{V \to \infty} f\{H(C)\}$$

exist, then:

1. *This limit function has an absolute minimum is the space of all points C, which is realized at certain points*

$$C = \bar{C} \in \mathcal{M}(2M_1).$$

2. *The inequalities*

$$f_\infty\{H(C)\} = \lim_{V \to \infty} f\{H(C)\}$$

are valid, where

$$\varepsilon\left(\frac{1}{V}\right) \to 0, \quad \delta_V \to 0 \quad \text{as} \quad V \to \infty,$$

$$\delta_V = \max_{(C \in \mathcal{M}(2M_1))} \left| f\{H(C)\} - f_\infty\{H(C)\} \right|.$$

4.   Construction of Asymptotic Relations for the Free Energy

We shall now make a special study of those cases when the operators T and J_α in the Hamiltonian (44) have the form (17)

As can easily be shown, the conditions of Theorem 1 will be fulfilled in such cases if

(a) $\dfrac{1}{V} \sum_p |T(p)\lambda_\alpha(p, \sigma)| \leq Q_0,$

(b) $\dfrac{1}{V} \sum_p |\lambda_\alpha(p, \sigma)| \leq Q_1,$

$$\text{(c)} \quad \frac{1}{V} \sum_p |\lambda_\alpha(p, \sigma)|^2 \leq Q_2. \tag{52}$$

Here, $\alpha = 1, \ldots, s$; $\sigma = \pm 1/2$, Q_0, Q_1, $Q_2 = $ const. Then, for example, in the inequalities (1) we can put

$$M_1 = Q_1, \quad M_3 = Q_2, \quad M_2 = 2Q_0.$$

Here, let the functions $\lambda_\alpha(p, \sigma)$ satisfy, in addition to the inequalities (52), the following conditions:

$$|\lambda_\alpha(p, \sigma)| \leq \bar{Q}, \quad \bar{Q} = \text{conts.} \tag{53}$$

The set of the discontinuities of the functions $\lambda_\alpha(p, \sigma)$ is a set

of measure zero in the space E. $\tag{54}$

We shall show that in this situation the conditions of Theorem 2 are also fulfilled.

Before proceeding to this problem, we note that the inequalities (52) and (53) are not independent. In fact, (52c) follows form the inequalities (52b) and (52a). Also, (52b) follows from (52a) and (53). Thus, all the inequalities imposed here on the λ_α are fulfilled if the inequalities (52a) and (53) are true.

We note further that (52a) and (53) hold if λ_α satisfy the inequalities

$$|\lambda_\alpha(p, \sigma)| \leq \frac{\mathcal{K}}{(p^2 + a)^3}, \quad \mathcal{K}, \ a = \text{const.} \tag{55}$$

We turn now to the question of the fulfillment of the conditions of Theorem 2. Since, in the situation being studied, the conditions of Theorem 1 are fulfilled, we need only show that for any fixed complex quantities $C_1, \ldots, C_s$ the limit

$$f_\infty\{H(C)\} = \lim_{(V \to \infty)} f_V\{H(C)\} \tag{56}$$

exists. For this, we write the operator form of the trial Hamiltonian as follows:

$$H(C) = \sum_f T(f) a_f^\dagger a_f - \frac{1}{2} \sum_f \{\Lambda(f) a_f^\dagger a_{-f}^\dagger + \Lambda^*(f) a_{-f} a_f\}$$

$$+ 2V \sum_{\alpha} g_\alpha C_\alpha^* C_\alpha, \tag{57}$$

where

$$\lambda(f) = 2 \sum_{\alpha} g_\alpha C_\alpha^* \lambda_\alpha(f). \tag{58}$$

Going over to the Fermi amplitudes α_f, $\alpha_f^\dagger$, which are related to the old a_f, $a_f^\dagger$ by the transformation

$$a_f = u(f)\alpha_f - v(f)\alpha_{-f}^\dagger,$$
$$a_{-f}^\dagger = u(f)\alpha_{-f}^\dagger + v^*(f)\alpha_f,$$

with

$$u(f) = \frac{1}{\sqrt{2}}\sqrt{1 + \frac{T(f)}{E(f)}}, \quad v(f) = -\frac{\Lambda(f)}{\sqrt{2}|\Lambda(f)|}\sqrt{1 - \frac{T(f)}{E(f)}},$$

we diagonalize the form (57) and obtain

$$H(C) = \sum_{f} E(f)\alpha_f^\dagger \alpha_f + V\left\{2\sum_{\alpha} g_\alpha C_\alpha^* C_\alpha - \frac{1}{V}\sum_{f}[E(f) - T(f)]\right\}, \tag{59}$$

where

$$E(f) = \sqrt{T^2(f) + |\Lambda(f)|^2}.$$

Hence, we find

$$f_V\{H(C)\} = -\frac{\theta}{V}\ln \operatorname{Tr} e^{-H(C)/\theta} = 2\sum_{\alpha} g_\alpha C_\alpha^* C_\alpha$$
$$- \frac{1}{C}\sum_{f}[E(f) - T(f)] - \frac{\theta}{V}\sum_{f}\ln(1 + e^{-E(f)/\theta}),$$

or, separating the indices p and σ in the summation,

$$f_V\{H(C)\} = 2\sum_{\alpha} g_\alpha C_\alpha^* C_\alpha - \sum_{\sigma}\frac{1}{2V}\sum_{p}\left\{E(p,\sigma) - \left(\frac{p^2}{2m} - \mu\right)\right\}$$
$$- \theta\sum_{\sigma}\frac{1}{V}\sum_{p}\ln(1 + e^{-E(p,\sigma)/\theta}). \tag{60}$$

On the other hand, it is not difficult to see that if we have some bounded function $F(\boldsymbol{p})$, defined everywhere on the space E, whose discontinuities form a set of measure zero, then

$$\frac{1}{V} \sum_{p \in S_r} F(\boldsymbol{p}) \rightarrow \frac{1}{(2\pi)^3} \int_{S_r} F(\boldsymbol{p}) \, d\boldsymbol{p} \tag{61}$$

for any sphere S_r with arbitrary fixed radius r. In fact, such a function will be Riemann-integrable in the region S_r. For the summation points

$$\boldsymbol{p} = \left(\frac{2\pi n_1}{L}, \frac{2\pi n_2}{L}, \frac{2\pi n_3}{L}\right)$$

we have

$$\Delta p_x \Delta p_y \Delta p_z = \left(\frac{2\pi}{L}\right) = \frac{(2\pi)^3}{V},$$

so that

$$\frac{(2\pi)^3}{V} \sum_{p \in S_r} F(\boldsymbol{p})$$

will be the Riemann sum for the integral $\int_{S_r} F(\boldsymbol{p}) \, d\boldsymbol{p}$. We note also that if

$$\frac{1}{V} \sum_{p} |F(\boldsymbol{p})| \leq A = \text{const},$$

then

$$\frac{1}{V} \sum_{p \in S_r} |F(\boldsymbol{p})| \leq A.$$

Hence, passing to the limit $V \rightarrow \infty$, we have

$$\frac{1}{(2\pi)^3} \int_{S_r} F(\boldsymbol{p}) \, d\boldsymbol{p} \leq A.$$

Because of the arbitrariness of the radius r, we see that $F(\boldsymbol{p})$ is an absolutely integrable function in the whole space, such that

$$\frac{1}{(2\pi)^3} \int F(\boldsymbol{p}) \, d\boldsymbol{p} \leq A.$$

For a given function $F(\boldsymbol{p})$, let the following inequality be valid:

$$\frac{1}{V} \sum_{\boldsymbol{p} \in E - S_r} |F(\boldsymbol{p})| \neq \eta_r,$$

where $(E - S_r)$ denotes the set of the points of E lying outside the sphere S_r, does not depend on V, and

$$\eta_r \to 0, \quad (r \to \infty).$$

Then, obviously

$$\frac{1}{V} \sum_{\boldsymbol{p}} F(\boldsymbol{p}) \to \frac{1}{(2\pi)^3} \int F(\boldsymbol{p}) \, d\boldsymbol{p}. \tag{62}$$

In fact, we shall fix an arbitrary small number $\varepsilon > 0$ and choose $r = r_0$ such that

$$\eta_{r_0} \leq \frac{\varepsilon}{4}.$$

In view of (61), we can find a number V_0 such that, for $V \geq V_0$, the inequality

$$\left| \sum_{\boldsymbol{p} \in S_{r_0}} F(\boldsymbol{p}) - \frac{1}{(2\pi)^3} \int_{S_{r_0}} F(\boldsymbol{p}) \, d\boldsymbol{p} \right| \leq \frac{\varepsilon}{2}$$

holds. We have, therefore

$$\left| \sum_{\boldsymbol{p}} F(\boldsymbol{p}) - \frac{1}{(2\pi)^3} \int F(\boldsymbol{p}) \, d\boldsymbol{p} \right|$$

$$\leq \left| \sum_{\boldsymbol{p} \in S_{r_0}} F(\boldsymbol{p}) - \frac{1}{(2\pi)^3} \int_{S_{r_0}} F(\boldsymbol{p}) \, d\boldsymbol{p} \right|$$

$$+ \sum_{\boldsymbol{p} \in E - S_{r_0}} |F(\boldsymbol{p})| - \frac{1}{(2\pi)^3} \int_{E - S_{r_0}} |F(\boldsymbol{p})| \, d\boldsymbol{p}$$

$$\leq \frac{\varepsilon}{2} + \frac{\varepsilon}{4} + \frac{\varepsilon}{4} = \varepsilon$$

for any $V \geq V_0$, and this establishes the validity of (62).

After these trivial remarks, we turn to the expression (60). It is clear from (58) and the conditions imposed above on the λ_α that

$$|\Lambda(\boldsymbol{p}, \sigma)| \leq \Lambda_0 = \text{const},$$

$$\frac{1}{V} \sum_p |\Lambda(\boldsymbol{p}, \sigma)| \leq \Lambda_1 = \text{const}.$$

We see that the discontinuities of the function $\Lambda(\boldsymbol{p}, \sigma)$, are consequently also of the function

$$E(p, \sigma) - \left(\frac{p^2}{2m} - \mu\right) = \sqrt{\left(\frac{p^2}{2m} - \mu\right)^2 + |\Lambda(\boldsymbol{p}, \sigma)|^2} - \left(\frac{p^2}{2m} - \mu\right)$$

form a set of measure zero in the space E. Further, for

$$p^2 \geq 4m\mu,$$

we have

$$\frac{p^2}{2m} - \mu \geq \frac{p^2}{4m},$$

$$0 \leq E(p, \sigma) - \left(\frac{p^2}{2m} - \mu\right) \leq \frac{|\Lambda(\boldsymbol{p}, \sigma)|^2}{2\frac{p^2}{4m}} = \frac{2m}{p^2}|\Lambda(\boldsymbol{p}, \sigma)|^2.$$

Hence,

$$\frac{1}{V} \sum_{\substack{p \in E - S_r \\ (p^2 \geq 4m\mu)}} \left\{E(p, \sigma) - \left(\frac{p^2}{2m} - \mu\right)\right\} \leq \frac{2m}{p^2} \Lambda_1.$$

Thus, taking into account the remarks made above, we see that

$$\frac{1}{V} \sum_p \left\{E(p, \sigma) - \left(\frac{p^2}{2m} - \mu\right)\right\}$$

$$\rightarrow \frac{1}{(2\pi)^3} \int \left\{E(p, \sigma) - \left(\frac{p^2}{2m} - \mu\right)\right\} d\boldsymbol{p}.$$

Further, we have

$$\ln\{1 + e^{-E(p,\sigma)/\theta}\} < e^{-E(p,\sigma)/\theta} \leq \text{const } e^{-p^2/2m\theta}.$$

Since this function decreases sufficiently rapidly as $\boldsymbol{p} \rightarrow \infty$ and its discontinuities form a set of measure zero, we also obtain

$$\frac{1}{V} \sum_p \ln\{1 + e^{-E(p,\sigma)/\theta}\} \rightarrow \frac{1}{(2\pi)^3} \int \ln\{1 + e^{-E(p,\sigma)/\theta}\} d\boldsymbol{p}.$$

This also proves the validity of the property (56). We have here

$$f_\infty\{H(C)\} = 2\sum_\alpha g_\alpha C_\alpha C_\alpha^* - \frac{1}{2}\sum_\sigma \frac{1}{(2\pi)^3}\int\left\{E(p,\sigma) - \left(\frac{p^2}{2m} - \mu\right)\right\}d\boldsymbol{p}$$
$$- \theta\sum_\sigma \frac{1}{(2\pi)^3}\int \ln\{1 + e^{-E(p,\sigma)/\theta}\}\,d\boldsymbol{p},$$

or, more compactly,

$$f_\infty\{H(C)\} = 2\sum_\alpha g_\alpha C_\alpha C_\alpha^* - \frac{1}{2(2\pi)^3}\int\{E(f) - T(f)\}\,df$$
$$- \frac{\theta}{(2\pi)^3}\int \ln\{1 + e^{-E(f)/\theta}\}\,df. \tag{63}$$

Here, the integration

$$\int(\ldots)\,df \quad \text{implies the operation} \quad \sum_\sigma\int(\ldots)\,d\boldsymbol{p}.$$

Thus, in the case under investigation, if the conditions (52a), (53) and (54) are fulfilled, the condition of Theorem 2 are satisfied. As was noted above, in the proof of this theorem, the convergence

$$f_V\{H(C)\} - f_\infty\{H(C)\} \to 0, \quad (V \to \infty) \tag{64}$$

is uniform on any *bounded* set of points C.

5. On the Uniform Convergence with Respect to θ of the Free Energy Function and on the Bounds for the Quantities δ_V

We shall show that, in the above case, the convergence (64) is also uniform with respect to θ in the interval $(0 < \theta \le \theta_0)$, where θ_0 is any fixed temperature. It can be seen that this property will be established once we have shown that

$$\left|\frac{\partial}{\partial\theta}(f_V\{H(C)\} - f_\infty\{H(C)\})\right| \le X = \text{const} \quad (0 < \theta \le \theta_0) \tag{65}$$

and this will be our aim now. We have

$$\frac{\partial}{\partial\theta} f_V\{H(C)\} = -\frac{1}{V}\sum_f \ln\{1 + e^{-E(f)/\theta}\} - \frac{1}{\theta V}\sum_f \frac{E(f)e^{-E(f)/\theta}}{1 + e^{-E(f)/\theta}}.$$

Since

$$\frac{E}{\theta} e^{-E/2\theta} \leq \frac{2}{e},$$

we can write

$$\left| \frac{\partial}{\partial \theta} f_V\{H(C)\} \right| \leq \frac{1}{V} \sum_f e^{-E(f)/\theta} + \frac{2}{e} \frac{1}{V} \sum_f e^{-E(f)/\theta}$$

$$\leq \left(1 + \frac{2}{e}\right) \sum_f e^{-E(f)/\theta_0}.$$

Completely analogously, we find

$$\left| \frac{\partial}{\partial \theta} f_\infty\{H(C)\} \right| \leq \left(1 + \frac{2}{e}\right) \frac{1}{(2\pi)^3} \int e^{-E(f)/\theta_0}\, df$$

and, therefore

$$\left| \frac{\partial}{\partial \theta} (f_V\{H(C)\} - f_\infty\{H(C)\}) \right|$$

$$\leq \left(1 + \frac{2}{e}\right) \left\{ \frac{1}{V} \sum_f e^{-E(f)/\theta} + \frac{1}{(2\pi)^3} \int e^{-E(f)/\theta_0}\, df \right\}.$$

In view of the rapid falling off of $e^{-E(f)/2\theta_0}$ as $|p| \to \infty$, the integral

$$\int e^{-E(f)/2\theta_0}\, df$$

has a finite value, and

$$\frac{1}{V} \sum_f e^{-E(f)/\theta} \to \frac{1}{(2\pi)^3} \int e^{-E(f)/\theta_0}\, df.$$

Inequality (65) is thus established, and the uniformity of the convergence (64) with respect to θ in the interval $(0 < \theta \leq \theta_0)$ is thereby also proved. Therefore, in the case under consideration, in Theorem 2 the relation

$$\delta_V \to 0 \quad (V \to \infty)$$

holds uniformly with respect to θ in this interval.

Thus, putting

$$\varepsilon\left(\frac{1}{V}\right) + \delta_V = \bar{\delta}_V,$$

we can formulate paragraph 2 of Theorem 2 in the form

$$|f_\infty\{H(C)\} - f(H)| \le \bar{\delta}_V, \tag{66}$$

$\bar{\delta}_V \to 0$ $(V \to \infty)$ uniformly with respect to θ in the interval $(0 < \theta \le \theta_0)$.

An explicit expression for $\varepsilon(1/V)$ has been obtained. (See footnote on page 269.) It would also not be difficult to obtain an explicit expression for the bound δ_V of the difference

$$f_V\{H(C)\} - f_\infty\{H(C)\} \tag{67}$$

if we impose on $\lambda_\alpha(f)$ the appropriate conditions of smoothness and falling of as $|\boldsymbol{p}| \to \infty$. In fact, as we have seen, (67) is the difference between a Riemann sum and the corresponding integral, so that here we can make use of well-known technique form the theory of the approximate calculations of three dimensional integrals.

Thus, we can show, for example, that if the functions λ_α of the point $(\boldsymbol{p})$ are continuous and differentiable everywhere, with the possible exception of certain sufficiently smooth discontinuity surfaces, and go to zero, together with $\partial\lambda_\alpha/\partial p$, sufficiently rapidly as $p \to \infty$, then

$$\delta_V \le \frac{\text{const}}{L} = \frac{\text{const}}{V^{1/3}}.$$

In the case when λ_α are everywhere continuous, possess derivatives of second order with respect to $(\boldsymbol{p})$, and, as $\boldsymbol{p} \to \infty$, go to zero along with their derivatives of up to and including second order, we can obtain the strong bound:

$$\delta_V \le \frac{\text{const}}{V^{2/3}}.$$

6. Properties of Partial Derivatives of the Free Energy Function. Theorem 3

We shall now study the partial derivatives of the function

$$f_\infty\{H(C)\} = 2\sum_\alpha g_\alpha C_\alpha^* C_\alpha$$

$$-\frac{1}{2(2\pi)^{3/2}}\int\{E(f)-T(f)+2\theta\ln(1+e^{-E(f)/\theta})\}\,df \qquad (68)$$

with respect to the variables $C_1,\ldots,C_s,\,C_1^*,\ldots,C_s^*$. We have

$$U = \frac{\partial}{\partial C_\alpha}\{E(f)-T(f)+2\theta\ln(1+e^{-E(f)/\theta})\}$$

$$= \left\{1-\frac{2}{1+e^{E(f)/\theta}}\right\}\frac{\partial E(f)}{\partial C_\alpha} = 2\frac{\tanh[E(f)/2\theta]}{E(f)}\left\{\sum_\beta g_\beta C_\beta^*\lambda_\beta(f)\right\}g_\alpha\lambda_\alpha^*(f).$$

But, by virtue of (53) and the inequality

$$0 \le \frac{\tanh x}{x} \le 1,$$

we see that U is a bounded function of $(\boldsymbol{p})$ in E:

$$|U| \le \frac{1}{\theta}\sum_\beta |C_\beta|\bar{Q}^2 g_\alpha.$$

It is clear also that U is a continuous and differentiable function of C in the whole space of points (C). On the other hand, since $|\tanh x| \le 1$, we also have

$$|U| \le \frac{2}{E(f)}\sum_\beta g_\alpha g_\beta |C_\beta|\cdot|\lambda_\beta(f)\lambda_\alpha^*(f)|$$

$$\le \frac{1}{\left|\dfrac{p^2}{2m}-\mu\right|}\sum_\beta g_\alpha g_\beta |C_\beta|\cdot|\lambda_\beta(f)\lambda_\alpha^*(f)|$$

Hence, it is not difficult to see that $U(p)$ is absolutely integrable in E, and

$$\left|\int U\,d\boldsymbol{p}-\int_{S_r} U\,d\boldsymbol{p}\right| = \left|\int_{E-S_r} U\,d\boldsymbol{p}\right| \le \int_{E-S_r}|U|\,d\boldsymbol{p}$$

$$\le \frac{4mg_\alpha}{r^2}\sum_\beta g_\beta|C_\beta|\sqrt{\int|\lambda_\beta(f)|^2\,d\boldsymbol{f}\int|\lambda_\alpha(f)|^2\,d\boldsymbol{f}}$$

for $r^2 \ge 4m\mu$. Consequently

$$\int_{S_r} U\,d\boldsymbol{p} \to \int U\,d\boldsymbol{p} \quad (r\to\infty)$$

uniformly with respect to C on any bounded set of points (C).

Thus, the expression (68) can be differentiated with respect to C_α (or C_α^*) under the integral sign, and the corresponding derivatives

$$\frac{\partial f_\infty\{H(C)\}}{\partial C_\alpha} = 2g_\alpha C_\alpha^* - \frac{g_\alpha}{(2\pi)^3}\int\frac{\tanh[E(f)/2\theta]}{E(f)}\Big\{\sum_\beta g_\beta C_\beta^*\lambda_\beta(f)\Big\}g_\alpha\lambda_\alpha^*(f)\,df,$$

$$\frac{\partial f_\infty\{H(C)\}}{\partial C_\alpha^*} = 2g_\alpha C_\alpha - \frac{g_\alpha}{(2\pi)^3}\int\frac{\tanh[E(f)/2\theta]}{E(f)}\Big\{\sum_\beta g_\beta C_\beta\lambda_\beta^*(f)\Big\}g_\alpha\lambda_\alpha(f)\,df$$

$$(69)$$

will be continuous functions of C in the whole space of points (C).

It is not difficult to see that an analogous treatment is valid for partial derivatives of $f_\infty\{H(C)\}$ of any order with respect to the variables $C_1,\ldots,C_s$, $C_1^*,\ldots,C_s^*$.

In fact, on further differentiation of the expression for U, in addition to the factor $\{\tanh[E(f)/2\theta]\}/E(f)$, the expressions

$$\Big\{\frac{1}{E}\frac{\partial}{\partial E}\Big(\frac{\tanh[E/2\theta]}{E}\Big)\Big\}_{E=E(f)},$$

$$\Big\{\frac{1}{E}\frac{\partial}{\partial E}\Big(\frac{1}{E}\frac{\partial}{\partial E}\frac{\tanh[E/2\theta]}{E}\Big)\Big\}_{E=E(f)},$$

also appear. These are bounded functions of E (since $\{\tanh[E/2\theta]\}/E$ can, for small E, be expanded in a Taylor series in *even* powers of E), and, as $E\to\infty$, fall off like

$$\frac{1}{E^2}\sim\frac{\text{const}}{p^4},\qquad\frac{1}{E^3}\sim\frac{\text{const}}{p^6},\qquad\ldots.$$

Moreover, on differentiation of U with respect to the variables C, there appear further polynomials in C_α, C_α^* and $\lambda_\alpha(f)$, which also fail to invalidate the above arguments.

Returning to the expressions (69) for the first derivatives, we see that, since they are continuous functions of C at the points $C=\bar{C}$ at which the absolute minimum of the function $f_\infty\{H(C)\}$ occurs, we have

$$2g_\alpha C_\alpha^* - \frac{g_\alpha}{(2\pi)^3}\int\frac{\tanh[E(f)/2\theta]}{E(f)}\Big\{\sum_\beta g_\beta C_\beta^*\lambda_\beta(f)\Big\}g_\alpha\lambda_\alpha^*(f)\,df = 0,$$

$$2g_\alpha C_\alpha - \frac{g_\alpha}{(2\pi)^3} \int \frac{\tanh[E(f)/2\theta]}{E(f)} \left\{ \sum_\beta g_\beta C_\beta \lambda_\beta^*(f) \right\} g_\alpha \lambda_\alpha(f) \, df = 0. \qquad (70)$$

Thus, summarizing the results just obtained, we see that the following theorem holds:

Theorem 3. *If in the Hamiltonian (44) the operators T and J_α have the form (17) and the functions $\lambda_\alpha(f)$ satisfy the conditions (52a), (53) and (54) then:*

1.

$$|f_V\{H(C)\} - f_\infty\{H(C)\}| \leq \delta_V \qquad (71)$$

   for

$$|C_\alpha| \leq 2M_1, \quad \alpha = 1, \ldots, s$$

   where $\delta_V \to 0$ uniformly with respect to θ in the interval $(0 < \theta < \theta_0)$.

Here, $f_\infty\{H(C)\}$ is given by the expression (68) and possess continuous partial derivatives of all orders with respect to the variables $C_1, \ldots, C_s$, $C_1^*, \ldots, C_s^*$ for all complex values of these variables.

This function has an absolute minimum in the space of all the points (C), which is realized at certain points $C = \bar{C}$:

$$\min_{(C)} f_\infty\{H(C)\} = f_\infty\{H(\bar{C})\},$$

satisfying the equation (70)

2. *The inequality*

$$|f_V\{H(C)\} - f_\infty\{H(\bar{C})\}| \leq \bar{\delta}_V \qquad (72)$$

   holds, where $\bar{\delta}_V = (\varepsilon(1/V) + \delta_V) \to \infty$ uniformly with respect to θ in the interval $(0 < \theta < \theta_0)$.

We shall now add a rider to this theorem.

7. Rider to Theorem 3 and Construction of an Auxiliary Inequality

The point $C = \bar{C}$ at which the function $f_\infty\{H(C)\}$ attains an absolute minimum is, in general, not unique. However, in the particular case when the absolute minimum is realized at the point $C = 0$, the uniqueness property holds.

In other words, if

$$\min_{(C)} f_\infty\{H(C)\} = f_\infty\{H(0)\},$$

then

$$f_\infty\{H(C)\} > f_\infty\{H(0)\}$$

for $C \neq 0$ (i.e. for C such that at last one of the components $C_\alpha \neq 0$). To establish this property of the free energy (68) taken for the trial Hamiltonian, we shall assume that the opposite is true.

Then there exists a point $\bar{C} \neq 0$ such that

$$f_\infty\{H(\bar{C})\} = f_\infty\{H(0)\}. \tag{73}$$

We put

$$C = \sqrt{\tau}\,\bar{C}, \quad \tau > 0,$$

and consider the function

$$\phi(\tau) = f_\infty\{H(\sqrt{\tau}\,\bar{C})\}.$$

Then, because of (73),

$$\phi(1) = \phi(0). \tag{74}$$

Making use of the expression (68) and differentiating, we find

$$\frac{d\phi(\tau)}{d\tau} = 2\sum_\alpha g_\alpha |\bar{C}_\alpha|^2 - \frac{1}{(2\pi)^3}\int \frac{\tanh[E(f)/2\theta]}{E(f)}\left|\sum_\beta g_\beta \bar{C}_\beta \lambda_\beta^*(f)\right|^2 df,$$

$$\frac{d^2\phi(\tau)}{d\tau^2} = \frac{4}{(2\pi)^3}\int \frac{e^{E/\theta}}{(1+e^{E/\theta})}\frac{\sinh(E/\theta)-E/\theta}{E^3}\left|\sum_\beta g_\beta \bar{C}_\beta \lambda_\beta^*(f)\right|^4 df.$$

Since

$$\frac{\sinh(E/\theta)-E/\theta}{E^3} > 0,$$

$d^2\phi(\tau)/d\tau^2$ can go to zero only if, for all f,

$$\left|\sum_\beta g_\beta \bar{C}_\beta \lambda_\beta^*(f)\right| = 0$$

identically. But in this case,

$$\int \frac{\tanh[E(f)/2\theta]}{E(f)} \left| \sum_\beta g_\beta \bar{C}_\beta \lambda_\beta^*(f) \right|^2 d = 0$$

also, so that

$$\frac{d\phi(\tau)}{d\tau} = 2\sum_\alpha g_\alpha |\bar{C}_\alpha|^2 > 0, \quad \tau \geq 0.$$

But this inequality contradicts (74). Consequently,

$$\frac{d^2\phi(\tau)}{d\tau^2} > 0. \tag{75}$$

On the other hand, since $C = 0$ gives the absolute minimum of $f_\infty\{H(C)\}$, we have

$$\phi(\tau) \geq \phi(0), \quad \tau > 0.$$

Therefore $(d\phi(\tau)/d\tau)_{\tau=0}$ cannot be negative:

$$\left(\frac{d\phi(\tau)}{d\tau} \right)_{\tau=0} \geq 0.$$

Hence, it follows from (75) that

$$\frac{d\phi(\tau)}{d\tau} > 0 \quad \text{for} \quad \tau > 0$$

and, consequently,

$$\phi(1) > \phi(0),$$

which again contradicts (74). Our rider is thus proved.

To conclude this section, which contains preliminary results relating to the properties of the free energies $f_V(H)$, $f_V\{H(C)\}$ and $f_\infty\{H(C)\}$, we shall prove one more equality, which we shall use frequently in the following discussions.

We shall consider systems defined by a Hamiltonian which depends linearly on some parameter τ:

$$H_\tau = \Gamma_0 + \tau\Gamma_1.$$

We shall formally define the expression

$$f_V(H_\tau) = -\frac{\theta}{V} \ln \text{Tr } e^{-H_\tau/\theta},$$

which we shall call the free energy per unit volume for the model system H_τ. Differentiating this expression, we have

$$\frac{d}{d\tau}\, f_V(H_\tau) = \frac{1}{V}\, \frac{\operatorname{Tr}\Gamma_1\, e^{-H_\tau/\theta}}{\operatorname{Tr}\, e^{-H_\tau/\theta}} = \frac{1}{V}\langle\Gamma_1\rangle_{H_\tau} \tag{76}$$

and

$$\frac{d^2 f_V(H_\tau)}{d\tau^2} = -\frac{1}{\theta}\int\limits_0^1 \frac{\operatorname{Tr}\{\bar\Gamma_1\, e^{-(H_\tau/\theta)\xi}\bar\Gamma_1\, e^{-(H_\tau/\theta)(1-\xi)}\}\, d\xi}{\operatorname{Tr}\, e^{-H_\tau/\theta}}$$

where

$$\bar\Gamma_1 = \Gamma_1 - \langle\Gamma_1\rangle_{H_\tau}.$$

But, as we have shown in section 1 of this chapter

$$\frac{d^2 f_V(H_\tau)}{d\tau^2} \le 0, \tag{77}$$

in view of which

$$\left\{\frac{df_V(H_\tau)}{d\tau}\right\}_{\tau=1} \le \frac{df_V(H_\tau)}{d\tau} \le \left\{\frac{df_V(H_\tau)}{d\tau}\right\}_{\tau=0} \quad (0 \le \tau \le 1)$$

and, therefore, for the difference

$$f_V(\Gamma_0 + \Gamma_1) - f_V(\Gamma_0) = \int\limits_0^1 \frac{d}{d\tau}\, f_V(H_\tau)\, d\tau$$

we obtain the inequality

$$\left\{\frac{df_V(H_\tau)}{d\tau}\right\}_{\tau=1} \le f_V(\Gamma_0 + \Gamma_1) - f_V(\Gamma_0) \le \left\{\frac{df_V(H_\tau)}{d\tau}\right\}_{\tau=0}.$$

Thus, on the basis of (76) we have established the following important inequality

$$\frac{1}{V}\langle\Gamma_1\rangle_{\Gamma_0+\Gamma_1} \le f_V(\Gamma_0 + \Gamma_1) - f_V(\Gamma_0) \le \frac{1}{V}\langle\Gamma_1\rangle_{\Gamma_0}. \tag{78}$$

We shall make use of these inequalities later when we specify concrete model and trial systems and choose the source terms in an appropriate way.

8.   On the Difficulties of Introducing Quasi-Averages

We shall now study the question of the determination of quasi-averages. Let $\mathscr{A}$ be some operator of the type for which the limit Theorems were formulated in [13, 14], e.g. a product of Fermi amplitudes, field functions or similar operators. Then the quasi-average

$$\prec \mathscr{A} \succ_H$$

of such an operator will be defined, for the Hamiltonian (44) under consideration, as the limit

$$\prec \mathscr{A} \succ_H = \lim_{\nu \to 0} \left(\lim_{V \to \infty} \langle \mathscr{A} \rangle_\Gamma \right) \tag{79}$$

of an ordinary average

$$\langle \mathscr{A} \rangle_\Gamma$$

taken over a Hamiltonian Γ obtained from H by adding "source terms" to it:

$$\Gamma = H - V \sum_\alpha (\nu_\alpha J_\alpha^\dagger + \nu_\alpha^* J^\dagger)$$

$$= T - 2V \sum_\alpha g_\alpha J_\alpha J_\alpha^\dagger - V \sum_\alpha (\nu_\alpha J_\alpha^\dagger + \nu_\alpha^* J). \tag{80}$$

We now wish to call attention to certain difficulties associated with the definition (79). Thus, in the definition given, there is no indication in which region the parameters ν must lie, or how they must tend to zero in order to ensure convergence in the definition (79).

We shall show that, even in the simplest cases, if $|\nu|$ tends to zero arbitrary, the limit $\lim_{\eta \to 0}$ *may not exist.*

We shall take, as an example, the Hamiltonian (11)

$$H = T - 2VgJJ^\dagger,$$

which we have examined in a number of papers;[ll] the basic results of these were summarized briefly in [13, 14].

We recall that here T and J are given by the formulae (2), and the function $\lambda(f)$ satisfies all the conditions imposed in §1 (conditions **I** and **I′**).

[ll]See, for example, [4, 6] and [7].

For Γ, we took a Hamiltonian with real positive ν:

$$\Gamma = \Gamma_\nu = T - 2VgJJ^\dagger - \nu V(J + J^\dagger), \quad (\nu > 0). \tag{81}$$

As we have shown, Γ reduces to the form (11)

$$\Gamma = \Gamma_a - 2Vg\big(J - C(\nu)\big)\big(J^\dagger - C(\nu)\big),$$

$$\Gamma_a = T - \frac{1}{2}\sum_f \Lambda(f)\{a_f^\dagger a_{-f}^\dagger + a_{-f}a_f\} + 2gVC^2,$$

$$\Lambda(f) = 2g\lambda(f)\Big\{C(\nu) + \frac{\nu}{2g}\Big\}.$$

Hence,

$$C(\nu) + \frac{\nu}{2g} > 0$$

and the quantity $C = C(\nu)$ realizes the absolute minimum of the function $f_\infty\{\Gamma(C)\}$:

$$\min_{(C)} f_\infty\{\Gamma(C)\} = f_\infty\{\Gamma(C(\nu))\}$$

Moreover (see (13)),

$$\langle\big(J - C(\nu)\big)\big(J^\dagger - C(\nu)\big)\rangle_\Gamma \le \varepsilon_V \to 0, \quad (V \to \infty). \tag{82}$$

It can be seen that the Hamiltonian belongs to the class (14), and, by virtue of the inequality (82) and the conditions imposed on $\lambda(f)$, the conditions **I** and **I'** of §1 are fulfilled.

Because of this, we can make use of the above-mentioned limit theorems and establish the existence of limit of the type

$$\lim_{V\to\infty} \langle\mathscr{A}\rangle_\Gamma = \lim_{V\to\infty} \langle\mathscr{A}\rangle_{\Gamma_a}.$$

We note further that, as has already been pointed out (see (7)),

$$C(\nu) \to C(0) = \bar{C} \quad (\nu > 0, \ \nu \to 0). \tag{83}$$

On the other hand, the expression

$$\lim_{V\to\infty} \langle\mathscr{A}\rangle_{\Gamma_a}$$

can be expanded in explicit form using the rules of Bloch and de Dominics, and it can be proved in a completely elementary way that the passage to the limit $\nu \to 0$ $(\nu > 0)$ can be made and reduces simply to replacing $C(\nu)$ by $\bar{C}$ in this expression, i.e. to replacing the averaging over Γ_a by averaging over $H(\bar{C})$.

In this way we can establish the existence of the quasi-averages

$$\prec A \succ_H = \lim_{\substack{(\nu \to 0, \\ \nu > 0)}} \lim_{V \to \infty} \langle \mathscr{A} \rangle_\Gamma = \lim_{V \to \infty} \langle \mathscr{A} \rangle_{H(\bar{C})}. \tag{84}$$

With the definition (79), and for the case when

$$\bar{C} \neq 0, \tag{85}$$

we shall now examine how the situation changes when we go over to complex values of ν and, in place of the Hamiltonian (81), we take

$$\Gamma_{\nu, \nu^*} = T - 2VgJJ^\dagger - V(\nu J^\dagger + \nu^* J). \tag{86}$$

Here, we shall put $\nu = |\nu| e^{i\varphi}$ and note that Γ_{ν, ν^*} can be reduced to the form $\Gamma = \Gamma_{|\nu|}$ (i.e. to the Hamiltonian (81) with $|\nu|$ in place of ν) by means of the gauge transformation

$$a_f \to a_f\, e^{i(\varphi/2)}, \quad a_f^\dagger \to a_f^\dagger\, e^{-i(\varphi/2)}.$$

Thus, we obtain, for example,

$$\langle a_f^\dagger(t) a_{-f}^\dagger(\tau) \rangle_{\Gamma_{\nu, \nu^*}} = e^{-i\varphi} \langle a_f^\dagger(t) a_{-f}^\dagger(\tau) \rangle_{\Gamma_{|\nu|}}$$
$$= \frac{\nu^*}{|\nu|} \langle a_f^\dagger(t) a_{-f}^\dagger(\tau) \rangle_{\Gamma_{|\nu|}}. \tag{87}$$

The limit

$$\lim_{|\nu| \to 0} \lim_{V \to \infty} \langle a_f^\dagger(t) a_{-f}^\dagger(\tau) \rangle_{\Gamma_{|\nu|}} \tag{88}$$

obviously exists and is given by the formula (8) in which $\bar{C}$ replaces in the expression for $u(f)$, $v(f)$ and $E(f)$. Then, in the case (85) under consideration, the expression (88) does not go identically to zero.

Consequently, although

$$\lim_{V \to \infty} \langle a_f^\dagger(t) a_{-f}^\dagger(\tau) \rangle_{\Gamma_{\nu, \nu^*}}$$

always exists for $|\nu| > 0$, the limit

$$\lim_{\nu \to 0} \lim_{V \to \infty} \langle a_f^\dagger(t) a_{-f}^\dagger(\tau) \rangle_{\Gamma_{\nu,\nu^*}} \tag{89}$$

does not, for the trivial reason that the ratio $\nu/|\nu|$ does not tend to any limit as $\nu \to 0$.

The limit (89) exists only when ν tends to zero in such a way that the ratio $\nu/|\nu|$ is finite.

In the general case (80), with the passage to the limit $\nu \to 0$ the situation is found, naturally, to be even more complicated. Apart from gauge invariance (due to the gauge group), other groups of transformations can also occur, e.h. the rotation group.

We shall now direct our attention to a difficulty which is specific for $s > 1$.

We take the Hamiltonian

$$H = T - 2Vg J_1 J_1^\dagger - 2Vg J_2 J_2^\dagger,$$
$$\Gamma = H - V\{\nu_1(J_1 + J_1^\dagger) + \nu_2(J_2 + j_2^\dagger)\}.$$

Here, we take ν_1 and ν_2 to be *real* and *positive*.

We put here

$$J_1 = \frac{J}{\sqrt{2}}, \quad J_2 = -\frac{J}{\sqrt{2}},$$

where the operators J and T have the same form as in the Hamiltonian (1), and (81).

In the given case, H will, in this way, be the same Hamiltonian (1) that we have just considered.

We take $\nu_1 = \nu_2$. Then the source will drop out completely:

$$\Gamma = H$$

and, since the operator H conserves the number of particles, we have, identically,

$$\langle a_f^\dagger a_{-f}^\dagger \rangle_\Gamma = 0.$$

It can be seen that in such a situation we cannot define a quasi-average correctly at all.

9. A New Method of Introducing Quasi-Averages

In order to avoid difficulties of the above type, *we propose that ν be taken proportional to $\bar{C}$ with positive proportionality coefficients:*

$$\nu_a = r_\alpha \bar{C}_\alpha, \quad r_\alpha > 0, \quad \alpha = 1, 2, \ldots, s. \tag{90}$$

In such a case, we shall consider the trial Hamiltonian

$$\Gamma_\alpha = T - 2V \sum_\alpha g_\alpha \{\bar{C}_\alpha J_\alpha^\dagger + \bar{C}_\alpha^* J_\alpha\}$$

$$- V \sum_\alpha r_\alpha \{\bar{C}_\alpha J_\alpha^\dagger + \bar{C}_\alpha^* J_\alpha\} + \text{const.} \tag{91}$$

Here we shall not write out the constant term, since it affect neither the calculations of the averages $\langle \ldots \rangle_{\Gamma_\alpha}$ nor the equation of motion. It can be seen that we have obtained a trial Hamiltonian for H with the transformed parameters

$$g_\alpha \rightarrow g_\alpha + \frac{r_\alpha}{2},$$

$$H_r = T - 2V \sum_\alpha \left(g_\alpha + \frac{r_\alpha}{2}\right) J_\alpha J_\alpha^\dagger.$$

In order that Γ_a from (91) be a trial Hamiltonian not for H_r but for the original H, we must replace g_α by $g_\alpha - r_\alpha/2$ in the expression (80) for Γ (in which ν is taken in accordance with (90)), thereby obtaining,

$$\Gamma = T - 2V \sum_\alpha \left(g_\alpha + \frac{r_\alpha}{2}\right) J_\alpha J_\alpha^\dagger - V \sum_\alpha r_\alpha \{J_\alpha^\dagger \bar{C}_\alpha + J_\alpha \bar{C}_\alpha^*\}.$$

It is clear that here, apart from adding "sources", we have performed a "renormalization" of the parameters g_α.

We can add any constant term to this expression for Γ since it will not affect either the average $\langle \ldots \rangle_\Gamma$ or the equation of motion.

As such a constant term, we shall take

$$V \sum_\alpha r_\alpha \bar{C}_\alpha^* \bar{C}_\alpha.$$

Then the Hamiltonian Γ will be represented by the form

$$\Gamma = T - 2V \sum_\alpha \left(g_\alpha + \frac{r_\alpha}{2}\right) J_\alpha J_\alpha^\dagger - V \sum_\alpha r_\alpha \{J_\alpha^\dagger \bar{C}_\alpha + J_\alpha \bar{C}_\alpha^*\}$$

$$+ V \sum_\alpha r_\alpha \bar{C}_\alpha^* \bar{C}_\alpha = H + V \sum_\alpha r_\alpha (J_\alpha - \bar{C}_\alpha)(J_\alpha^\dagger - \bar{C}_\alpha^*). \qquad (92)$$

We emphasize that here, as always, $\bar{C}$ denotes the point at which the absolute minimum of the function $f_\infty\{H(C)\}$ (see (63)) is attained. For notational convenience, in (92) we put

$$r_\alpha = 2\tau_\alpha g_\alpha, \quad \text{where} \quad \tau_\alpha > 0.$$

Thus, we shall be concerned with a Hamiltonian having the form

$$\Gamma = H + 2V \sum_\alpha \tau_\alpha g_\alpha (J_\alpha - \bar{C}_\alpha)(J_\alpha^\dagger - \bar{C}_\alpha^*) = T - 2v \sum_\alpha g_\alpha J_\alpha J_\alpha^\dagger$$

$$+ 2V \sum_\alpha \tau_\alpha g_\alpha (J_\alpha - \bar{C}_\alpha)(J_\alpha^\dagger - \bar{C}_\alpha^*)$$

$$\tau_\alpha > 0, \quad \alpha = 1, 2, \ldots, s. \qquad (93)$$

We shall show that, with the above choice of Γ, no difficulties will now arise in the definition of the quasi-averages

$$\preceq \mathscr{A} \succ_H = \lim_{\tau \to 0} \lim_{V \to \infty} \langle \mathscr{A} \rangle_\Gamma, \quad \tau_\alpha > 0, \quad \alpha = 1, \ldots, s.$$

For this, we note first of all that, for

$$\tau_1 = 1, \quad \ldots, \quad \tau_s = 1,$$

from (93) we shall have:

$$\Gamma = T - 2V \sum_\alpha g_\alpha (J_\alpha^\dagger \bar{C}_\alpha + J_\alpha \bar{C}_\alpha^*) + 2V \sum_\alpha g_\alpha \bar{C}_\alpha^* \bar{C}_\alpha = H(\bar{C})$$

(cf. also formula (99)). Since

$$(J_\alpha - \bar{C}_\alpha)(J_\alpha^\dagger - \bar{C}_\alpha^*) \geq 0,$$

we see that

$$H(\bar{C}) - \Gamma \geq 0 \quad \text{and} \quad \Gamma - H \geq 0 \quad \text{for} \quad 0 < \tau_\alpha < 1. \qquad (94)$$

Consequently, the inequalities

$$f_V\{H(C)\} \geq f_V(\Gamma) \geq f_V(H) \quad (\text{for } 0 < \tau_\alpha < 1) \qquad (95)$$

are valid. However,

$$0 \leq f_V\{H(\bar{C})\} - f_V(H) \leq \left| f_\infty\{H(\bar{C})\} - f_V(H) \right| + \left| f_\infty\{H(\bar{C})\} - f_V\{H(\bar{C})\} \right|.$$

Therefore, on the basis of Theorem 3, we obtain

$$0 \leq f_V\{H(\bar{C})\} - f_V(H) \leq \bar{\delta}_V + \delta_V.$$

Hence, taking (95) into account, we find

$$0 \leq f_V(\Gamma) - f_V(H) \leq \bar{\delta}_V + \delta_V,$$
$$0 \leq f_V\{H(\bar{C})\} - f_V(\Gamma) \leq \bar{\delta}_V + \delta_V. \tag{96}$$

We make use now of inequality (78) and substitute into it

$$\Gamma_0 = H, \quad \Gamma_1 = \Gamma - H = 2V \sum_\alpha \tau_\alpha g_\alpha (J_\alpha - \bar{C}_\alpha)(J_\alpha^\dagger - \bar{C}_\alpha^*).$$

Then, form the first of the inequalities (96), we obtain

$$2 \sum_\alpha \tau_\alpha g_\alpha \langle (J_\alpha - \bar{C}_\alpha)(J_\alpha^\dagger - \bar{C}_\alpha^*) \rangle_\Gamma \leq \bar{\delta}_V + \delta_V. \tag{97}$$

We have thus proved the following theorem:

Theorem 4. *Let the conditions of Theorem 3 be fulfilled, and let Γ be represented by the expression* (93) *in which*

$$0 < \tau_\alpha < 1, \quad \alpha = 1, 2, \ldots, s.$$

Then the following inequalities hold:

$$0 \leq f_V(\Gamma) - f_V(H) \leq \bar{\delta}_V + \delta_V \to 0 \quad as \quad V \to \infty,$$
$$\sum_\alpha g_\alpha \langle (J_\alpha - \bar{C}_\alpha)(J_\alpha^\dagger - \bar{C}_\alpha^*) \rangle_\Gamma \leq \frac{\bar{\delta}_V + \delta_V}{2\tau_0} \to 0 \quad as \quad V \to \infty, \tag{98}$$

where τ_0 is the smallest of the quantities $\tau_1, \tau_2, \ldots, \tau_s$.

RIDER TO THEOREM 4

We shall consider the more general case when, in the expression

$$H = T - 2V \sum_\alpha g_\alpha J_\alpha J_\alpha^\dagger,$$

the operators T and J_α are not represented by (17) and satisfy only the conditions of Theorem 2.

Then, replacing Theorem 3 by Theorem 2 in the arguments carried through above, we see that Theorem 4 remains true.

We note, further, that in the case when the operators have the specific form (17) and the conditions of Theorem (3) are fulfilled, the Theorem 4 proved above enables us to transform the Hamiltonian Γ directly to the form (14), (15), and the conditions $\mathbf{I}$ and $\mathbf{I}'$ of §1 are found to be fulfilled.

In fact, we have

$$H = T - 2V \sum_\alpha g_\alpha(J_\alpha^\dagger \bar{C}_\alpha + J_\alpha \bar{C}_\alpha^*) + 2V \sum_\alpha g_\alpha \bar{C}_\alpha^* \bar{C}_\alpha$$
$$- 2V \sum_\alpha g_\alpha(J_\alpha - \bar{C}_\alpha)(J_\alpha^\dagger - \bar{C}_\alpha^*),$$

and, therefore

$$\Gamma = H(\bar{C}) - 2V \sum_\alpha g_\alpha(1 - \tau_\alpha)(J_\alpha - \bar{C}_\alpha)(J_\alpha^\dagger - \bar{C}_\alpha^*), \tag{99}$$

where

$$H(\bar{C}) = T - 2V \sum_\alpha g_\alpha(J_\alpha^\dagger \bar{C}_\alpha + J_\alpha \bar{C}_\alpha^*) + 2V \sum_\alpha g_\alpha \bar{C}_\alpha^* \bar{C}_\alpha.$$

It can be seen that this Hamiltonian has the form of the Hamiltonian (14), (15) in which we put

$$\Gamma_\alpha = H(\bar{C}),$$
$$\Lambda(f) = 2 \sum_\alpha g_\alpha \lambda_\alpha(f) \bar{C}_\alpha^*,$$
$$K = 2V \sum_\alpha g_\alpha \bar{C}_\alpha^* \bar{C}_\alpha,$$
$$G_\alpha = 2g_\alpha(1 - \tau_\alpha) > 0, \quad C_\alpha = \bar{C}_\alpha. \tag{100}$$

By virtue of Theorem 4, the following inequality is fulfilled:

$$\sum_{\alpha} G_{\alpha} \langle (J_{\alpha} - \bar{C}_{\alpha})(J_{\alpha}^{\dagger} - \bar{C}_{\alpha}^{*}) \rangle_{\Gamma} \leq \varepsilon_{V}, \tag{101}$$

where

$$\varepsilon_{V} = \frac{\delta_{V} + \bar{\delta}_{V}}{\tau_{0}} \quad (V \to \infty)$$

uniformly with respect to the temperature θ in any interval of the form $(0 < \theta \leq \theta_{0})$.

The validity of paragraph 3 of condition I is thereby also established.

The remaining paragraphs of conditions **I** and **I'** follow trivially form the inequalities (52) and (53), the condition (54), the fact that s is finite in the sum over α, and the fact that the quantities C are independent of V.

We can therefore make use of all the limit theorems proved in [13, 14].

Since $\Gamma_{a} = H_{\infty}(\bar{C})$, we write the theorems on the existence of the limits

$$\lim_{V \to \infty} \langle \mathscr{A} \rangle_{\Gamma} = \lim_{V \to \infty} \langle \mathscr{A} \rangle_{\Gamma_{a}}$$

in the form

$$\lim_{V \to \infty} \langle \mathscr{A} \rangle_{\Gamma} = \lim_{V \to \infty} \langle \mathscr{A} \rangle_{H(\bar{C})}.$$

But $H(C)$ is independent of the parameters τ in the case under consideration when

$$0 < \tau_{\alpha} < 1, \quad \alpha = 1, 2, \ldots, s. \tag{102}$$

Therefore, the expression

$$\lim_{V \to \infty} \langle \mathscr{A} \rangle_{\Gamma}$$

is also independent of the τ lying in the region (102)

Consequently, when all the $\tau_{1}, \ldots, \tau_{s}$ tend to zero while remaining positive, we have, trivially

$$\lim_{\tau \to 0} \lim_{V \to \infty} \langle \mathscr{A} \rangle_{\Gamma} = \lim_{V \to \infty} \langle \mathscr{A} \rangle_{H(\bar{C})}.$$

We can define the quasi-average in this situation by the relations

$$\prec \mathscr{A} \succ_{H} = \lim_{V \to \infty} \langle \mathscr{A} \rangle_{\Gamma} = \lim_{V \to \infty} \langle \mathscr{A} \rangle_{H(\bar{C})} \tag{103}$$

in which the τ can take any values from the range (102).

We emphasize again that the most important point in our arguments was the establishment of the inequality (97), based on the inequality (96).

It can be seen that from the inequality (96) there follows the asymptotic relation

$$\lim_{V\to\infty} f_V(\Gamma) = \lim_{V\to\infty} f_V(H). \tag{104}$$

10. The Question of the Choice of Sign for the Source-Terms

We note, in passing, that the above asymptotic relation ceases, in general, to be true for negative values of τ. In fact, for example,

$$\tau_\alpha = -w_\alpha; \quad w_\alpha > 0 \quad (\alpha = 1, 2, \ldots, s). \tag{105}$$

Then

$$\Gamma = \Gamma_w = H - 2V \sum_\alpha w_\alpha g_\alpha (J_\alpha - \bar{C}_\alpha)(J_\alpha^\dagger - \bar{C}_\alpha^*).$$

We make use of the inequality (78), substituting in it

$$\Gamma_0 = \Gamma, \quad \Gamma_1 = 2V \sum w_\alpha g_\alpha (J_\alpha - \bar{C}_\alpha)(J_\alpha^\dagger - \bar{C}_\alpha^*).$$

Then in (78) the Hamiltonian will be

$$\Gamma_0 + \Gamma_1 = H$$

and

$$f_V(H) - f_V(\Gamma_w) \geq 2 \sum_\alpha w_\alpha g_\alpha \langle (J_\alpha - \bar{C}_\alpha)(J_\alpha^\dagger - \bar{C}_\alpha^*) \rangle_H. \tag{106}$$

But, obviously,

$$\langle a_f^\dagger a_{-f}^\dagger \rangle_H = \langle a_{-f} a_f \rangle_H = 0,$$

and therefore

$$\langle J_\alpha \rangle_H = 0, \quad \langle J_\alpha^\dagger \rangle_H = 0.$$

We have, consequently

$$\langle (J_\alpha - \bar{C}_\alpha)(J_\alpha^\dagger - \bar{C}_\alpha^*) \rangle_H = \langle J_\alpha J_\alpha^\dagger \rangle_H + |\bar{C}_\alpha|^2,$$

whence, by virtue of (106), we find:

$$f_V(H) - f_V(\Gamma_w) \geq 2 \sum w_\alpha g_\alpha |\bar{C}_\alpha|^2,$$

and, passing to the limit,

$$\lim_{V \to \infty} f_V(H) - \lim_{V \to \infty} f_V(\Gamma_\omega) \geq 2 \sum w_\alpha g_\alpha |\bar{C}_\alpha|^2. \tag{107}$$

Thus, if $\bar{C} \neq 0$, the equality (104) is not true for negative τ (105).

It was to take this fact into account that we required that the proportionality coefficients r_α in our choice (90) of parameters ν_α characterizing the sources included in the Hamiltonian be positive.

11. The Construction of Upper-Bound Inequalities in the Case $\bar{C} = 0$

We give here a separate treatment of the case when $\bar{C} = 0$, i.e. when

$$\bar{C}_1 = \bar{C}_2 = \ldots = \bar{C}_s = 0. \tag{108}$$

In this case,

$$H(\bar{C}) = T,$$

and, therefore,

$$f_v(H) - f_V(T) \to 0 \quad \text{as} \quad V \to \infty.$$

Thus, the interaction terms

$$-2V \sum_{1 \leq \alpha \leq s} g_\alpha J_\alpha J_\alpha^\dagger$$

of the Hamiltonian H are asymptotically (as $V \to \infty$) ineffective in the calculation of the free energy.

Further, we have (compare with (93))

$$\Gamma = \Gamma_\tau = H + 2V \sum_\alpha g_\alpha \tau_\alpha J_\alpha J_\alpha^\dagger = T - 2V \sum_\alpha g_\alpha (1 - \tau_\alpha) J_\alpha J_\alpha^\dagger, \tag{109}$$

and, in view of our earlier proofs, we can write an upper bound for a correlation average constructed on the basis of this Hamiltonian:

$$\left\langle \sum_\alpha g_\alpha J_\alpha J_\alpha^\dagger \right\rangle_\Gamma \leq \frac{\delta_V + \bar{\delta}_V}{2\tau_0} \to 0 \quad \text{as} \quad V \to \infty. \tag{110}$$

We shall show that, in the given case (108), we have also

$$\left\langle \sum_\alpha g_\alpha J_\alpha J_\alpha^\dagger \right\rangle_\Gamma \leq \zeta_V \to 0 \quad (V \to \infty). \tag{111}$$

For this, we shall take the Hamiltonian

$$H_\omega = T - 2V(1+\omega) \sum_\alpha g_\alpha J_\alpha J_\alpha^\dagger \quad (1 > \omega > 0), \tag{112}$$

and formulate a trial Hamiltonian

$$H_\omega(C) = T - 2V(1+\omega) \sum_\alpha g_\alpha(C_\alpha J_\alpha + C_\alpha^* J_\alpha^\dagger)$$
$$+ 2V(1+\omega) \sum_\alpha g_\alpha C_\alpha^* C_\alpha. \tag{113}$$

We denote by $\bar{C}^{(\omega)}$ the point C giving the absolute minimum of the function $f_\infty\{H_\infty(C)\}$. If for any positive ω, however small,

$$\bar{C}^{(\omega)} = 0, \tag{114}$$

the proof of the relation (111) is trivial. We need only replace H by H_ω in the equality (110) and for τ_α take

$$\tau_\alpha = \frac{\omega}{1+\omega}$$

in the Hamiltonian Γ_τ. Then the Hamiltonian Γ in (110) coincides with H. It remains, therefore, to consider the case when (114) is not true for some positive value of ω, however small.

We note that the value $C = C^{(\omega)}$ must satisfy the equations

$$\frac{\partial f_\infty\{H_\omega(C)\}}{\partial C_\alpha^*} = 0 \quad (0 \leq \alpha \leq s);$$

i.e. from (69),

$$\bar{C}_\alpha^{(\omega)} = \frac{1}{2(2\pi)^3} \int \frac{\tanh[E_\omega(f)/2\theta]}{E_\omega(f)} \left\{ \sum_\beta (1+\omega) g_\beta \bar{C}_\beta^{(\omega)} \lambda_\beta^*(f) \right\} \lambda_\alpha(f) \, df,$$

where

$$E_\omega^2(f) = \left(\frac{p^2}{2m} - \mu\right)^2 + 4(1+\omega)^2 \left|\sum_\beta g_\beta \bar{C}_\beta^{(\omega)} \lambda_\beta^*(f)\right|^2.$$

Hence, it follows that

$$|\bar{C}_\alpha^{(\omega)}| \le \frac{1}{4(2\pi)^3} \int |\lambda_\alpha(f)|\, df \le \frac{\bar{Q}_1}{2}. \tag{115}$$

We now put

$$(1+\omega)C_\alpha = X_\alpha, \tag{116}$$

and note that

$$f_\infty\{H_\omega(C)\} = f_\infty\{H(X)\} - 2\frac{\omega}{1+\omega}\sum_\alpha g_\alpha |X_\alpha|^2. \tag{117}$$

Thus, for

$$X_\alpha^{(\omega)} = (1+\omega)\bar{C}_\alpha^{(\omega)},$$

the expressions on the right-hand side of (117) attains an absolute minimum.
Therefore

$$f_\infty\{H(X^{(\omega)})\} - 2\frac{\omega}{1+\omega}\sum_\alpha g_\alpha |X_\alpha^{(\omega)}|^2 \le f_\infty\{H(0)\}. \tag{118}$$

On the other hand,

$$f_\infty\{H(0)\} = \min_{(X)} f_\infty\{H(X)\},$$

in view of which,

$$f_\infty\{H(X^{(\omega)}\} \ge f_\infty\{H(0)\}.$$

Hence,

$$0 \le f_\infty\{H(X^{(\omega)}\} - f_\infty\{H(0)\} \le 2\frac{\omega}{1+\omega}\sum_\alpha g_\alpha |X_\alpha^{(\omega)}|^2. \tag{119}$$

We shall show now that

$$X^{(\omega)} \to 0 \quad (\omega \to 0). \tag{120}$$

In fact, we assume the opposite to be the case. Then, since, by virtue of (115), $X^{(\omega)}$ is bounded:

$$|X_\alpha^{(\omega)}| \le (1+\omega)\frac{\bar{Q}_1}{2} \le \bar{Q}_1,$$

we can always choose a sequence of positive $\omega' \to 0$ such that

$$X^{(\omega')} \to \tilde{X},$$

with

$$\tilde{X} \neq 0. \tag{121}$$

Putting $\omega = \omega'$ in (119) and passing to the limit, we find

$$f_\infty\{H(\tilde{X})\} = f_\infty\{H(0)\}. \tag{122}$$

But, as we have seen, if the point $C = 0$ gives the absolute minimum of the function $f_\infty\{H(C)\}$, no other points realizing the absolute minimum of this function exists; in view of this, (122) and (121) are inconsistent and we have arrived at a contradiction.

Thus, the relation (120) is proved. Noting that

$$f_\infty\{H(0)\} = f_\infty(T),$$

form (117) and (119) we obtain

$$-2\frac{\omega}{1+\omega}\sum_\alpha g_\alpha|X_\alpha^{(\omega)}|^2 \le f_\infty\{H_\omega(\bar{C}^{(\omega)})\} - f_\infty(T) \le 0;$$

i.e.;

$$0 \le f_\infty(T) - f_\infty\{H_\omega(\bar{C}^{(\omega)})\} \le \omega\xi(\omega), \tag{123}$$

where

$$\xi(\omega) = \frac{2}{1+\omega}\sum_\alpha g_\alpha|X_\alpha^{(\omega)}|^2 \to 0 \quad \text{as} \quad \omega \to 0. \tag{124}$$

We now invoke Theorem 3. Since, in the case under consideration, $H(\bar{C}) = H(0) = T$, we can write

$$|f_V(H) - f_\infty(T)| \le \bar{\delta}_V \to 0, \quad (V \to \infty).$$

For the Hamiltonian H_ω, we also have

$$|f_V(H_\omega) - f_\infty\{H_\omega(\bar{C}^{(\omega)})\}| \le \bar{\delta}_V(\omega) \to 0, \quad \text{as} \quad V \to \infty.$$

Here, $\bar{\delta}_V(\omega)$ denotes $\bar{\delta}_V$ for H_ω. On the other hand,

$$0 \leq f_V(H) - f_V(H_\omega) = f_V(H) - f_\infty(T) + f_\infty(T)$$
$$- f_\infty\{H_\omega(\bar{C}^{(\omega)})\} + f_\infty\{H_\omega(\bar{C}^{(\omega)})\} - f_V(H_\omega),$$

and, therefore,

$$0 \leq f_V(H) - f_V(H_\omega) \leq \bar{\delta}_V + \bar{\delta}_V(\omega) + \omega\xi(\omega). \tag{125}$$

We shall now make use of the inequality (78); we substitute in it

$$\Gamma_0 = H_\omega, \quad \Gamma_1 = 2V\omega \sum_\alpha g_\alpha J_\alpha J_\alpha^\dagger,$$

$$\Gamma_0 + \Gamma_1 = H.$$

Then, from (125) we obtain

$$\left\langle \sum_\alpha g_\alpha J_\alpha J_\alpha^\dagger \right\rangle_H \leq \frac{\bar{\delta}_V + \bar{\delta}_V(\omega)}{2\omega} + \frac{1}{2}\xi(\omega). \tag{126}$$

However, this inequality is valid for any value of ω in the interval $(0 < \omega < 1)$, and its left-hand side is entirely independent of ω. Consequently, the left-hand side of (126) will not exceed the lower bound of the right-hand side in the given interval:

$$\left\langle \sum_\alpha g_\alpha J_\alpha J_\alpha^\dagger \right\rangle_H \leq \zeta_V,$$

$$\zeta_V = \inf_{(0<\omega<1)} \left\{ \frac{\bar{\delta}_V + \bar{\delta}_V(\omega)}{2\omega} + \frac{1}{2}\xi(\omega) \right\}.$$

It only remains for us to show that ζ_V goes to zero as $V \to \infty$.

We shall fix an arbitrary small number ϱ. On the basis of (124), we can, in the interval $(0 < \omega < 1)$ under consideration, fix a number ω_0 such that

$$\xi(\omega_0) \leq \varrho.$$

We see then that

$$\zeta_V \leq \frac{\bar{\delta}_V + \bar{\delta}_V(\omega)}{2\omega} + \frac{\varrho}{2}.$$

But, since ω_0 is fixed, we have

$$\frac{\bar{\delta}_V + \bar{\delta}_V(\omega)}{2\omega} \to 0 \quad \text{as} \quad V \to \infty.$$

We can find a value V_0 such that

$$\frac{\bar{\delta}_V + \bar{\delta}_V(\omega)}{2\omega} \leq \frac{\varrho}{2} \quad \text{for} \quad V \geq V_0.$$

Thus,

$$\zeta_V \leq \varrho \quad \text{for} \quad V \geq V_0;$$

i.e., $\zeta_V \to 0$ as $V \to \infty$, and the relation (111) is proved.

It is now clear that H is of the type (1.14), (1.15), with

$$\Gamma_a = T, \quad \Lambda(f) = 0, \quad C_\alpha = 0, \quad G_\alpha = 2g_\alpha.$$

By virtue of (111), paragraph 3 of condition **I** is fulfilled; the remaining paragraphs of conditions **I** and **I′** are trivial in the given case.

We shall, therefore, make use of the limit theorems proved in [13,14]. We see that, for the operators $\mathscr{A}$ with which these theorems are concerned, we can write

$$\lim_{V \to \infty} \langle \mathscr{A} \rangle_H = \lim_{V \to \infty} \langle \mathscr{A} \rangle_T.$$

Applying (103), we see that, in the case under investigation ($\bar{C} = 0$), we have

$$\prec \mathscr{A} \succ_H = \lim_{V \to \infty} \langle \mathscr{A} \rangle_H = \lim_{V \to \infty} \langle \mathscr{A} \rangle_T.$$

Thus, the quasi-averages and the usual averages of the operators considered above are asymptotically equal to the corresponding averages taken over the Hamiltonian T. Moreover, the interaction terms in H turn out to have *no effect* here.

Some applications for Approximating Hamiltonian Method (AHM) were also done in [3,5,6,16,18–28,30]. It is also possible to consider Hamiltonians with coupling constants of different signs and formulate on this basis minimax principle for the Hamiltonian

$$H = T_0 + 2V \sum_{(1 \leq \alpha \leq r)} g_\alpha J_\alpha J_\alpha^\dagger - 2V \sum_{(r+1 \leq \alpha \leq r+s)} g_\alpha J_\alpha J_\alpha^\dagger,$$

$$T_0 = \sum_f \left(\frac{p^2}{2m} - \mu \right) a_f^\dagger a_f, \quad J_\alpha = \frac{1}{2V} \sum_f \lambda_\alpha(f) a_f^\dagger a_{-f}^\dagger,$$

$$\lambda_\alpha(-f) = -\lambda_\alpha(f), \quad g_\alpha > 0.$$

For more details, the reader is referred to the works by the author [10,16].

References

1. Bogolyubov, N. N., Zubarev, D. N., and Tserkovnikov, Yu. A. *Docl. Acad. Nauk. SSSR* **117**, 788 (1957) [*Sov. Phys. Doclady* **2**, 535 (1957)].

2. Bogolyubov, N. N., Zubarev, D. N., and Tserkovnikov, Yu. A. *Zh. Eksp. Teor. Phys.* **39**, 120 (1960) [*Sov. Phys. JETP* **12**, 88 (1960)].

3. Bogolyubov, N.N., Jr. *Ukrainsk. Matematichesk. Journal* **17**, 3 (1965) [in Russian].

4. Bogolyubov, N. N., Jr. *Vestn. Mosc. Univ. Fiz. Atsron.* **21**(1), 94 (1966) [*Moscow Univ. Phys. Bull.* **21**(1), 67 (1966)].

5. Bogolyubov, N. N., Jr. *Physica* **32**, 933 (1966).

6. Bogolyubov, N. N., Jr. *Docl. Acad. Nauk. SSSR* **168**, 766 (1966) [*Sov. Phys. Doclady* **11**, 482 (1966)].

7. Bogolyubov, N. N., Jr. ITPK (Institute for Theoretical Physics, Kiev) Preprint 67-1 (1967).

8. Bogolyubov, N. N., Jr. JINRD Preprint P4-4184 (1968).

9. Bogolyubov, N. N., Jr. *Docl. Acad. Nauk. SSSR* **182**, 797 (1968) [*Sov. Phys. Doclady* **13**, 1111 (1969)].

10. Bogoliubov, N. N., Jr. *Yad. Fiz.* **10**, 425 (1969) [*Sov. J. Nucl. Phys.* **10**, 243 (1970)].

11. Bogoliubov, N. N., Jr. *Physica* **41**, 601 (1969).

12. Bogolyubov, N. N., Jr. and Shumovski, A. S. *Int. J. Pure Appl. Phys.* **8**, 121 (1970).

13. Bogolyubov, N. N., Jr. *Theoret. and Math. Phys.* **4**, 929 (1970).

14. Bogolyubov, N. N., Jr. *Theoret. and Math. Phys.* **5**, 1038 (1970).

15. Bogolyubov, N. N., Jr. and Shumovski, A. S. *Phys. Lett.* **35A**, 380 (1971).

16. Bogolyubov, N. N., Jr. *A Method for Studying Model Hamiltonians,* Pergamon Press (1972).

17. Bogolyubov, N. N., Jr. *Theoret. and Math. Phys.* **13**, 1032 (1972).

18. Bogolyubov, N. N., Jr. and Plechko V.N. *Physica* **82**A, (163) (1975); *Docl. Acad. Nauk. SSSR* **288**, 1061 (1976).

19. Bogolyubov, N. N., Jr. and Plechko V.N., Repnikov N. F. *Theoret. and Math. Phys.* **24**, 886 (1975).

20. Bogolyubov, N. N., Jr., Shumovski A.S. *Tr. Mat. Inst., Akad. Nauk SSSR* **136** 351 (1975) (in Russian).

21. Bogolyubov, N. N., Jr., Brankov J.G., Zagrebnov V.A., Kurbatov A.M., Tonchev N.S. *The Approximating Hamiltonian Method in Statistical Physics*. Publ. House Bulg. Acad. Sci., Sofia, 1981 [in Russian].

22. Bogolyubov, N. N., Jr., Plechko V.N., Shumovski A. S. *Fiz. Elem. Chastits At. Yadra* **14**, 1443 (1983) [in Russian].

23. Bogolyubov, N. N., Jr., Brankov J.G., Zagrebnov V.A., Kurbatov A.M., Tonchev N.S. *Russian Math. Surveys* **39**, 1 (1984).

24. Bogolyubov, N. N., Jr. and Bogolyubova, E. N. *Ukrainian Journal of Physics* **45**, 4 (2000).

25. Bogolubov, N.N., Jr., Bogolubova, E.N. and Kruchinin, S.P. *Calculation of Correlation Functions for Superconductivity Models* in New Trends in Superconductivity, NATO Science Series **67**, 277 (2002).

26. Brankov, J.G., Danchev, D. M., Tonchev N.S. *Theory of Critical Phenomena in Finite-Size Systems: Scaling and Quantum Effects*, Series in Modern Condensed Matter Physics v. 9, World Scientific, Singapore, 2000.

27. Brankov, J.G. and Tonchev, N.S. *Cond. Matter Physics* **14**, 13003 (2011)

28. Brankov J.G. and Tonchev N.S. *Phys. Rev. E* **85**, 031115 (2012).

29. Maison, H.D. Preprint Th. 1299-CERN, Geneva, (1971).

30. Plechko V. N. *Theoret. and Math. Phys.* **28** 677 (1976).

31. Shilov, G. E. and Gurevich, V. L. *Integral, Measure and Derivative. A Unified Approach*, Nauka, Moscow, 1964 (Translation published by Prentice-Hall, Inc., N. Y., 1966)

32. Shumovski, A. S. ITPK Preprint 71-57E, 71-56P (1971).

33. Thirring, W. and Wehrl, A. *Commun. Math. Phys.* **4**, 303 (1967).

34. Thirring, W. Preprint, Institute of Theoretical Physics, University of Vienna (1968).

35. Wentzel, G. *Helv. Phys. Acta.* **33**, 859 (1960).

Additional References

I. GENERAL REFERENCES

1. N. N. Bogoliubov, *Collection of Scientific Papers*, In 12 vol. ed. A.D.Sukhanov (Moscow, Nauka, 2005-2009).

2. N. N. Bogoliubov and N. N. Bogoliubov, Jr., *Introduction to Quantum Statistical Mechanics*, 2nd ed. (World Scientific, Singapore, 2009).

3. N. N. Bogoliubov, Jr., B. I. Sadovnikov, A. S. Schumovsky, *Mathematical Methods for Statistical Mechanics of Model Systems*, (Nauka, Moscow, 1989) [in Russian].

4. N. N. Bogoliubov Jr. and D. P. Sankovich, *N. N. Bogoliubov and Statistical Mechanics*, Usp. Mat. Nauk. **49**, 21 (1994).

5. D. Ya. Petrina, *Mathematical Foundations of Quantum Statistical Mechanics*, (Kluwer Academic Publ., Dordrecht, 1995).

6. P.A. Martin and F. Rothen. *Many-body Problems and Quantum Field Theory*, (Springer, Berlin, 2004).

7. D. V. Shirkov, *60 Years of Broken Symmetries in Quantum Physics (from the Bogoliubov theory of superfluidity to the Standard Model)*, Usp. Fiz. Nauk **179**, 581 (2009).

8. A.L. Kuzemsky, *Bogoliubov's Vision: Quasiaverages and Broken Symmetry to Quantum Protectorate and Emergence*, Int. J. Mod. Phys. **B24**, 835-935 (2010).

II. BOGOLIUBOV TRANSFORMATIONS

1. W. Witschel, *On the General Linear (Bogoliubov) Transformation for Bosons*, Z.Physik **B21**, 313 (1975)

2. S. N. M. Ruijsenaars, *On Bogoliubov Transformations. The General Case.* Annals of Physics **116**, 105 (1978).

3. S. N. M. Ruijsenaars, *Integrable Quantum Field Theories and Bogoliubov Transformations*, Annals of Physics **132**, 328 (1981).

4. Nguyen Ba An, *A Step-by-step Bogoliubov Transformation Method for Diagonalising a Kind of non-Hermitian Effective Hamiltonian*, J. Phys. C: Solid State Phys. **21**, L1209 (1988).

5. F. J. W. Hanle and A. Klein, *Generalization of the Quantized Bogoliubov-Valatin Transformation*, Phys.Lett. B **229**, 1 (1989).

6. P. C. W. Fung, C. C. Lam and P. Y. Kwok, *U-Matrix Theory, Bogoliubov Transformation and BCS Trial Wave Function*, Science in China **32**, 1072 (1989).

7. W.-S. Liu and X.-P. Li, *Time-dependent formulation of the Bogoliubov transformation and time-evolution operators for time-dependent quantum oscillators*, Europhys. Lett. **58**, 639 (2002).

8. J.-W. van Holten and K. Scharnhorst, *Nonlinear Bogolyubov Valatin transformations and quaternions*, J. Phys. A: Math. Gen. **38** 10245 (2005).

9. L. Bruneaua and J. Dereziski, *Bogoliubov Hamiltonians and one-parameter groups of Bogoliubov transformations*, J. Math. Phys. **48**, 022101 (2007).

10. A. I.Vdovin, A. A. Dzhioev, *Thermal Bogoliubov Transformation in Nuclear Structure Theory*, Physics of Particles and Nuclei **41**, 2093 (2010).

11. K. Scharnhorst and J.-W. van Holten, *Nonlinear Bogolyubov-Valatin transformations: Two modes*, Annals of Physics **326**, 2868 (2011).

12. J. Katriel, *A Nonlinear Bogoliubov Transformation*, Phys. Lett. A **307**, 1 (2003).

13. K.Takayanagi, *Utilizing Group Property of Bogoliubov Transformation*, Nucl. Phys. A **808**, 17 (2008).

14. S. Caracciolo, F. Palumbo and G. Viola, *Bogoliubov Transformations and Fermion Condensates in Lattice Field Theories*, Annals of Physics **324**, 584 (2009).

III. QUASIAVERAGES

1. N. N. Bogoliubov, Jr., *Method of Calculating Quasiaverages*, J. Math. Phys. **14**, 79 (1973).

2. N. N. Bogoliubov and N. N. Bogoliubov, Jr., *Introduction to Quantum Statistical Mechanics*, 2nd ed. (World Scientific, Singapore, 2009).

3. D. Ya. Petrina, *Mathematical Foundations of Quantum Statistical Mechanics*, (Kluwer Academic Publ., Dordrecht, 1995).

4. S. V. Peletminskii, A. I. Sokolovskii, *Flux Operators of Physical Variables and the Method of Quasiaverages*, Theor. Math. Phys. **18**, 121 (1974).

5. V. I. Vozyakov, *On an Application of the Method of Quasiaverages in the Theory of Quantum Crystals*, Theor. Math. Phys. **39**, 129 (1979).

6. D. V. Peregoudov, *Effective Potentials and Bogoliubov's Quasiaverages*, Theor. Math. Phys. **113**, 149 (1997).

7. N. N. Bogoliubov, Jr., D. A. Demyanenko, M. Yu. Kovalevsky, N. N. Chekanova, *Quasiaverages and Classification of Equilibrium States of Condensed Media with Spontaneously Broken Symmetry*, Physics of Atomic Nuclei **72**, 761 (2009).

8. A. L. Kuzemsky, *Symmetry Breaking, Quantum Protectorate and Quasiaverages in Condensed Matter Physics*, Physics of Particles and Nuclei **41**, 1031 (2010).

9. A. L. Kuzemsky, *Quasiaverages, Symmetry Breaking and Irreducible Green Functions Method*, Condensed Matter Physics (http://www.icmp.lviv.ua/journal), **13**, N4, p.43001: 1-20 (2010).

IV. BOGOLIUBOV'S INEQUALITIES

1. A. Ishihara, *The Gibbs-Bogoliubov inequality*, J. Phys. A **2**, 539 (1968).

2. N. D. Mermin, *Some Applications of Bogoliubov's Inequality in Equilibrium Statistical Mechanics*, J. Phys. Soc. Japan **26** Supplement, 203 (1969).

3. S. Okubo. *Some General Inequalities in Quantum Statistical Mechanics*, J. Math. Phys. **12**, 1123 (1971)

4. B. I. Sadovnikov, V. K. Fedyanin, *N. N. Bogoliubov's Inequalities in Systems of Many Interacting Particles with Broken Symmetry*, Theor. Math. Phys. **16**, 368 (1973).

5. J. C. Garrison, J. Wong, *Bogoliubov Inequalities for Infinite Systems*, Commun. Math. Phys. **26**, 1 (1972).

6. L. Pitaevskii, S. Stringari, *Uncertainty Principle, Quantum Fluctuations, and Broken Symmetries*, J. Low Temp. Phys. **85**, 377 (1991).

7. A. V. Soldatov, *Generalization of the Peierls-Bogoliubov Inequality by Means of a Quantum-Mechanical Variational Principle*, Physics of Particles and Nuclei **31**, 138 (2000).

8. A. L. Kuzemsky, *Bogoliubov's Vision: Quasiaverages and Broken Symmetry to Quantum Protectorate and Emergence*. Int. J. Mod. Phys. B **24**, 835 (2010).

V. SUPERFLUIDITY, BOSE GAS

1. V.V. Tolmachev, *Theory of Bose Gas*. (Izd-vo Mosk. Un-ta, Moscow, 1969) [in Russian].

2. V. A. Zagrebnov, *Bogoliubov's Theory of a Weakly Imperfect Bose Gas and its Modern Development*, in: N. N. Bogoliubov, *Collection of Scientific Papers* In 12 vol. ed. A. D. Sukhanov, vol.8. (Moscow, Nauka, 2007) p.576.

3. E.H. Lieb, *The Bose Gas: a Subtle Many-Body Problem*, The proceedings of the XIII International Congress of Mathematical Physics, London, July 18-24, 2000. `arXiv:math-ph/0009009v1`

4. V. A. Zagrebnov and J.-B. Bru, *The Bogoliubov Model of Weakly Imperfect Bose Gas*, Phys. Rep., **350**, 291 (2001).

5. M. Corgini and D. P. Sankovich, *Study of a Non-Iinteracting Boson Gas*, Int. J. Mod. Phys. B **16**, 497 (2002).

6. V.I. Yukalov, *Self-consistent Theory of Bose-condensed Systems*. Phys. Lett., A **359**, 712 (2006).

7. V. I. Yukalov and H. Kleinert, *Gapless Hartree-Fock-Bogoliubov Approximation for Bose Gases*, Phys. Rev., A **73**, 063612 (2006).

8. N. N. Bogoliubov, Jr. and D. P. Sankovich, *Bogoliubov's Approximation for Bosons*, Ukr. J. Phys., **55**, 104 (2010).

9. D. P. Sankovich, *Bogoliubov's Theory of Superfluidity, Revisited*, Int. J. Mod. Phys. B **24**, 5327 (2010).

10. V. I. Yukalov, *Basics of Bose-Einstein Condensation*. Physics of Particles and Nuclei, **42**, 460 (2011).

11. L. Pitaevskii, S. Stringari, *Bose-Einstein Condensation*, (Oxford University Press, Oxford, 2003).

12. C. J. Pethick, H. Smith, *Bose-Einstein Condensation in Dilute Gases*, (Cambridge University Press, Cambridge, 2002).

13. A. Griffin, T. Nikuni, E. Zaremba, *Bose-Condensed Gases at Finite Temperatures*, (Cambridge University Press, Cambridge, 2009)

14. A. Griffin, *New light on the intriguing history of superfluidity in liquid 4He*. J. Phys.: Condens. Matter **21**, 164220 (2009).

15. A. Verbeure, *Many-Body Boson Systems: Half a Century Later*, (Springer, Berlin, 2011).

16. C. W. Gardiner, *Particle-number-conserving Bogoliubov method which demonstrates the validity of the time-dependent Gross-Pitaevskii equation for a highly condensed Bose gas*, Phys. Rev., A **56**, 1414 (1997).

17. A. Brunello, F. Dalfovo, L. Pitaevskii, S. Stringari, *How to Measure the Bogoliubov Quasiparticle Amplitudes in a Trapped Condensate*, Phys. Rev. Lett. **85**, 4422 (2000).

18. J. M. Vogels, K. Xu, C. Raman, J. R. Abo-Shaeer, W. Ketterle, *Experimental Observation of the Bogoliubov Transformation for a Bose-Einstein Condensed Gas*, Phys. Rev. Lett. **88**, 060402 (2002).

19. M. W. J. Romans and H. T. C. Stoof, *Bogoliubov theory of Feshbach molecules in the BEC-BCS crossover*, Phys. Rev., A **74**, 053618 (2006).

20. E. H. Lieb, R. Seiringer, J. Yngvason, *Bose-Einstein Condensation and Spontaneous Symmetry Breaking*, Rep. Math. Phys. **59**, 389 (2007).

21. P. Shygorin and A. Svidzynsky, *Modified Bogolyubovs Derivation of the Two-Fluid Hydrodynamics*, Ukr. J. Phys. **55**, 109 (2010).

22. A. A. Svidzinsky and M. O. Scully, *Condensation of N bosons: Microscopic approach to fluctuations in an interacting Bose gas*, Phys. Rev., A **82**, 063630 (2010).

VI. SUPERCONDUCTIVITY

1. N. N. Bogoliubov, *To the Question of Model Hamiltonian in Superconductivity Theory*, Sov. J. Part. Nucl. **1**, 1 (1971).

2. V. V. Tolmachev, *Theory of Fermi Gas.* (Izd-vo Mosk. Un-ta, Moscow, 1973) [in Russian].

3. C. T. Chen-Tsai, *On the Bogoliubov-Zubarev-Tserkovnikov Method in the Theory of Superconductivity*, Chinese J. Phys. **3**, 22 (1965).

4. D. H. Kobe, *Green's Functions for Fermion Systems with Pairing Correlations*, Ann. Phys. **20**, 279 (1962).

5. D. H. Kobe, *Green's Functions Derivation of the Method of Approximate Second Quantization*, Ann. Phys. **25**, 121 (1963).

6. D. H. Kobe, *Self-Energy for the Bogoliubov Quasiparticle*. Ann. Phys. **28**, 400 (1964).

7. D. H. Kobe, *Best Energy Criterion and the Principle of Compensation of Dangerous Diagrams*, Ann. Phys. **40**, 395 (1966).

8. D. H. Kobe, *Derivation of the Principle of Compensation of Dangerous Diagrams*, J. Math. Phys. **8**, 1200 (1967).

9. G. A. Raggio and R.F. Werner, *The Gibbs Variational Principle for General BCS-type Models*, Europhys. Lett. **9**, 633 (1989).

10. D. Waxman, *Fredholm Determinant for a Bogoliubov Hamiltonian*. Phys. Rev. Lett. **72**, 570 (1994).

11. N. N. Bogoliubov, V. A. Moskalenko, *On Question of the Existence of Superconductivity in the Hubbard Model*, Theor. Math. Phys. **86**, 16 (1991).

12. V.V. Tolmachev, *Superconducting BoseEinstein condensates of Cooper pairs interacting with electrons*, Phys. Lett. A **266**, 400 (2000).

13. H. Matsui, T. Sato, T. Takahashi, S.-C. Wang, H.-B. Yang, H. Ding, T. Fujii, T. Watanabe, A. Matsuda, *BCS-Like Bogoliubov Quasiparticles in High-Tc Superconductors Observed by Angle-Resolved Photoemission Spectroscopy*, Phys. Rev. Lett. **90**, 217002 (2003).

14. M. de Llano and V. V. Tolmachev, *A Generalized BoseEinstein Condensation Theory of Superconductivity Inspired by Bogolyubov*. Ukr. J. Phys. **55**, 79 (2010).

15. J. Batle, M. Casas, M. Fortes, F. J. Sevilla, M. A. Solis, M. de Llano, O. Rojo and V. V. Tolmachev, *BCS and BEC Finally Unified: a Brief Review*, in: Condensed Matter Theories, volume 18 (Nova Science Publishers, Inc., 2003) p.111.

Tee for two, on the country, dacha near Moscow N.N. Bogolubov and E.A. Bogolubova (Pirashkova)
(family collection)

Printed in the United States
By Bookmasters